Organic Reactions

Organic Reactions

VOLUME 24

JOHN WILEY & SONS, INC.

NEW YORK · LONDON · SYDNEY · TORONTO

SPECIAL PREFACE TO VOLUME 24

In the SPECIAL PREFACE to Volume 22, it was pointed out that *Organic Reactions* was initiating a program of updating earlier reviews. In the present volume, the chapter covering Selenium Dioxide Oxidation is an update of the original review that appeared in Volume 5 and the chapter covering the Meerwein Arylation Reaction is an update of the original review that appeared in Volume 11.

The Editorial Board, which selected the first chapters to be revised, welcomes your suggestions about which earlier reactions warrant a current coverage. Your recommendations should list the developments that have been made in the area so that a basis for review can be evaluated. Send your suggestions directly to the Editor-in-Chief. The Editorial Board also welcomes comments on the updated reviews in this volume. Such comments may serve as a guide in the Board's future plans.

WILLIAM G. DAUBEN
EDITOR-IN-CHIEF

September 1976
Department of Chemistry
University of California
Berkeley, California

PREFACE TO THE SERIES

In the course of nearly every program of research in organic chemistry the investigator finds it necessary to use several of the better-known synthetic reactions. To discover the optimum conditions for the application of even the most familiar one to a compound not previously subjected to the reaction often requires an extensive search of the literature; even then a series of experiments may be necessary. When the results of the investigation are published, the synthesis, which may have required months of work, is usually described without comment. The background of knowledge and experience gained in the literature search and experimentation is thus lost to those who subsequently have occasion to apply the general method. The student of preparative organic chemistry faces similar difficulties. The textbooks and laboratory manuals furnish numerous examples of the application of various syntheses, but only rarely do they convey an accurate conception of the scope and usefulness of the processes.

For many years American organic chemists have discussed these problems. The plan of compiling critical discussions of the more important reactions thus was evolved. The volumes of *Organic Reactions* are collections of chapters each devoted to a single reaction, or a definite phase of a reaction, of wide applicability. The authors have had experience with the processes surveyed. The subjects are presented from the preparative viewpoint, and particular attention is given to limitations, interfering influences, effects of structure, and the selection of experimental techniques. Each chapter includes several detailed procedures illustrating the significant modifications of the method. Most of these procedures have been found satisfactory by the author or one of the editors, but unlike those in *Organic Syntheses* they have not been subjected to careful testing in two or more laboratories.

Each chapter contains tables that include all the examples of the reaction under consideration that the author has been able to find. It is inevitable, however, that in the search of the literature some examples will be missed, especially when the reaction is used as one step in an extended synthesis. Nevertheless, the investigator will be able to use the tables and their accompanying bibliographies in place of most or all of the literature search so often required.

Because of the systematic arrangement of the material in the chapters and the entries in the tables, users of the books will be able to find information desired by reference to the table of contents of the appropriate chapter. In the interest of economy the entries in the indices have been kept to a minimum, and in particular, the compounds listed in the tables are not repeated in the indices.

The success of this publication, which will appear periodically, depends upon the cooperation of organic chemists and their willingness to devote time and effort to the preparation of the chapters. They have manifested their interest already by the almost unanimous acceptance of invitations to contribute to the work. The editors will welcome their continued interest and their suggestions for improvements in *Organic Reactions*.

Chemists who are considering the preparation of a manuscript for submission to Organic Reactions are urged to write either secretary before they begin work.

CONTENTS

Organic Reactions

CHAPTER 1

HOMOGENEOUS HYDROGENATION CATALYSTS IN ORGANIC SYNTHESIS

A. J. Birch and D. H. Williamson*

Australian National University, Canberra, Australia

CONTENTS

* We are greatly indebted to Mrs. Bev Cooper for expert typing of a demanding manuscript.

1

INTRODUCTION

During recent years, studies of the activation of hydrogen by soluble catalysts have been intensively pursued. Reviews of certain aspects of the subject have appeared.[1] The scope of this chapter has been limited to those catalysts utilizing hydrogen gas in the reduction of organic substrates. This has led to omission of reference to reduction of inorganic ions, to stoichiometric reductions by metal hydride complexes generated in other ways, and to most catalysts utilizing, for example, water, mineral acids, phenols, alcohols, hydrazine, sodium borohydride, and alkanes as sources of hydrogen.[2] The approach is based on experimental procedure and the distinction is rather artificial in some cases, since hydrogenations often

---

[1] (a) M. Orchin, *Adv. Catal.*, **5,** 385 (1953); (b) H. W. Sternberg and I. Wender, *Int. Conf. Coord. Chem.*, London, **1959** [*Chem. Soc. Special Publ.*, [13] 35 (1959); (c) J. Halpern, *Proc. 3rd Int. Congr. Catal.*, Amsterdam, **1964;** (d) G. C. Bond, *Ann. Rep. Progr. Chem.*, **63,** 27 (1966); (e) J. A. Osborn, *Endeavour*, **26,** 144 (1967); (f) M. M. Taqui Khan, *Chem. Process. Eng.*, *Annual*, **1969,** 17; (g) B. R. James, *Inorg. Chim. Acta Rev.*, **4,** 73 (1970); (h) R. F. Evans, *Modern Reactions in Organic Synthesis*, C. J. Timmons, Ed., Van Nostrand Reinhold Co., London, 1970, p. 16; (i) R. Ugo, *Aspects of Homogeneous Catalysis*, Carlo Manfredi, Ed, Vol. 1, Milano, 1970; (j) J. E. Lyons, L. E. Rennick, and J. L. Burmeister, *Ind. Eng. Chem.*, *Prod. Res. Develop.*, **9,** 2 (1970); (k) W. Strohmeier, *Fortschr. Chem. Forsch.*, **25,** 71 (1972).

[2] (a) H. W. Sternberg, R. A. Friedel, R. Markby, and I. Wender, *J. Am. Chem. Soc.*, **78,** 3621 (1956); (b) H. W. Sternberg, R. Markby, and I. Wender, *ibid.*, **79,** 6116 (1957); (c) D. A. Brown, J. P. Hargaden, C. M. McMullin, N. Gogan, and H. Sloan, *J. Chem. Soc.*, **1963,** 4914; (d) C. E. Castro and R. D. Stephens, *J. Amer. Chem. Soc.*, **86,** 4358 (1964); (e) K. Isogai and Y. Hazeyama, *Nippon Kagaku Zasshi*, **86,** 869 (1965) [*C.A.*, **64,** 12578d (1966)]; (f) T. Suzuki and T. Kwan, *Nippon Kagaku Zasshi*, **87,** 926 (1966) [*C.A.*, **66,** 64817s (1967)]; (g) T. Suzuki and T. Kwan, *ibid.*, **88,** 440 (1967) [*C.A.* .**67,** 43325k (1967)]; (h) T. Mizuta and T. Kwan, *ibid.*, **88,** 471 (1967) [*C.A.*, **67,** 99537y (1967)]; (i) H. H. Brongersma, H. M. Buck, H. P. J. M. Dekkers, and L. J. Oosterhoff, *J. Catal.*, **10,** 149 (1968); (j) W. H. Dennis, D. H. Rosenblatt, R. R. Rickmond, G. A. Finseth, and G. T. Davis, *Tetrahedron Lett.*, **1968,** 1821; (k) M. Pereyre and J. Valade, *ibid.*, **1969,** 489; (l) H. B. Henbest and T. R. B. Mitchell, *J. Chem. Soc., C*, **1970,** 785; (m) T. Nishiguchi and K. Fukuzumi, *Chem. Commun.*, **1971,** 139; (n) Y. Sasson and J. Blum, *Tetrahedron Lett.*, **1971,** 2167; (o) M. E. Vol'pin, V. P. Kukolev, V. O. Chernyshev, and I. S. Kolomnikov, *ibid.*, **1971,** 4435; (p) R. Noyori, I. Umeda, and T. Ishigami, *J. Org. Chem.*, **37,** 1542 (1972).

occur via metal hydride intermediates which can be used stoichiometrically and which can sometimes be generated by other routes.[3]

The first hydrogenation of an organic molecule using a soluble catalyst, rather than classical divided metal catalysts, was reported by M. Calvin in 1938. He noted the reduction of copper(II) salts, and the hydrogenation of quinone, by a quinoline solution of copper(I) acetate in the presence of hydrogen at atmospheric pressure and 100°. This may, however, be an electron-transfer reaction with hydrogen the source of electrons.

No significant advances were made until the early 1950s, when studies of the industrial hydroformylation of olefins by carbon monoxide and hydrogen over the dicobaltoctacarbonyl catalyst revealed the presence of products of hydrogenation of the starting olefin with the aldehydes produced by the hydroformylation process. Subsequent alterations of reaction conditions were found to suppress hydroformylation and to permit hydrogenation.

In the early 1960s aqueous solutions of the pentacyanocobaltate(II) ion were shown to activate molecular hydrogen for the reduction of various organic substrates, a conjugated π system being necessary for reaction.

These catalytic systems are not ideally suited to many organic syntheses under usual laboratory conditions, in that they require either an unsuitable solvent (water) or elevated temperatures and pressures.

In 1963 a complex platinum-tin chloride catalyst in methanol solution was reported to be effective in the hydrogenation of ethylene and acetylene at ambient pressure and temperature. This discovery was followed in 1965 by that of chlorotris(triphenylphosphine)rhodium(I) which catalyzes the hydrogenation of olefins and acetylenes under similar conditions.

These catalysts, efficient in hydrogenating a range of organic compounds under mild conditions, promised far-reaching developments which, indeed, are still in progress.

Homogeneous and heterogeneous catalysts employ a similar range of metals, but the soluble complex catalysts are uniform and therefore show more clearly defined activity and selectivity. Variation of ligands in a soluble catalyst, resulting in a range of properties, may be likened to

[3] (a) I. Wender, H. W. Sternberg, and M. Orchin, *J. Amer. Chem. Soc.*, **75**, 3041 (1953); (b) J. H. Flynn and H. M. Hulburt, *ibid.*, **76**, 3393 (1954); (c) N. Kelso King and M. E. Winfield, *ibid.*, **83**, 3366 (1961); (d) M. Murakami, *Proc. 7th Int. Conf. Coord. Chem.*, Sweden, **1962**, 268; (e) R. W. Goetz and M. Orchin, *J. Org. Chem.*, **27**, 3698 (1962); *J. Amer. Chem. Soc.*, **85**, 2782 (1963); (f) M. Murakami, J.-W. Kang, H. Itatani, S. Senoh, and N. Matsusato, *Nippon Kagaku Zasshi*, **84**, 48, 51, 53 (1963) [*C.A.*, **59**, 15207f (1963)]; (g) J.-W. Kang, *ibid.*, **84**, 56 (1963) [*C.A.*, **59**, 15208d (1963)]; (h) A. Misono, Y. Uchida, K. Tamai, and M. Hidai, *Bull. Chem. Soc. Jap.*, **40**, 931 (1967); (i) A. Miyake and H. Kondo, *Angew. Chem., Int. Ed. Engl.*, **7**, 631, 880 (1968); (j) K. Tarama and T. Funabiki, *Bull. Chem. Soc. Jap.*, **41**, 1744 (1968).

poisoning of metal surfaces; but the use of different ligands in theory makes possible precise control of a range of properties in soluble catalysts. Ligands employed may exert both electronic and steric effects to influence the catalytic activity; however, the practical results of these influences depend on individual mechanisms and are not predictable over the whole range of catalysts. Optically active ligands have been used in asymmetric hydrogenations. A theoretical advantage of homogeneous catalysts is their use with bulky molecules, particularly polymers, where surface adsorptions in suitable orientations on solid catalysts may be difficult. Polymers are in fact often readily hydrogenated in this way. One drawback to date is lack of a range of homogeneous catalysts completely general for difficult reductions, such as hydrogenation of aromatic compounds.

The basic mechanistic similarity between homogeneous and heterogeneous catalysts combined with the greater ease of conducting kinetic studies in homogeneous media gives hope that such studies may throw further light on the mode of action of the kinetically complicated heterogeneous systems.[4] In the present context, mechanisms are discussed only to the extent that they may assist in understanding the scope and use of a particular catalyst. It is noted that many hydrogenation systems have been reported where the exact catalytic intermediate and its mode of reaction are not known.

Some dissimilarities are observed between the types of functional groups attacked in catalytic hydrogenation and in reduction by other methods used widely in organic chemistry: dissolving metals and complex metal hydrides. Dissolving metal reductions require the substrate molecule to accept an electron or electrons to form charged species; therefore only conjugated and aromatic systems or ones containing an appropriate heteroatom are readily reduced. Complex metal hydrides rely on an initial polar addition reaction of anionic hydride and therefore cause reduction only of highly polar unsaturated bonds. The main requirement of the substrate in catalytic hydrogenation is for π electrons capable of forming a bond to the metal. Whether these electrons are capable of donation by an isolated or conjugated bond with a high or low dipole moment (e.g., C=C or C=O) is a function of the metal employed and, in the case of a soluble catalyst, of the ligands coordinated to it.

Many developments are hopefully yet to come, and with them an increasing utility of soluble catalysts in general organic synthesis. In this context we note the use of polymer-supported phosphines to make analogs

[4] (a) J. Halpern, *Adv. Catal.*, **11**, 301 (1959); (b) S. Carra and R. Ugo, *Inorg. Chim. Acta Rev.*, **1**, 49 (1967); (c) I. Jardine and F. J. McQuillin, *Tetrahedron Lett.*, **1968**, 5189; (d) I. Jardine, R. W. Howsam, and F. J. McQuillin, *J. Chem. Soc.*, *C*, **1969**, 260; (e) R. L. Augustine and J. Van Peppen, *Ann. N.Y. Acad. Sci.*, **158**, 482 (1969).

of "soluble" catalysts.[5] This principle could give a system combining the uniformity of active centers and potential control of selectivity of a homogeneous catalyst with the practical advantages of a heterogeneous catalyst. In this case the catalyst is also selective on the basis of molecular size.

Many of the catalysts mentioned are, in their present form, not very useful to the synthetic chemist, but it is likely that most have been insufficiently investigated from this viewpoint, often by those whose primary interests were only in the structure and mode of action of the catalyst.

In this chapter the literature has been covered up to about the end of 1974.

THE CATALYSTS

Chlorotris(triphenylphosphine)rhodium—RhCl(PPh₃)₃—and Related Catalysts

Independently reported by three groups in 1965,[6, 7] but mainly investigated by Wilkinson and his collaborators, this catalyst has proved to be amongst the most generally applicable discovered to date. Its utility and mechanistic details have been studied in some detail.

The catalyst has been modified in several ways, with variation of the ligands and the central metal atom.

(a) The metal from rhodium to iridium.
(b) The halide from chloride to bromide, iodide, and nitrosyl.
(c) The group V donor atom from phosphorus to arsenic and antimony.
(d) The triphenylphosphine to other tertiary phosphines.

Increased rates of hydrogenation have received most attention, although catalysts showing greater specificity may emerge. The modifications and their effects are discussed on p. 22.

Preparation

Reaction of rhodium trichloride trihydrate with excess triphenylphosphine in boiling ethyl alcohol gives dark-red crystals of RhCl(PPh₃)₃ in 88% yield, the phosphine serving both to reduce rhodium(III) and to complex the resulting rhodium(I).[8] Excess triphenylphosphine also

[5] (a) R. H. Grubbs and L. C. Kroll, *J. Amer. Chem. Soc.*, **93**, 3062 (1971); (b) R. H. Grubbs, L. C. Kroll, and E. M. Sweet, *J. Macromol. Sci. Chem.* **A7**, 1047 (1973); (c) M. Capka, P. Suoba, M. Cerny, and J. Hetfleje, *Tetrahedron Lett.*, **1971**, 4787; (d) J. P. Collman, L. S. Hegedus, M. P. Cooke, J. R. Norton, G. Dolcetti, and D. N. Marquardt, *J. Amer. Chem. Soc.*, **94**, 1789 (1972).

[6] (a) F. H. Jardine, J. A. Osborn, G. Wilkinson, and J. F. Young, *Chem. Ind.* (London), **1965**, 560; (b) ICI Ltd., Brit. Pat. 1121642 (1965) [*C.A.*, **66**, 10556y (1967)].

[7] M. A. Bennett and P. A. Longstaff, *Chem. Ind.* (London), **1965**, 846.

[8] J. A. Osborn, F. H. Jardine, J. F. Young, and G. Wilkinson, *J. Chem. Soc.*, A, **1966**, 1711.

prevents formation of the chloro-bridged dimer [RhCl(PPh$_3$)$_2$]$_2$. A second orange crystalline form is precipitated if insufficient ethyl alcohol is employed; continued refluxing causes conversion into the red form.

The bromo and iodo analogs are most conveniently prepared by addition of excess lithium bromide or iodide to solutions of the chloro compound or to the reaction mixture used in its preparation. Direct reactions of rhodium tribromide and rhodium triiodide with triphenylphosphine are also possible.[7-9] Orange and deep-brown crystalline forms of RhBr(PPh$_3$)$_3$ have been described; the iodo complex is isolated as dark-red crystals.

Such direct preparations are successful only in formation of triphenylphosphine derivatives. A second more general method involves ligand exchange on preformed rhodium(I) complexes of the type [Rh(olefin)$_2$Cl]$_2$ or [Rh(diolefin)Cl]$_2$, the exchange being accompanied by splitting of the chloro-bridged dimers. Olefins used include ethylene, cyclooctene, and

$$[\text{Rh(olefin)}_2\text{Cl}]_2 + 6\,\text{PR}_3 \rightarrow 2\,\text{RhCl(PR}_3)_3 + 2\,(\text{olefin})$$

1,5-hexadiene.

The triphenylarsine and triphenylstibine analogs, RhCl(AsPh$_3$)$_3$ and RhCl(SbPh$_3$)$_3$, are prepared from [Rh(C$_2$H$_4$)$_2$Cl]$_2$ in the manner above.[10] They are mentioned only for the sake of completeness: as catalysts they fall far short of RhCl(PPh$_3$)$_3$.

Catalysts incorporating a wide range of phosphines have been prepared by this second method. The reaction is carried out under an inert atmosphere in benzene solution using a stoichiometric quantity of phosphine. Isolation of pure complex is simply accomplished by evaporation of the solvent under vacuum and washing the residue with pentane or hexane.[11] However, isolation is not strictly necessary because addition of phosphine to a solution of the rhodium(I) olefin complex constitutes a convenient *in situ* preparation of catalyst solution.[12] Such *in situ* preparations allow variation of the ratio phosphine:rhodium, sometimes giving substantial changes in rates of hydrogenation. Use of the readily available ethylene complex [Rh(C$_2$H$_4$)$_2$Cl]$_2$[10, 13] rather than complexes of higher olefins should prevent any contamination of hydrogenation products.

The related nitrosyl complex Rh(NO)(PPh$_3$)$_3$ is prepared by bubbling nitric oxide through a tetrahydrofuran solution of rhodium trichloride at 65° in the presence of excess triphenylphosphine and granulated zinc.[14]

[9] G. C. Bond and R. A. Hillyard, *Discuss. Faraday Soc.*, **46**, 20 (1968).

[10] J. T. Mague and G. Wilkinson, *J. Chem. Soc., A*, **1966**, 1736.

[11] Y. Chevallier, R. Stern, and L. Sajus, *Tetrahedron Lett.*, **1969**, 1197.

[12] R. Stern, Y. Chevallier, and L. Sajus, *C.R. Acad. Sci., Ser. C*, **264**, 1740 (1967).

[13] R. D. Cramer, *Inorg. Chem.*, **1**, 722 (1962).

[14] J. P. Collman, N. W. Hoffman, and D. E. Morris, *J. Amer. Chem. Soc.*, **91**, 5659 (1969).

Properties

Solid $RhCl(PPh_3)_3$ is stable indefinitely in air at room temperature; at 25° it is moderately soluble in benzene, chloroform, and dichloromethane, slightly soluble in acetic acid, acetone, and other ketones, methanol, ethanol, and other alcohols. Light petroleum and cyclohexane are poor solvents. Its solutions are unstable; decomposition leads to slow formation of the insoluble dimer, $[RhCl(PPh_3)_2]_2$.

$$2\ RhCl(PPh_3)_3 \rightarrow [RhCl(PPh_3)_2]_2 + 2\ PPh_3$$

In solution, reactions with carbon monoxide, hydrogen, oxygen, peroxides, and ethylene cause displacement of triphenylphosphine and formation of new complexes. Complexing solvents (L) such as pyridine, dimethyl sulfoxide, or acetonitrile similarly displace triphenylphosphine, giving complexes $RhCl(PPh_3)_2L$. Reactions with chlorinated solvents under hydrogen produce catalytically inactive $RhCl_2H(PPh_3)_2$.

The bromo and iodo compounds, as well as those containing differently substituted phosphines, have similar properties. Lower stability toward oxygen in some cases is perhaps the most notable variation so far as their preparation and manipulation are concerned; they are all safely handled in nitrogen and other inert atmospheres.

Mechanism

A general mechanistic picture has emerged for hydrogenations over complex metal catalysts.[15, 16] The overall pathways can be broken into several distinct steps: hydrogen activation, substrate activation, and hydrogen transfers.

Activation of molecular hydrogen at the metal center may result in heterolytic or homolytic cleavage of the hydrogen. Heterolytic cleavage is often accompanied by ligand displacement at the metal atom as shown in Eq. 1.

$$[Ru^{III}Cl_6]^{3-} + H_2 \rightarrow [Ru^{III}HCl_5]^{3-} + H^+ + Cl^- \qquad \text{(Eq. 1)}$$

Homolytic cleavage of hydrogen involves oxidative addition to the metal atom. Two types of homolytic cleavage may be distinguished: (a) addition of one hydrogen atom to each of two metal atoms,

$$2\ [Co^{II}(CN)_5]^{3-} + H_2 \rightarrow 2\ [Co^{III}H(CN)_5]^{3-}$$

$$Co_2(CO)_8 + H_2 \rightarrow 2\ Co^IH(CO)_4$$

[15] (a) J. Halpern, *Chem. Eng. News*, **44**, 68 (Oct. 31, 1966); (b) J. P. Collman, *Trans. N.Y. Acad. Sci.*, Ser. 2, **30**, 479 (1967–68); (c) J. P. Collman, *Accounts Chem. Res.*, **1**, 138 (1968); (d) L. Vaska, *Accts. Chem. Res.*, **1**, 335 (1968).

[16] J. Halpern, *Quart. Rev.*, **10**, 463 (1956); *J. Phys. Chem.*, **63**, 398 (1959); *Ann. Rev. Phys. Chem.*, **16**, 103 (1965).

and (b) addition of both hydrogen atoms to a single metal atom, giving a dihydride. The latter mechanism has been demonstrated in many

$$Ir^{I}Cl(CO)(PPh_3)_2 + H_2 \rightarrow Ir^{III}H_2Cl(CO)(PPh_3)_2$$

of the more recently discovered transition metal complexes, including $RhCl(PPh_3)_3$.

$$Rh^{I}Cl(PPh_3)_3 + H_2 \rightarrow Rh^{III}H_2Cl(PPh_3)_3$$

Formation of a π complex between an unsaturated organic molecule and the metal center serves to bring the substrate into a suitable environment for addition of hydrogen atoms. Since the syntheses and structures of stable π complexes are well documented, this aspect of the reaction needs little discussion.

Successful hydrogenation thus requires initially a coordinatively saturated complex able to lose ligands by dissociation or, alternatively, a coordinatively unsaturated complex. In either case the metal atom must be able to bind hydrogen and substrate simultaneously. Coordination of both species is a necessary but not sufficient condition for hydrogenation. Which step occurs first has not always been elucidated in mechanistic studies. Kinetic results show that in some cases rate is limited by hydrogen concentration, in others by both hydrogen and substrate.

Details of hydrogen and substrate activation by $RhCl(PPh_3)_3$ have been the subject of some controversy. The original cryoscopic and osmotic pressure measurements of Wilkinson and others indicated almost complete dissociation of the complex in solution according to Eq. 2. Absorption of

$$RhCl(PPh_3)_3 \rightleftharpoons RhCl(PPh_3)_2 + PPh_3 \qquad \text{(Eq. 2)}$$

hydrogen by solutions of the complex results in consumption of one mole of hydrogen per mole of rhodium; white $RhH_2Cl(PPh_3)_2$ is isolated from chloroform solution. This result seems to give chemical evidence for the dissociation in Eq. 2 and points to an obvious pathway for hydrogen and substrate activation.

$$RhCl(PPh_3)_3 \rightarrow RhCl(PPh_3)_2 \rightarrow RhH_2Cl(PPh_3)_2 \rightarrow$$

$$RhH_2Cl(PPh_3)_2 \text{(unsaturate)}$$

It is now clear that this picture is not entirely exact. More recent work has provided spectrophotometric and nuclear magnetic resonance measurements as well as chemical evidence showing that such dissociation does not occur to a large extent, and that addition of hydrogen to $RhCl(PPh_3)_3$ in solution forms octahedral $RhH_2Cl(PPh_3)_3$ with the configuration shown in **1**.[17–21a–d] [31]P nuclear magnetic resonance spectra of **1** and of Rh(NO)(PPh$_3$)$_3$ indicate rapid exchange between the phosphine ligands and free phosphine; competitive exchange with an unsaturated molecule may be the pathway which allows the substrate activation necessary for

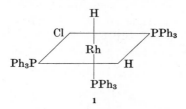

1

hydrogenation to take place.[21d, e] The competitive nature of phosphine displacement is confirmed by the increased rates of hydrogenation shown by catalysts prepared *in situ* with a deficiency of phosphine (less than three moles per mole of rhodium) in solution, and by the inactivity of the complex derived from a chelating triphosphine.[21b] Participation of the dimers [RhCl(PPh$_3$)$_2$]$_2$ and H$_2$[RhCl(PPh$_3$)$_2$]$_2$ in the hydrogenation system has also been demonstrated.[21d]

The processes described above serve to generate a metal atom bearing the π-bonded unsaturated ligand and one or two hydride ligands. Insertion of the organic molecule into a metal-hydrogen bond leads to formation of a σ-bonded metal alkyl derivative. That this step is frequently reversible is shown by the olefin isomerization observed over many catalysts.

In dihydride-type intermediates, completion of the reaction involves transfer of the second hydrogen with irreversible decomposition to the saturated product and regeneration of the catalyst. Theoretically, distinc-

$$RhH_2Cl(PPh_3)_2(\text{unsaturate}) \rightarrow RhHCl(PPh_3)_2(\sigma\text{-alkyl})$$

$$\rightarrow RhCl(PPh_3)_2 + \text{saturated substrate}$$

tion can be made between stepwise and synchronous hydrogen transfers, although there undoubtedly exists a continuous spectrum of situations between the two extremes. Synchronous transfer should lead to stereospecific *cis* addition of hydrogen to substrate; stepwise addition gives the

[17] D. R. Eaton and S. R. Suart, *J. Amer. Chem. Soc.*, **90**, 4170 (1968).

[18] T. H. Brown and P. J. Green, *J. Amer. Chem. Soc.*, **92**, 2359 (1970).

[19] (a) D. D. Lehman, D. F. Shriver, and I. Wharf, *Chem. Commun.*, **1970**, 1486; (b) R. W. Mitchell, J. D. Ruddick, and G. Wilkinson, *J. Chem. Soc., A*, **1971**, 3224.

[20] H. Arai and J. Halpern, *Chem. Commun.*, **1971**, 157.

[21] (a) P. Meakin, J. P. Jesson, and C. A. Tolman, *J. Amer. Chem. Soc.*, **94**, 3240 (1972); (b) T. E. Nappier, D. W. Meek, R. M. Kirchner, and J. A. Ibers, *ibid.*, **95**, 4194 (1973); (c) Y. Demortier and I. de Aguirre, *Bull. Soc. Chim. Fr.*, **1974**, 1614 and 1619; (d) C. A. Tolman, P. Z. Meakin, D. L. Lindner, and J. P. Jesson, *J. Amer. Chem. Soc.*, **96**, 2762 (1974); (e) K. G. Caulton, *Inorg. Chem.*, **13**, 1774 (1974).

possibility of olefin isomerization and nonstereospecific hydrogen addition to the unsaturated bond. Hydrogenations over $RhCl(PPh_3)_3$ appear to approach synchronous hydrogen transfer, specific *cis* addition occurring in many of the cases studied.

Where the σ-alkylmetal intermediate does not contain a second co-ordinated hydrogen, cleavage to saturated product and regenerated catalyst requires a further hydrogen activation step. Hydrogen addition to the metal displaces the product and regenerates the initial metal hydride.

$$CoH(CO)_4 \rightarrow CoH(CO)_4(\text{unsaturate})$$

$$\rightarrow Co(CO)_4(\sigma\text{-alkyl})$$

$$\xrightarrow{H_2} CoH(CO)_4 + \text{saturated substrate}$$

In this case, hydrogen transfers must be stepwise. Consequently there always exists the possibility of nonstereospecific hydrogen addition. Failure of this second activation and subsequent reversal of the reaction account for the role of hydrogen in the olefin isomerization (not accompanied by hydrogenation) observed over many transition metal complexes. Often there is no reaction in the absence of hydrogen.

Solvent molecules undoubtedly play an active role in the overall process, principally by occupying vacant sites on the metal atom at different stages during the catalytic cycle. The process is critically dependent upon the metal-hydrogen, π-substrate-metal, and σ-alkyl-metal bonds all having suitable energies for the sequence of steps to be carried to completion. Too weak a metal-hydrogen or metal-substrate bond results in the reaction not being initiated; too stable a bond at any stage does not allow the final irreversible step to be reached.

The strength of bonding to hydrogen, unsaturated substrate, and alkyl groups is determined by the residual ligands on the metal. The influence of different ligands through electronic and steric effects has been reviewed.[22] Suitable ligands must also be present to prevent reduction of the complex catalyst to the free metal.[23] Figure 1 summarizes the possible hydrogenation pathways.

Hydrogen activation, substrate activation, or hydrogen transfer could be rate-determining in this sequence. The low deuterium and tritium isotope effects observed in hydrogenations over $RhCl(PPh_3)_3$ indicate that hydrogen is not directly involved in the slow step for this catalyst.[24, 25] Increased rates of hydrogenations carried out using a deficiency of phos-

[22] G. Henrici-Olive and S. Olive, *Angew. Chem., Int. Ed. Engl.*, **10**, 105 (1971).

[23] B. R. James, F. T. T. Ng, and G. L. Rempel, *Inorg. Nucl. Chem. Lett.*, **4**, 197 (1968).

[24] H. Simon and O. Berngruber, *Tetrahedron*, **26**, 1401 (1970).

[25] S. Siegel and D. W. Ohrt, *Chem. Commun.*, **1971**, 1529.

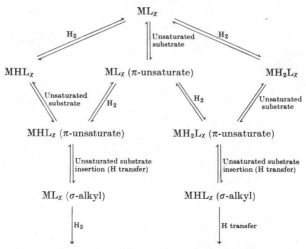

M = metal; L_χ = generalized ligands (χ may change during the reaction sequence)

FIGURE 1

phine support the conclusion that substrate coordination is the rate-determining step.

Scope and Limitations

Only olefinic and acetylenic bonds can normally be hydrogenated. Groups such as keto, hydroxy, cyano, nitro, chloro, azo, ether, ester, aldehyde, and carboxylic acid are not affected.

Mono- and di-substituted olefins are hydrogenated rapidly over $RhCl(PPh_3)_3$ at ambient temperatures and atmospheric pressure. Rates of reaction compare favorably with those obtained over common heterogeneous catalysts. Tri- and especially tetra-substituted bonds react far more slowly; cyclohexene, for example, reacts at more than 50 times the rate of 1-methylcyclohexene.[26a] Even greater differentiation in the rates of hydrogenation of these two compounds has been found using catalysts prepared from piperidylphosphines.

Terminal olefins are reduced more rapidly than internal olefins (cyclic olefins reacting at intermediate rates) and *cis* olefins more rapidly than the corresponding *trans* isomers. Table I shows rate constants for hydrogenations of some representative olefinic bonds.

[26] (a) F. H. Jardine, J. A. Osborn, and G. Wilkinson, *J. Chem. Soc., A,* **1967,** 1574; (b) W. Strohmeier and R. Endres, *Z. Naturforsch., B,* **25,** 1068 (1970).

TABLE I. RATE CONSTANTS FOR HYDROGEN-
ATIONS OVER RhCl(PPh$_3$)$_3$[26]
[In benzene solution; 1.25×10^{-9} M RhCl(PPh$_3$)$_3$]

Substrate	$k^1 \times 10^2$ (l mol^{-1})
Cyclopentene	34.3
Cyclohexene	31.6
1-Methylcyclohexene	0.6
Cycloheptene	21.8
1-Hexene	29.1
1-Dodecene	34.3
2-Methyl-1-pentene	26.6
cis-2-Pentene	23.2
cis-4-Methyl-2-pentene	9.9
trans-4-Methyl-2-pentene	1.8

Considerable selectivity can thus be achieved in the hydrogenation of compounds containing two or more differently substituted double bonds. Several noteworthy applications in the steroid and natural product fields are shown in Chart 1.[27–30]

Sterically unhindered conjugated olefins and chelating nonconjugated diolefins (e.g., 1,5-cyclooctadiene, norbornadiene) are hydrogenated only slowly under a hydrogen pressure of 1 atm. The formation of stable rhodium-substrate complexes undoubtedly causes this inhibition. More rapid reaction occurs at elevated pressure (60 atm).[8, 26a] Some rather sterically hindered dienes (e.g., ergosterol) are rapidly hydrogenated.

Acetylenes usually undergo complete hydrogenation to saturated compounds. This reaction is a sequence of two separate reductions, acetylene to olefin followed by olefin to saturated compound.[8] The use of acidic alcohols (2,2,2-trifluoroethyl alcohol or phenol—see Table II) as co-solvents gives some selectivity between the two stages, reduction of alkene being slowed relative to the initial alkyne hydrogenation.[31a] This technique deserves further exploration. Selective reduction of cyclic and acyclic allenes to fair yields of monoolefins has been achieved.[31b]

Specific deuterations are possible using RhCl(PPh$_3$)$_3$, little or no scrambling of deuterium being observed in many examples.[8, 27, 28, 32]

---

[27] C. Djerassi and J. Grutzwiller, *J. Amer. Chem. Soc.*, **88**, 4537 (1966).

[28] A. J. Birch and K. A. M. Walker, *J. Chem. Soc.*, C, **1966**, 1894.

[29] J. F. Biellmann and H. Liesenfelt, *Bull. Soc. Chim. Fr.*, **1966**, 4029.

[30] M. Brown and L. W. Piszkiewicz, *J. Org. Chem.*, **32**, 2013 (1967).

[31] (a) J. P. Candlin and A. R. Oldham, *Discuss. Faraday Soc.*, **46**, 60 (1968); (b) M. M. Bhagwat and D. Devaprabhakara, *Tetrahedron Lett.*, **1972**, 1391.

[32] (a) A. J. Birch and K. A. M. Walker, *Tetrahedron Lett.*, **1966**, 4939; (b) J. R. Morandi and H. B. Jensen, *J. Org. Chem.*, **34**, 1889 (1969).

$(CH_3)_2C\!\!=\!\!CH(CH_2)_2C(OH)(CH_3)CH\!\!=\!\!CH_2 \xrightarrow{RhCl(PPh_3)_3,\,H_2}$

$(CH_3)_2C\!\!=\!\!CH(CH_2)_2C(OH)(CH_3)CH_2CH_3$
(80%, 96% pure)

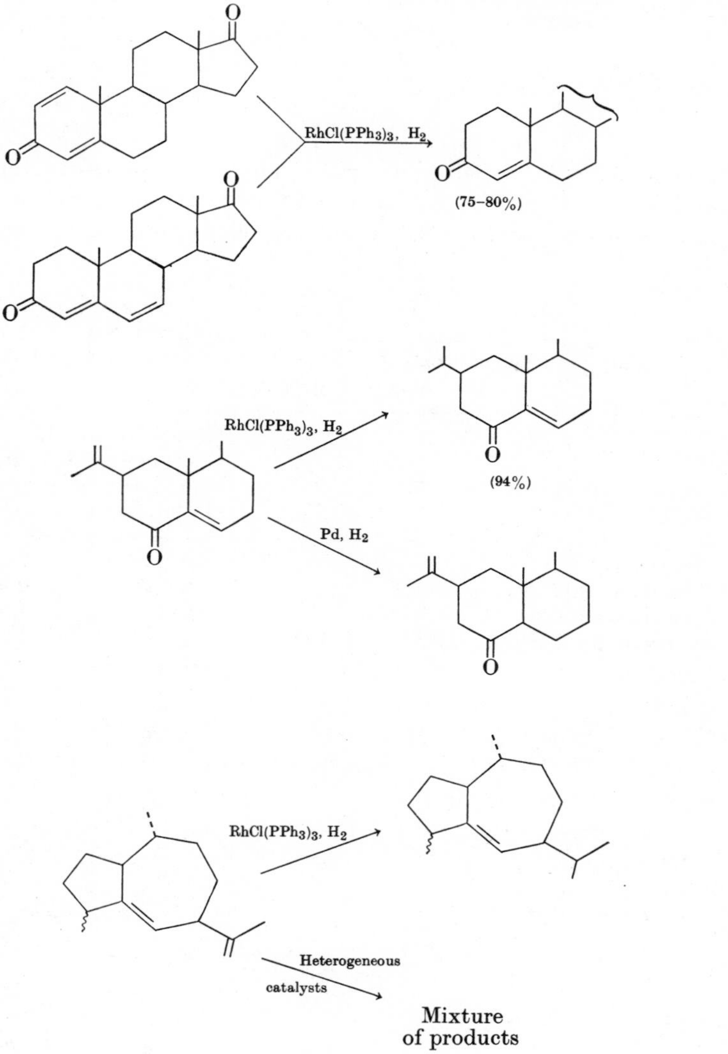

(75–80%)

(94%)

Mixture
of products

Chart 1

Exclusive *cis* addition of deuterium has been proved in a number of cases. Such reactions contrast with most deuterations conducted over heterogeneous catalysts where scrambling of olefinic and allylic hydrogen atoms

(68%)

is common.[33,34a] In an industrially important process, RhCl)PPh₃)₃ catalyzes stereospecific hydrogen addition to the 6-methylene group of methacycline to give the antibiotic doxycycline (α-6-deoxytetracycline).[34b]

TABLE II.   RELATIVE RATES OF HYDROGENATION IN "NORMAL" AND ACIDIC SOLVENTS[31a]
(Catalyst concentration, $10^{-2} M$; substrate concentration, $0.5 M$; 1 atm of hydrogen at 22°)

Solvent	Rates (Relative to 1-Octene in Benzene = 1.0)	
	1-Octene	1-Hexyne
Benzene	1.0	0.9
Benzene-ethyl alcohol (1:1)	0.9–1.7	0.9
Benzene-phenol (1:1)	0.9–1.0	1.7–2.4
Benzene-2,2,2-trifluoroethyl alcohol (1:1)	0.9	>12

[33] N. Dinh-Nguyen and R. Ryhage, *Acta Chem. Scand.*, **13**, 1032 (1959).
[34] (a) H. Budzikiewiez, C. Djerassi, and D. H. Williams, *Structure Elucidation of Natural Products by Mass Spectrometry, Vol. 1, Alkaloids*, Holden-Day, San Francisco, 1964, p. 24; (b) German Patent Application, OS 2,308, 227 (1974).

$$CH_3(CH_2)_7(CHD)_2(CH_2)_7CO_2CH_3$$

RhCl(PPh₃)₃, D₂

cis-$CH_3(CH_2)_7CH{=}CH(CH_2)_7CO_2CH_3$

Adams catalyst, D₂

Mixture d_2- to d_{18}-methyl stearates

RhCl(PPh₃)₃, D₂

Pd/C, D₂

d_1 to d_4-Cholestanes

Instances of olefin isomerization and deuterium scrambling in hydrogenations over $RhCl(PPh_3)_3$ have been noted, however.[35–40a, b] (See examples at top of p. 18.) These reactions must be the consequence of reversible stepwise transfer of hydrogen from rhodium to alkene, with rearrangement of the intermediate alkylrhodium. Modification of ligands can perhaps be useful in this respect; the catalysts $RhX(PPh_3)_3$ where $X = Br$, I, NO all exhibit greater specificity than $RhCl(PPh_3)_3$.

Control of the stereochemistry of hydrogen addition by pre-coordination of the substrate to rhodium is achieved in the novel hydrogenation shown on the bottom of p. 18.[40c]

The presence of oxygen and use of benzene-ethyl alcohol rather than pure benzene as solvent have been shown to promote isomerization.[37,41] (See example at top of p. 19.) The formation of peroxo-rhodium species

[35] A. S. Hussey and Y. Takeuchi, *J. Org. Chem.*, **35**, 643 (1970).

[36] L. Horner, H. Buthe, and H. Siegel, *Tetrahedron Lett.*, **1968**, 4023.

[37] G. V. Smith and R. J. Shuford, *Tetrahedron Lett.*, **1970**, 525.

[38] J. J. Sims, V. K. Honward, and L. H. Selman, *Tetrahedron Lett.*, 1969, 87.

[39] (a) C. H. Heathcock and S. R. Poulter, *Tetrahedron Lett.*, **1969**, 2755; (b) J. F. Biellman and M. J. Jung, *J. Amer. Chem. Soc.*, **90**, 1673 (1968).

[40] (a) A. S. Hussey and Y. Takeuchi, *J. Amer. Chem. Soc.*, **91**, 672 (1969); (b) A. L. Odel, J. B. Richardson, and W. R. Roper, *J. Catal.*, **8**, 393 (1967); (c) H. W. Thompson and E. McPherson, *J. Amer. Chem. Soc.*, **96**, 6232 (1974); (d) C. W. Dudley, G. Reid, and P. J. C. Walker, *J. Chem. Soc. (Dalton)*, **1974**, 1926; (e) A. A. Blanc, H. Arzoumanian, E. J. Vincent, and J. Metzger, *Bull. Soc. Chim. Fr.*, **1974**, 2175.

[41] (a) R. L. Augustine and J. F. Van Peppen, *Chem. Commun.*, **1970**, 495, 571; (b) *ibid.*, **497**.

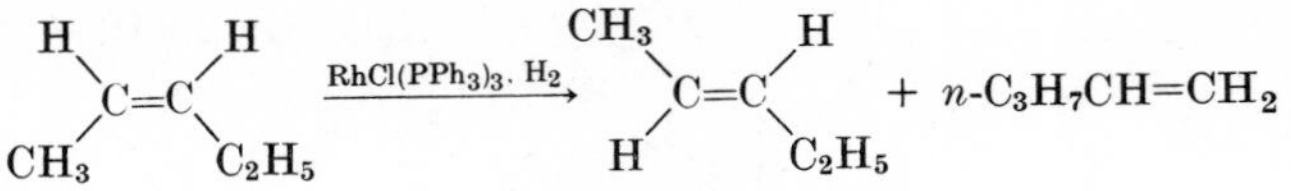
H
H
CH3
C2H5
RhCl(PPh3)3, H2
CH3
H
H
C2H5
+ n-C3H7CH=CH2

(Isolated at intermediate stages in the hydrogenation)

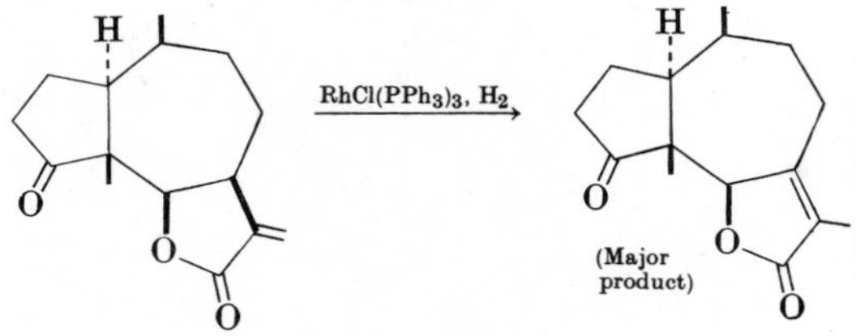
RhCl(PPh3)3, H2
(Major product)

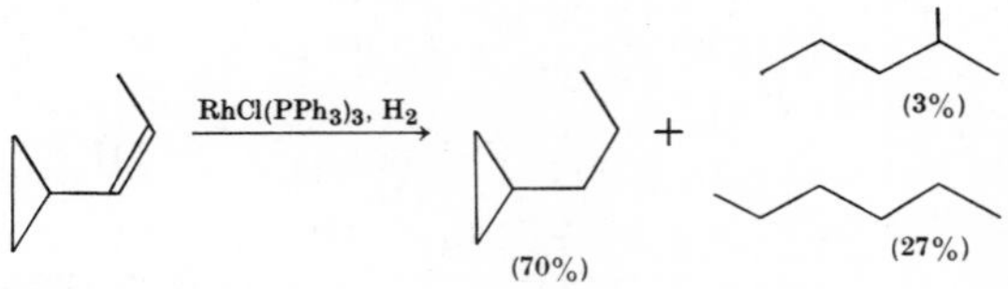
RhCl(PPh3)3, H2
(3%)
(27%)
(70%)

has been demonstrated under these conditions.[21d, 40d] Though the part played by oxygen is not entirely clear, well-deoxygenated solvents and prehydrogenated catalyst solutions appear to limit isomerization and exchange reactions.

Controlled admission of oxygen or hydrogen peroxide in small amounts activates the catalyst in hydrogenation of cyclohexene.[42] Limited oxidation of triphenylphosphine to triphenylphosphine oxide has the effect of lowering the phosphine:rhodium ratio, giving a consequent increase in rate. In the presence of excess oxygen terminal olefins are oxidized to methyl ketones, and cyclohexene to its hydroperoxide.[40d, e]

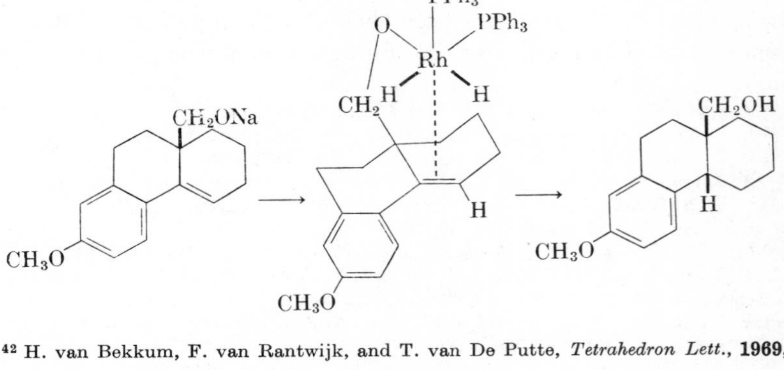
PPh3
PPh3
O
Rh
H
H
CH2
CH2ONa
H
CH2OH
H
CH3O
CH3O
CH3O

[42] H. van Bekkum, F. van Rantwijk, and T. van De Putte, *Tetrahedron Lett.*, **1969**, 1.

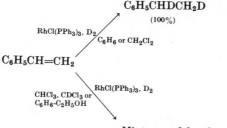

$$C_6H_5CH=CH_2 \begin{array}{l} \xrightarrow[\text{RhCl(PPh}_3)_3, D_2]{C_6H_6 \text{ or } CH_2Cl_2} C_6H_5CHDCH_2D \quad (100\%) \\ \\ \xrightarrow[\text{CHCl}_3, \text{CDCl}_3 \text{ or } C_6H_6\text{-}C_2H_5OH]{\text{RhCl(PPh}_3)_3, D_2} \text{Mixtures of } d_0\text{-}, d_1\text{-}, d_2\text{-ethylbenzenes} \end{array}$$

Heated chloroform or benzene solutions of the catalyst equilibrate 1,4- to 1,3-dienes in the absence of hydrogen.[43a] Allyl ethers are isomerized to 1-propenyl ethers.[43b] The mechanism here has not been defined but some oxygenated complex is probably present. Avoidance of isomerization by oxygen-free conditions is again indicated. Catalytic dimerization and trimerization of norbornadiene also occurs under rather forcing conditions (5 days at 90°).[44] In contrast, 1,4-dihydrotetralin is not isomerized in benzene-ethyl alcohol at room temperature over several days.[38]

Most functional groups do not interfere with hydrogenation; aldehydes and some primary alcohols are significant exceptions (see below). Successful reduction of unsaturated compounds containing keto, ester, lactone, amide, p-toluenesulfonamide, sulfide, ethylene ketal, hydroxy, methoxy, nitrile, fluoro, nitro, and tertiary amine groups have been recorded. Examples include acrylonitrile and the p-toluenesulfonamide shown

$$CH_2=CHCN \xrightarrow{\text{RhCl(PPh}_3)_3, H_2} CH_3CH_2CN$$

$$p\text{-}CH_3C_6H_4SO_2NHCH_2\text{-}\langle\bigcirc\rangle \xrightarrow[\text{H}_2]{\text{RhCl(PPh}_3)_3} p\text{-}CH_3C_6H_4SO_2NHCH_2\text{-}\langle\bigcirc\rangle$$

below. In spite of poisoning of the catalyst by chlorinated solvents and pyridine, it is interesting to note the hydrogenation of compounds such functions.[28, 45, 46] (See examples at top of p. 20.)

Decarbonylation of the aldehyde group in attempted hydrogenation of unsaturated aldehydes leads to formation of inactive $RhCl(CO)(PPh_3)_2$

[43] (a) A. J. Birch and G. S. R. Subba Rao, *Tetrahedron Lett.*, **1968**, 3797; (b) E. J. Corey and J. W. Suggs, *J. Org. Chem.*, **38**, 3224 (1973).

[44] N. Acton, R. H. Roth, T. J. Katz, J. K. Frank, C. A. Maier, and I. C. Paul, *J. Amer. Chem. Soc.*, **94**, 5446 (1972).

[45] A. J. Birch and K. A. M. Walker, *Aust. J. Chem.*, **24**, 513 (1971).

[46] A. J. Birch and H. H. Mantsch, *Aust. J. Chem.*, **22**, 1103 (1969).

and consequent loss of catalytic activity. Use of low aldehyde and high catalyst and hydrogen concentrations may overcome this problem to some extent; higher yields of the saturated aldehyde are obtained from

the unsaturated compound under the modified conditions. The procedure is not entirely satisfactory, however; some catalyst poisoning and reduction of the carbonyl group occur under these conditions. Primary allylic alcohols behave similarly. The decarbonylation reaction, while placing a restriction on hydrogenations, can be used as an organic synthetic procedure.[48,49]

[47] F. H. Jardine and G. Wilkinson, *J. Chem. Soc., C*, **1967**, 270.
[48] J. Tsuji and K. Ohno, *Tetrahedron Lett.*, **1965**, 3969.
[49] J. Blum, *Tetrahedron Lett.*, **1966**, 1605.

Ketones do not undergo carbon monoxide abstraction. However α,β-unsaturated ketones are converted into the corresponding saturated dimethyl ketals during hydrogenations in benzene-methyl alcohol solution.[50] Ketal formation, which occurs to a lesser extent with higher alcohols, could be useful or otherwise according to circumstances, but can obviously be avoided by not using alcoholic solvents. Similar conversion of an enol-ether into a ketal has been reported.[45]

A particularly useful aspect of hydrogenations catalyzed by RhCl(PPh$_3$)$_3$ is the lack of hydrogenolysis of groups normally cleaved during reactions over heterogeneous catalysts.[28, 45, 51]

Similarly, disproportionation of 1,4-dihydroaromatic compounds which occurs readily over heterogeneous catalysts is almost absent over

[50] W. Voelter and C. Djerassi, *Chem. Ber.*, **101**, 1154 (1968).
[51] (a) J. F. Biellmann and H. Liesenfelt, *C.R. Acad. Sci., Ser. C*, **263**, 251 (1966); (b) A. C. Day, J. Nabney, and A. I. Scott, *J. Chem. Soc.*, **1961**, 4067.

RhCl(PPh$_3$)$_3$.[38, 45] This feature has led to simple syntheses of some hitherto less easily prepared compounds, e.g., the tricyclic diene.

Modification of the Catalyst

As indicated by the examples cited, the majority of wide-ranging applications of this family of catalysts to organic synthesis have concentrated on the parent complex. In most situations, RhCl(PPh$_3$)$_3$ offers ease of preparation and handling combined with general utility. For difficult reductions and especially where there are interfering side reactions it may be helpful to use modified catalysts. The following discussion indicates the effects of different variations.

Attempted asymmetric hydrogenations using catalysts prepared from optically active phosphines have met with limited but encouraging success. Optical yields up to 90% have been obtained. They are described in the section on Asymmetric Hydrogenation (p. 74).

In situ preparations of catalyst solutions have brought to light variation in rates of hydrogenation with the molar ratio phosphine:rhodium. (Table III). Substantial increases in rates (sometimes ten- or twenty-fold) can be achieved; maximum rate generally occurs with phosphine: rhodium about 2.[11, 36, 52a] At this ratio, formation of a coordinatively unsaturated complex, RhCl(PPh$_3$)$_2$ (most probably a dimer of this formula), leads to efficient coordination and subsequent hydrogenation of the substrate. Phosphine in excess of this concentration has an inhibiting effect by competing with substrate for the coordination site on rhodium. The efficiency of such coordinatively unsaturated complexes formed from diphosphines depends both on the length of the carbon chain linking the two phosphorus atoms and on the compound being hydrogenated.[52b] While early experiments indicated the iridium analog IrCl(PPh$_3$)$_3$ to be inactive as a hydrogenation catalyst, subsequent *in situ* preparations utilizing phosphine:iridium ratios of 1 or 2 have shown surprisingly high catalytic activity: up to ten times that of the corresponding rhodium system.[53-55] However, rapid isomerization of olefins accompanies hydrogenation, detracting from practical application of the iridium complex.

[52] (a) S. Montelatici, A. van der Ent, J. A. Osborn, and G. Wilkinson, *J. Chem. Soc., A,* **1968,** 1054.

[52] (b) J.-C. Poulin, T.-P. Dang, and H. B. Kagan, *J. Organometal. Chem.,* **84,** 87 (1975).

[53] M. A. Bennett and D. L. Milner, *J. Amer. Chem. Soc.,* **91,** 6983 (1969).

[54] H. van Gaal, H. G. A. M. Cuppers, and A. van der Ent, *Chem. Commun.,* **1970,** 1694.

[55] J. Solodar, *J. Org. Chem.,* **37,** 1840 (1972).

TABLE III. VARIATION IN RATES OF HYDROGENATION
WITH PHOSPHINE:RHODIUM RATIO[a]
([A] Hydrogenation of 0.5 M 1-hexene in benzene; [Rh] $= 5 \times 10^{-3} M$)[36]
([B] Hydrogenation of styrene; [Rh] $= 10^{-2} M$)[11]

Phosphine	Phosphine: Rhodium	Rate
Ethyldiphenyl [A]	2	10 (ml H$_2$/min)
	3	0.2
Tri-(n-butyl) [A]	2.2	2.5
	3	0.2
Triphenyl [B]	2	0.44 (mole l^{-1} min^{-1})
	3	0.39
	4	0.05
Diphenylpyrrolidino [B]	2	0.67
	3	0.38
	3.5	0.23

[a] See also Table V, p. 25.

A detailed comparison of rates of hydrogenation and isomerization of 1-pentene and *cis*- and *trans*-2-pentene over RhCl(PPh$_3$)$_3$, RhBr(PPh$_3$)$_3$, and RhI(PPh$_3$)$_3$ has been made.[9] The bromo and iodo complexes give faster hydrogenation of terminal olefins and lower rates of isomerization than the chloro compound. Their use where isomerization is an interfering reaction with RhCl(PPh$_3$)$_3$ could be profitable; less deuterium scrambling might also be expected in consequence.

Tris(triphenylphosphine)nitrosylrhodium, Rh(NO)(PPh$_3$)$_3$, catalyzes the hydrogenation of a range of alkenes and alkynes.[56a, 56b] Reduction of cyclohexene with deuterium gives d$_2$-cyclohexane in greater than 99 % purity.[14] [Note, however, that RhCl(PPh$_3$)$_3$ also gives highly specific deuterium addition in this case.[28]] The corresponding carboxylate complexes, Rh(RCO$_2$)(PPh$_3$)$_3$, also catalyze hydrogenation of alkenes and alkynes.[19b]

Substitution of phosphines in RhCl(PPh$_3$)$_3$ has varying effects on rates of hydrogenation (Table IV). *Para* substituents on the aromatic rings of triphenylphosphine give increased rates when electron donating (N,N-dimethylamino, methyl, methoxy) and decreased rates when electron withdrawing (fluoro, chloro, acetyl).[36, 52a, 56c]

[56] (a) G. Dolcetti, *Inorg. Nucl. Chem. Lett.*, **9**, 705 (1973); (b) W. Strohmeier and R. Endres, *Z. Naturforsch.*, *B*, **27**, 1415 (1972); (c) L. Horner and H. Siegel, *Ann. Chem.*, **715**, 135 (1971); (d) T. Nishiguchi and K. Fukuzumi, *J. Amer. Chem. Soc.*, **96**, 1893 (1974), and references therein; (e) T. Nishiguchi, K. Tachi, and K. Fukuzumi, *J. Org. Chem.*, **40**, 237 (1975).

TABLE IV. EFFECT OF SUBSTITUENTS IN
TRIPHENYLPHOSPHINE ON RATE OF HYDROGENATION[36]
(Hydrogenation of 0.5 M 1-hexene in benzene;
[Rh] $= 5 \times 10^{-3} M$)

Phosphine	Relative Rate
Triphenyl	16
Diphenyl-p-dimethylaminophenyl	~50
Tri(p-tolyl)	29
Tri(p-methoxyphenyl)	33
Tri(p-chlorophenyl)	<1.7
Tri(α-naphthyl)	<0.1

Steric effects of bulky phosphines, e.g., tri-(α-naphthyl)phosphine, appear to limit olefin coordination and decrease the rate of hydrogenation.

Successive replacement of phenyl groups in triphenylphosphine by alkyl results in successively lower catalytic activities. Increasing basicity of the phosphines (P) drives the equilibrium in Eq. 3 to the left and also reduces the lability of the coordinated hydrogen, both effects producing lower rates of hydrogenation. Lower lability of the coordinated hydrogen

$$RhClH_2P_3 + \text{olefin} \rightleftharpoons RhClH_2P_2(\text{olefin}) + P \qquad \text{(Eq. 3)}$$

in $RhClH_2P_3$ when P = diethylphenylphosphine compared to P = triphenylphosphine has been noted.[52a] Catalysts formed from diethylphenylphosphine and triethylphosphine bring about only very slow hydrogenation of cyclohexene; those derived from phosphites, $P(OR)_3$ (e.g., trimethyl-, triethyl-, and triphenyl-phosphite) are inactive.

Catalytic systems utilizing a number of aminophosphines, e.g., phenylbis(dimethylamino)phosphine, diphenylmorpholinophosphine, have been prepared.[11, 12] Rates of hydrogenation vary widely. The catalyst containing phenyldipiperidylphosphine assists hydrogenation of styrene at three times the rate of $RhCl(PPh_3)_3$. Other derivatives, e.g., of tris(diethylamino)phosphine, are almost inactive.

Replacement of the phosphine by triphenylarsine or triphenylstibene produces catalysts of inferior properties. The activity of the arsine complex in hydrogenation of cyclohexenes is less than 1 % that of the phosphine complex.[10, 35] Hydrogenation of styrene proceeds at higher rates (Table V).[12]

A result which could have useful synthetic applications is that catalysts prepared from diphenylpiperidylphosphine and phenyldipiperidylphosphine, while more active than $RhCl(PPh_3)_3$ in hydrogenating the disubstituted double bond of cyclohexene, at 25° show no activity toward the

TABLE V. RATES OF HYDROGENATION OVER CATALYSTS CONTAINING VARIOUS PHOSPHINES, AND TRIPHENYL-ARSINE AND -STIBINE[12]
(Hydrogenation of 20 ml styrene in 5 ml benzene; catalyst, 0.04 M in benzene prepared from $[Rh(C_2H_4)_2Cl]_2$; hydrogen pressure 1 atm; temperature 40°)

| Phosphine | Rate (mol H_2 min^{-1} l^{-1}) | |
	Phosphine: Rhodium = 2	Phosphine: Rhodium = 3
Phenyldipiperidyl	1.12	1.11
Phenyldimorpholino	0.722	0.615
Diphenylpiperidyl	0.592	0.43
Diphenylmorpholino	0.414	0.372
Tripiperidyl	0.104	—
Triphenyl	0.381	0.267
Diphenylcyclohexyl	0.394	0.117
Phenyldicyclohexyl	0.0935	0.162
Tricyclohexyl	0.0244	—
Diphenylbornyl	—	0.021
Phenyldi-isobutyl	0.168	—
Diphenylbenzoyl	0.136	0.012
Trithienyl	0.0063	—
Tri(cyanoethyl)	0.0021	0.002
Triphenylarsine	0.395	0.344
Triphenylstibine	0.017	0.005

trisubstituted bond of 1-methylcyclohexene.[35] The latter is slowly hydrogenated over the triphenylphosphine complex. This increased specificity is attributed to steric crowding in the coordination sphere of the rhodium atom. Catalysts showing greater specificity than RhCl(PPh$_3$)$_3$ could perhaps evolve along these lines.

Experimental Conditions

Solvents generally employed in hydrogenations are benzene or benzene-ethyl alcohol (up to 50 % of the alcohol). The co-solvent leads to a significant increase in rate. Other solvents and/or co-solvents that have been used include methyl alcohol, t-butyl alcohol, isopropyl alcohol, phenol, glacial acetic acid, ethyl acetate, dimethylformamide, dioxane, dimethyl sulfoxide, dichloromethane, 1,2-dichloroethane, nitrobenzene, cyclohexanone, tetrahydrofuran, cyclohexanol, nitromethane, malonic ester, and 2,2,2-trifluoroethyl alcohol.[8, 31a, 35, 36] Some of them give reduced rates or slow catalyst poisoning but may be useful in overcoming solubility problems.

1,4-Dioxane and many other ethers, alcohols, amines, and hydroaromatics can act as hydrogen source for reduction of olefins.[56d, e] Their effect as solvents on selectivity and stereochemistry of hydrogenation has not been examined extensively.

Most reactions proceed satisfactorily under mild conditions of temperature and pressure; 25–40° and 1 atm. More difficult reductions may require higher hydrogen pressures. Higher temperatures are not advisable (see below).

Presaturation of the solvent with hydrogen is desirable for promotion of solution of $RhCl(PPh_3)_3$ by immediate formation of the more soluble $RhClH_2(PPh_3)_3$. Solubility of the catalyst in benzene is of the order of 10^{-2} mol per liter at 1 atm pressure.

In searching for optimum conditions for any hydrogenation the following points should be borne in mind.

1. Chlorinated solvents should be avoided wherever possible. Hydrogen transfer to the chloro compound liberates hydrogen chloride which forms inactive $RhCl_2H(PPh_3)_2$.

2. Abstraction of carbon monoxide from allyl alcohol, acetate ion, dimethylformamide and dioxane deactivates the catalyst by formation of $RhCl(CO)(PPh_3)_2$.[7, 8]

3. Solvents able to coordinate strongly with the metal inhibit catalysis by displacing triphenylphosphine to form inactive complexes $RhCl(PPh_3)_2$-(solvent). They include pyridine, dimethyl sulfoxide, and acetonitrile.

4. Purification of solvents to remove peroxides and hydroperoxides is desirable. They react with the catalyst, giving more rapid but less specific hydrogenation.

5. Use of benzene rather than benzene-ethyl alcohol and saturation of solvents with hydrogen before addition of the catalyst may be useful where it is desired to minimize isomerization and deuterium scrambling.[40a, 41]

6. Alcoholic co-solvents (especially methyl alcohol) react during hydrogenation of α,β-unsaturated ketones, and with saturated ketones, to form the saturated ketals.[50]

7. The use of acidic alcohols (notably 2,2,2-trifluoroethyl alcohol) as cosolvents can be advantageous in the partial reduction of alkynes to alkenes.[31a]

8. High hydrogen pressure (60 atm) assists hydrogenation of conjugated diolefins and chelating nonconjugated diolefins.[8, 26a]

9. High hydrogen pressure also allows hydrogenation of some unsaturated aldehydes, despite the deactivation caused by carbon monoxide abstraction at lower pressure.[47]

10. High temperatures (greater than 60°) lead to dimerization of the catalyst and a consequent loss of activity.

Rhodium Borohydride Complex—[RhCl(pyridine)$_2$(dimethyl-formamide)(BH$_4$)]Cl

Preparation and Properties

A saturated solution of trichloritris(pyridine)rhodium(III) in dimethyl-formamide is treated with one equivalent of sodium borohydride. Dilution of the solution with diethyl ether and recrystallization of the resulting precipitate from chloroform gives the dark-brown to red crystalline complex.[57] Its ionic nature is shown by conductance measurements. The borohydride is probably a bidentate ligand, coordinated through hydrogen bridges to rhodium. The complex is air-stable both in solution and in the solid state; isolation and storage thus present no problems.

In situ preparation using the same procedure is also possible: finely ground sodium borohydride and RhCl$_3$(pyridine)$_3$ are equilibrated by shaking in warm dimethylformamide under hydrogen. Introduction of substrate then initiates hydrogenation.[57, 58]

Mechanism

The detailed mechanism of hydrogenation by this catalyst has not been elucidated, but the following points which bear on practical applications may be noted.

1. Hydrogen added to the substrate is derived from both hydrogen gas and the borohydride ligand. Thus experiments using hydrogen/borodeuteride and deuterium/borohydride show considerable isotopic scrambling in products.[57] Dimethylformamide does not enter into the scrambling process.

2. Added pyridine has an inhibitory effect: dissociation of pyridine from rhodium is necessary for hydrogenation to proceed.[57, 58]

3. Dimethylformamide is present not only as solvent but also as a ligand during hydrogenation. Hydrogenations in the presence of optically active amides show that asymmetry induced in products is not due to simple asymmetric solvation.[59]

4. Hydrogen transfer from rhodium to substrate appears to be the rate-limiting step. Even at relatively low concentrations the rate of hydrogenation of olefins is independent of olefin concentration.[60]

[57] P. Abley, I. Jardine, and F. J. McQuillin, *J. Chem. Soc., C,* **1971,** 840.
[58] I. Jardine and F. J. McQuillin, *Chem. Commun.,* **1969,** 477.
[59] P. Abley and F. J. McQuillin, *J. Chem. Soc., C,* **1971,** 844.
[60] I. Jardine and F. J. McQuillin, *Chem. Commun.,* **1969,** 502.

Scope and Limitations

A variety of unsaturated bonds are reduced. They include C=C, C=N, N=N, and N=O linkages.

Straight chain terminal olefins from 1-pentene to 1-octene are readily hydrogenated, as are the cyclic olefins cyclopentene to cyclooctene, and norbornene. More hindered olefinic bonds are also hydrogenated, e.g., methyl 3-phenylbutenoate to methyl 3-phenylbutanoate.[57] Optically active

$$\underset{CH_3}{\overset{C_6H_5}{>}}C{=}CHCO_2CH_3 \longrightarrow \underset{CH_3}{\overset{C_6H_5}{>}}CHCH_2CO_2CH_3$$

products are obtained by substituting an optically active amide for dimethylformamide (see Asymmetric Hydrogenation, p. 74).

The steroidal 4-en-3-one system of 4-cholesten-3-one, testosterone, 17-methyltestosterone, and progesterone is also hydrogenated, the introduction of 5α and 5β hydrogens being influenced by substituents at the 17 position in a pattern similar to that observed over heterogeneous catalysts.[61]

R^1	R^2	5α (%)	5β (%)
C_8H_{17}	H	20	80
OH	H	80	20
OH	CH_3	25	75
$COCH_3$	H	80	20

Of possibly greater interest are applications to the reduction of unsaturated bonds containing heteroatoms. Slow saturation of the pyridine ligands occurs on stirring solutions of the complex under hydrogen. Bulk samples of pyridine are similarly reduced to piperidine, and quinoline to the 1,2,3,4-tetrahydro derivative.[62a] Isoquinoline and indole are not hydrogenated.

    [61] I. Jardine and F. J. McQuillin, *Chem. Commun.*, **1969**, 503.

    [62] (a) I. Jardine and F. J. McQuillin, *Chem. Commun.*, **1970**, 626; (b) C. J. Love and F. J. McQuillin, *J. Chem. Soc.* (*Perkin I*), **1973**, 2509.

Nitrocyclohexane, nitrobenzene, and substituted nitrobenzenes are cleanly hydrogenated to the corresponding amines; benzalaniline gives benzylaniline.[62]

$$RNO_2 \rightarrow RNH_2$$

$$R = C_6H_{11}, C_6H_5,$$

$$p\text{-}CH_3C_6H_4, p\text{-}HO_2CC_6H_4,$$

$$p\text{-}(CH_3)_2NC_6H_4$$

$$C_6H_5CH{=\!=}NC_6H_5 \rightarrow C_6H_5CH_2NHC_6H_5$$

Azobenzene is rapidly reduced to hydrazobenzene, which is subsequently slowly converted into aniline. Aromatic ketones (benzophenone, acetophenone, benzoin) are hydrogenated although slowly; aliphatic ketones are not affected.

Efficient hydrogenolysis of a variety of carbon-halogen bonds over the catalyst precludes its application to hydrogenation of unsaturated halogen derivatives.[62b]

Experimental Conditions

Dimethylformamide is most frequently used pure as a solvent.

An alternative procedure is advantageous when an amide is only available in limited quantities, e.g., optically active amides. Hydrogenations can be carried out in dilute solutions (5%) of the amide in diethylene glycol monoethyl ether (to which up to 10% of water can be added without affecting results).

Catalyst concentrations used are in the range 10^{-3} to 10^{-2} M. All hydrogenations proceed under 1 atm of hydrogen at 20°.

Vaska's Compound—IrCl(CO)(PPh$_3$)$_2$—and Related Complexes

Preparation and Properties

trans-IrCl(CO)(PPh$_3$)$_2$ can be prepared by several methods. The original preparation involves heating iridium trichloride trihydrate or ammonium chloroiridate with excess triphenylphosphine in an alcohol.[63] Carbon monoxide abstraction from the alcohol leads to the carbonyl complex. A modification of this method using diethylene glycol as solvent gives improved yields, as does a third procedure which replaces the alcohol by dimethylformamide.[64]

A fourth method utilizes carbon monoxide: a solution of sodium chloroiridite in diethylene glycol is heated under reflux in the presence of carbon monoxide to form a chlorocarbonyliridium species. Addition of two equivalents of triphenylphosphine completes the reaction. This method is economical in the quantity of phosphine required and is a convenient route to complexes other than the triphenylphosphine derivative.[65] In each case the product crystallizes when the reaction mixture is cooled.

The bromo and iodo analogs can be prepared from the respective iridium salts[64] or from the chloro complex by metathesis with lithium bromide or sodium iodide.[66]

The rhodium analog was originally prepared by the reaction of triphenylphosphine with $[RhCl(CO)_2]_2$.[67] Other phosphines react similarly. Direct

$$[RhCl(CO)_2]_2 + 4\,PPh_3 \rightarrow 2\,RhCl(CO)(PPh_3)_2 + 2\,CO$$

reaction of rhodium trichloride with triphenylphosphine in 2-methoxyethyl alcohol is also successful.[68a]

The complexes are bright yellow and air-stable as solids, soluble in aromatic hydrocarbons and chloroform and insoluble in alcohols. In solution they oxidatively add numerous compounds including oxygen itself which is therefore excluded during their preparation and use.

Mechanism

Rapid reversible addition of hydrogen to Vaska's compound gives an isolable dihydrido derivative; slower addition of olefins produces π-olefin complexes which exist only in solution and in the presence of excess olefin.[68b] Olefin addition is the rate-determining step in hydrogenation.

$$IrCl(CO)(PPh_3)_2 + H_2 \rightleftharpoons IrH_2Cl(CO)(PPh_3)_2$$

$$IrCl(CO)(PPh_3)_2 + C_2H_4 \rightleftharpoons Ir(C_2H_4)Cl(CO)(PPh_3)_2$$

It occurs to a small extent, giving a low concentration of nonhydrogenated

[63] L. Vaska and J. W. DiLuzio, J. Amer. Chem. Soc., 83, 2784 (1961).

[64] K. Vrieze, J. P. Collman, C. T. Sears, and M. Kubota, Inorg. Synth., 11, 101 (1968).

[65] W. Strohmeier and T. Onoda, Z. Naturforsch., B, 23, 1377 (1968).

[66] P. B. Chock and J. Halpern, J. Amer. Chem. Soc., 88, 3511 (1966).

[67] L. Vallarino, J. Chem. Soc., 1957, 2287.

[68] (a) J. Chatt and B. L. Shaw, Chem. Ind. (London), 1961, 290; (b) L. Vaska and R. E. Rhodes, J. Amer. Chem. Soc., 87, 4970 (1965).

complex in equilibrium with the dihydride. Dissociation of one phosphine

$$IrH_2Cl(CO)(PPh_3)_2 \overset{H_2}{\rightleftharpoons} IrCl(CO)(PPh_3)_2$$

$$\overset{olefin}{\rightleftharpoons} Ir(\pi\text{-olefin})Cl(CO)(PPh_3)_2$$

from the resulting π-olefin complex allows hydrogen activation, which is followed by stepwise hydrogen transfer to the substrate.[69–71]

$$Ir(\pi\text{-olefin})Cl(CO)(PPh_3)_2 \overset{-PPh_3}{\rightleftharpoons} Ir(\pi\text{-olefin})Cl(CO)(PPh_3)$$

$$\overset{H_2}{\rightleftharpoons} IrH_2(\pi\text{-olefin})Cl(CO)(PPh_3)$$

$$\rightleftharpoons IrH(\sigma\text{-alkyl})Cl(CO)(PPh_3)$$

$$\overset{PPh_3}{\longrightarrow} IrCl(CO)(PPh_3)_2 + alkane$$

Reversible stepwise transfer of hydrogen to substrate is indicated by extensive olefin isomerization observed over the catalyst, and by the dependence of this isomerization on the presence of hydrogen.[72–75]

Exchange of phosphines at the metal center has been attributed to dissociation of $IrCl(CO)(PPh_3)_2$ in solution at 20°, but an associative mechanism similar to olefin addition in the scheme above could explain this result.[76a] Increased rates observed in the presence of traces of oxygen presumably result from oxidation of triphenylphosphine to the phosphine oxide in a manner analogous to the similar activation of $RhCl(PPh_3)_3$.[69, 76b]

Scope and Limitations

Olefinic and activated acetylenic compounds are reduced to saturated compounds, rates of hydrogenation depending on steric effects similar to those observed for $RhCl(PPh_3)_3$ (terminal olefin > *cis* olefin > *trans* olefin, etc.). Rates for some representative olefins are given in Table VI.

[69] B. R. James and N. A. Memon, *Can. J. Chem.*, **46**, 217 (1968).

[70] W. Strohmeier and T. Onoda, *Z. Naturforsch., B*, **24**, 1493 (1969).

[71] M. G. Burnett, R. J. Morrison, and C. J. Strugnell, *J. Chem. Soc. (Dalton)*, **1973**, 701.

[72] W. Strohmeier and W. Rehder-Stirnweiss, *J. Organometal. Chem.*, **19**, 417 (1969).

[73] W. Strohmeier and R. Fleischmann, *J. Organometal. Chem.*, **42**, 163 (1972).

[74] W. Strohmeier, R. Fleischmann, and W. Rehder-Stirnweiss, *J. Organometal. Chem.*, **47**, C37 (1973).

[75] W. Strohmeier and W. Diehl, *Z. Naturforsch., B*, **28**, 207 (1973).

[76] (a) W. Strohmeier, W. Rehder-Stirnweiss, and G. Reischig, *J. Organometal. Chem.*, **27**, 393 (1971); (b) F. van Rantwijk, Th. G. Spec, and H. van Bekkum, *Rec. Trav. Chim., Pays-Bas*, **91**, 1057 (1972).

TABLE VI. RATES OF HYDROGENATION OVER
$IrCl(CO)(PPh_3)_2$[77]
(Catalyst 2×10^{-3} M, substrate 0.8 M in toluene; under
1 atm of hydrogen at 80°)

Substrate	Rate (mol l^{-1} min^{-1} $\times$ 10^{-3})
1-Heptene	8.93
cis-2-Heptene	0.97
trans-2-Heptene	0.55
trans-3-Heptene	0.72
Cycloheptene	4.50
Ethyl acrylate	10.0
Dimethyl maleate	0.44
Dimethyl fumarate	0
Styrene	6.66
ω-Bromostyrene	0
1-Hexyne	0
Phenylacetylene	0.66

In addition to hydrogenation of olefins the catalyst equilibrates *cis*, *trans*, and positional isomers in the series. This process is more rapid than hydrogenation and is most apparent in hydrogenations of terminal olefins when internal (less readily hydrogenated) isomers accumulate[73, 74] (Table VII).

TABLE VII. HYDROGENATION OF HEPTENES OVER $IrCl(CO)(PPh_3)_2$[73]
(Catalyst 2×10^{-3} M, substrate 0.8 M in toluene; products after 4 hours
at 80° under 470 mm hydrogen pressure.)

Substrate	Products (mol %)					
	Heptane	1-Heptene	cis-2-Heptene	trans-2-Heptene	cis-3-Heptene	trans-3-Heptene
1-Heptene	53	0	10	37	0	0
cis-2-Heptene	41	25	12	18	0	4
trans-2-Heptene	48	2	12	34	0	4
cis-3-Heptene	10	0	6	20	6	58

[77] W. Strohmeier, W. Rehder-Stirnweiss, and R. Fleischmann, *Z. Naturforsch.*, *B*, **25**, 1481 (1970).

The reversible nature of hydrogen transfer from iridium to olefin which accounts for isomerization also leads to considerable isotope scrambling in deuterations conducted over the catalyst.[75] Deuterium addition to ethylene gives a mixture of d_0- to d_4-ethanes, and recovered ethylene contains d_0- to d_2-isomers.[78] Hydrogen-deuterium exchange is also catalyzed.

Early work notes disproportionation of 1,4-cyclohexadiene into cyclohexene and benzene in a nitrogen atmosphere over $IrClCO(PPh_3)_2$, but subsequent hydrogenation studies show that clean reduction of 1,4- and 1,3-cyclohexadiene to cyclohexene is possible.[79a, b] Isotope scrambling is observed in deuterations.

The catalyst is activated by weak ultraviolet irradiation.[79c] The increase in rate upon photolysis depends on substrate; factors for several compounds are: ethyl acrylate (40 ×), 1,3-cyclohexadiene (10 ×), 1,4-cyclohexadiene (2.5 ×), cyclohexene (no change). Selective hydrogenation of 1,3-cyclohexadiene to cyclohexene, not possible under thermal conditions, is achieved in the photolytically activated hydrogenation.[79d]

Variation of the Catalyst

Influences of variations in the metal, halide, and phosphine have all been studied extensively.[72, 74, 77, 80]

Changing the central metal atom from iridium to rhodium results in lower rates of hydrogenation coupled with more rapid olefin isomerization. Different phosphines give catalysts showing different rates of hydrogenation; however, these rates do not correlate with the π-acceptor abilities of the phosphines and depend to some extent on individual substrates. Steric bulk appears to be the overriding factor. Rates of hydrogenation over catalysts incorporating different halogens decrease through the series chlorine > bromine > iodine.

Related hydrido complexes $MH(CO)(PPh_3)_3$ (where M = rhodium or iridium), $IrHX_2L_3$ (where X = halogen, L = triphenyl-phosphine, -arsine, or -stibine), and $IrH_2(CO)(PPh_3)_2[Ge(CH_3)_3]$ are also active for hydrogenation of olefins. Synthetic applications have not been widely investigated,

[78] G. G. Eberhardt and L. Vaska, *J. Catal.*, **8**, 183 (1967).

[79] (a) J. E. Lyons, *Chem. Commun.*, **1969**, 154; (b) J. E. Lyons, *J. Catal.*, **30**, 490 (1973); (c) W. Strohmeier and G. Csontos, *J. Organometal. Chem.*, **72**, 277 (1974); (d) W. Strohmeier and L. Weigelt, *J. Organometal. Chem.*, **82**, 417 (1974).

[80] (a) W. Strohmeier and T. Onoda, *Z. Naturforsch.*, *B*,**24**, 461, 515, (1969); (b) W. Strohmeier and F. J. Müller, *ibid.*, **24**, 931 (1969); (c) W. Strohmeier and R. Fleischmann, *ibid.*, **24**, 1217 (1969); (d) W. Strohmeier and W. Rehder-Stirnweiss, *ibid.*, **24**, 1219 (1969); (e) W. Strohmeier and W. Rehder-Stirnweiss, *J. Organometal. Chem.*, **18**, P28 (1969); (f) W. Strohmeier and W. Rehder-Stirnweiss, *Z. Naturforsch.*, *B*, **26**, 61 (1971); (g) W. Strohmeier, *J. Organometal. Chem.*, **32**, 137 (1971); (h) W. Strohmeier, R. Fleischmann, and T. Onoda, *ibid.*, **28**, 281 (1971); (i) W. Strohmeier and R. Fleischmann, *ibid.*, **29**, C39 (1971).

but reaction at lower temperatures than those necessary with IrCl(CO)-$(PPh_3)_2$ may be useful with thermally unstable substrates.[81–86a, b] $RhH(CO)(PPh_3)_3$ is activated by ultraviolet irradiation in the hydrogenation of ethyl acrylate.[86c]

TABLE VIII. RATES OF HYDROGENATION
OVER $MH(CO)(PPh_3)_3$[83a]
(Catalyst $2 \times 10^{-3} M$, substrate $0.8 M$ in toluene; under
1 atm of hydrogen at 25°)

| | Rate (mol l^{-1} min$^{-1} \times 10^{-3}$) | |
Substrate	M = Ir	M = Rh
1-Heptene	0.27	2.85
cis-2-Heptene	0.13	0.18
trans-2-Heptene	0.13	0.14
trans-3-Heptene	0.08	0.13
Cycloheptene	0.08	0.12
1,3-Cyclooctadiene	0.09	0.11
Ethyl acrylate	0.27	7.42
Dimethyl maleate	0.23	0.98
Dimethyl fumarate	0.05	0.11
Styrene	0.35	2.23
ω-Bromostyrene	0.01	0.03
1-Hexyne	0.13	1.00
Phenylacetylene	0.15	1.12

Experimental Conditions

Solvents for hydrogenation include toluene, dimethylformamide, and dimethylacetamide. Rates do not vary significantly between the different solvents. Catalyst concentration is generally about $10^{-3} M$, substrate concentration up to 1 M.

Hydrogenations over Vaska's compound proceed only slowly below 40°; temperatures of 70–80° give more satisfactory rates. At these temperatures a pressure of one atmosphere of hydrogen is sufficient. The effect of higher pressures has not been investigated.

[81] M. Yamaguchi, *Kogyo Kagaku Zasshi*, **70**, 675 (1967) [*C.A.*, **67**, 99542w (1967)].

[82] L. Vaska, *Inorg. Nucl. Chem. Lett.*, **1**, 89 (1967).

[83] (a) W. Strohmeier and S. Hohmann, *Z. Naturforsch.*, *B*, **25**, 1309 (1970); (b) W. Strohmeier and W. Rehder-Stirnweiss, *ibid.*, **26**, 193 (1971).

[84] F. Glocking and M. D. Wilbey, *J. Chem. Soc.*, *A*, **1970**, 1675.

[85] M. G. Burnett and R. J. Morrison, *J. Chem. Soc.*, *A*, **1971**, 2325.

[86] (a) M. G. Burnett and R. J. Morrison, *J. Chem. Soc.* (*Dalton*), **1973**, 632; (b) M. G. Burnett, R. J. Morrison, and C. J. Strugnell, *ibid.*, **1974**, 1663; (c) W. Strohmeier and G. Csontos, *J. Organometal. Chem.*, **67**, C27 (1974).

Further Complexes of Rhodium and Iridium

The Henbest Catalyst: Chloroiridic Acid-Trimethyl Phosphite

This catalyst has wide application for the stereospecific reduction of cyclic ketones to axial alcohols, in contrast with high proportions of equatorial alcohols produced by the majority of reduction procedures. It is included here because of its practical importance, in spite of not utilizing molecular hydrogen.

The catalytic system consists of chloroiridic acid or iridium tetrachloride and trimethyl phosphite, with the ketone to be reduced, in aqueous isopropyl alcohol. Reduction proceeds on refluxing the solution; exclusion of air is not necessary. Sodium chloroiridate or iridium trichloride can be used as an alternative source of iridium, and $RhCl(PPh_3)_3$ (in the presence of trimethyl phosphite) also gives highly stereospecific reductions. The complexes $IrH_3(PPh_3)_2$, $IrH_3(PPh_3)_3$, and $CoH_3(PPh_3)_3$ catalyze reduction in the absence of trimethyl phosphite but give high proportions of equatorial alcohols.[87a, b] However, a hydridoiridium complex isolated from the reaction of $[IrCl(C_8H_{14})_2]_2$ with dimethyl phosphite catalyzes reduction of 4-t-butylcyclohexanone with specificity equal to that of the chloroiridic acid/trimethyl phosphite system.[87c] The corresponding rhodium complex gives similar results.

Phosphorous acid or dimethyl sulfoxide can replace trimethyl phosphite; in the latter case, production of equatorial alcohols is again enhanced.[88, 89a]

One mole of phosphite per mole of substrate is necessary for reduction. In some cases it appears that the phosphite acts as reductant, little acetone being produced from isopropyl alcohol during the reaction.[87a] Proportions of reactants generally employed are 0.1 mol of iridium and 2 mol of phosphite per mol of substrate in 10% aqueous alcohol. The presence of at least 5% of water in the solvent prevents reductive etherification of the ketone: use of anhydrous alcohols as solvents leads to formation of ethers from the product alcohol.[88]

Results obtained using different co-catalysts (phosphites, dimethyl sulfoxide, amines) have been reviewed recently.[89b]

Practical use of the catalyst has concentrated on reduction of cyclohexanones, including steroidal derivatives. Simple cyclohexanones are

[87] (a) H. B. Henbest and T. R. B. Mitchell, *J. Chem. Soc., C*, **1970**, 785; (b) E. Malunowicz, S. Tyrlik, and Z. Lasocki, *J. Organometal. Chem.*, **72**, 269 (1974); (c) M. A. Bennett and T. R. B. Mitchell, *ibid.*, **70**, C30 (1974).

[88] Y. M. Y. Haddad, H. B. Henbest, J. Husbands, and T. R. B. Mitchell, *Proc. Chem. Soc.*, **1964**, 361.

[89] (a) M. Gulotti, R. Ugo, and S. Colonna, *J. Chem. Soc., C*, **1971**, 2652; (b) Y. M. Y. Haddad, H. B. Henbest, J. Husbands, T. R. B. Mitchell, and J. Trocha-Grimshaw, *J. Chem. Soc. (Perkin I)*, **1974**, 596.

$$\text{(R = ethyl, isopropyl, cyclopentyl)}$$

reduced to the corresponding alcohols in good yield; the proportions of axial alcohols produced are generally better than 95 %.[87a] Very sterically

hindered ketones are not readily reduced: 2,2-dimethylcyclohexanone gives only 60 % of alcohol whereas 2,2,6-trimethylcyclohexanone does not react.

The effect of steric hindrance is also apparent in steroidal ketones, where 2- and 3-keto groups are easily reduced, 17-keto groups are reduced less easily, and 4-, 6-, 11-, 17-, and 20-keto groups are unaffected. (See examples on p. 37.) High yields of the axial alcohols are again obtained, the only side reaction noted being some epimerization of the 17β-side chain in pregnane-20-one derivatives.[90, 91]

The catalyst has also been applied to alkaloid synthesis.[92a] (See p. 37.)

Other carbonyl groups, including aldehydes, and some activated carbon-carbon multiple bonds are reduced.[92b] Rates of reaction are, however, often low and the catalyst does not have obvious advantages in this direction. Four examples are shown at the top of p. 38.[87a]

90 P. A. Browne and D. N. Kirk, *J. Chem. Soc.*, *C*, **1969**, 1653.

91 J. C. Orr, M. Mersereau, and A. Sanford, *Chem. Commun.*, **1970**, 162.

92 (a) M. Hanaoka, N. Ogawa, and Y. Arata, *Tetrahedron Lett.*, **1973**, 2355; (b) H. B. Henbest and J. Trocha-Grimshaw, *J. Chem. Soc.* (*Perkin I*), **1974**, 601.

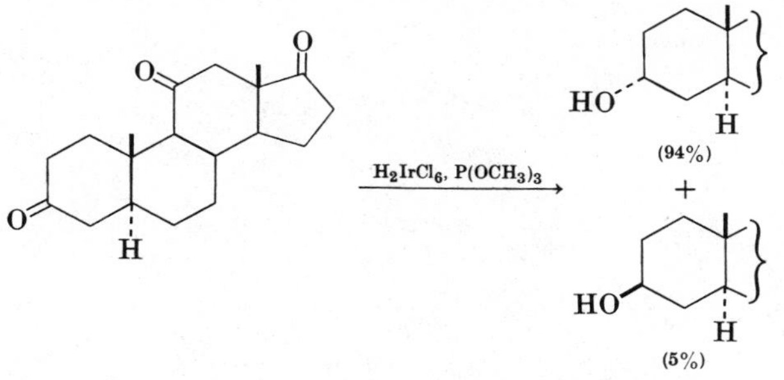
O
O
O
H
H_2IrCl_6, $P(OCH_3)_3$
HO
H
(94%)
+
HO
H
(5%)

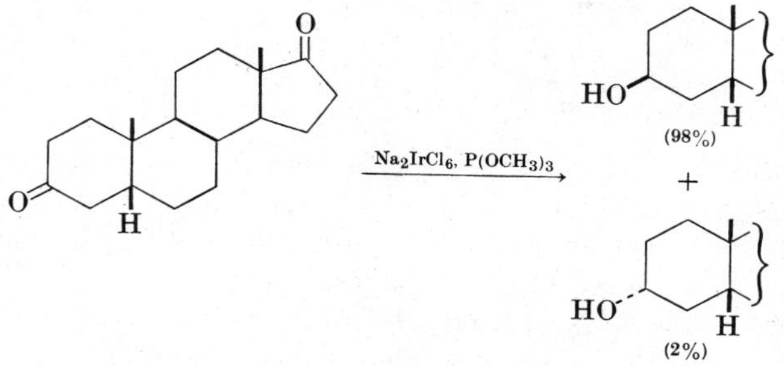
O
O
H
Na_2IrCl_6, $P(OCH_3)_3$
HO
H
(98%)
+
HO
H
(2%)

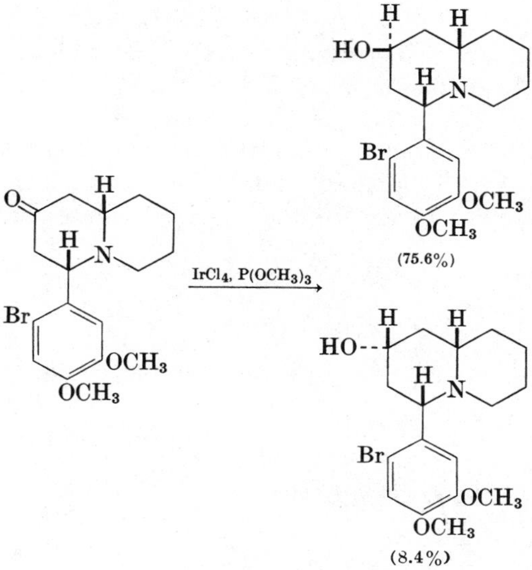
O
H
H
N
Br
OCH_3
OCH_3
$IrCl_4$, $P(OCH_3)_3$
H
H
HO
H
N
Br
OCH_3
OCH_3
(75.6%)
H
H
HO
H
N
Br
OCH_3
OCH_3
(8.4%)

$$CH_3CO(CH_2)_{16}CH_3 \xrightarrow{\text{IrCl}_4,\ \text{P(OH)}_3} CH_3CHOH(CH_2)_{16}CH_3$$
$$(98\%)$$

$$C_6H_5CHO \xrightarrow{\text{IrCl}_4,\ \text{HP(O)(OCH}_3)_2} C_6H_5CH_2OH$$
$$(95\%)$$

$$C_6H_5CO(CH{=}CH)_2C_6H_5 \xrightarrow{\text{H}_2\text{IrCl}_6,\ (\text{CH}_3)_2\text{SO}} C_6H_5CO(CH_2)_4C_6H_5$$
$$(75\%)$$

(30%)

Hydrogenation of Aldehydes and Ketones

Several rhodium and iridium complexes are effective for the homogeneous hydrogenation of aldehydes and ketones to alcohols.

Trihydridotris(triphenylphosphine)iridium, $IrH_3(PPh_3)_3$, in the presence of acetic acid (which reacts to form hydride acetate complexes) catalyzes hydrogenation of n-butyraldehyde to n-butyl alcohol under mild conditions (50°, 1 atm of hydrogen).[93] Under identical conditions only activated olefins (acrylic acid, methyl acrylate) are hydrogenated; octenes do not react. The catalyst thus should give selectivity in the hydrogenation of nonconjugated unsaturated aldehydes to the corresponding unsaturated alcohols.

Cationic rhodium and iridium complexes with a general formula $[MH_2(PPh_3)_2L_2]^+$ (M = rhodium or iridium, L = complexed solvent) as hexafluorophosphate or perchlorate catalyze hydrogenation of olefins and acetylenes.[94] Furthermore $[IrH_2(PPh_3)_2(\text{acetone})_2]^+PF_6^-$ in dioxane solution catalyzes the reduction of butyraldehyde. Substitution of basic tertiary phosphines for triphenylphosphine in either the rhodium or iridium complexes gives catalysts which allow hydrogenation of ketones under mild conditions (25°, 1 atm).[95] Other ketones reduced include acetone,

trans (86%)   cis (14%)

[93] R. S. Coffey, *Chem. Commun.*, **1967**, 923.

[94] J. R. Shapley, R. R. Schrock, and J. A. Osborn, *J. Amer. Chem. Soc.*, **91**, 2816 (1969).

[95] R. R. Schrock and J. A. Osborn, *Chem. Commun.*, **1970**, 567.

2-butanone, cyclohexanone, and acetophenone. The presence of a trace of water (1 % by volume) greatly increases the rate of reduction of ketones but inhibits the hydrogenation of olefins. Some selectivity might thus be achieved using this system. Reduction of acetone in the presence of deuterium produces isopropyl alcohol labeled specifically at the α-carbon atom; no β-deuterium incorporation is detected even in the presence of 1 % water.

Reduction of aldehydes to alcohols is also possible using $RhCl_3(PPh_3)_3$.[96] However, rather harsh conditions are required (110°, 50 atm).

Selective Hydrogenation of Acetylenes and Olefins

Cationic rhodium complexes $[Rh(diene)P_x]^+PF_6^-$ (where diene = norbornadiene or 1,5-cyclooctadiene, P = tertiary phosphine, x = 2 or 3) are briefly reported to allow reduction of internal alkynes to the corresponding cis olefins in virtually quantitative yield and with specificity greater than 95 %.[97] Details of the procedure were not given.

A potentially useful catalyst for selective hydrogenation of terminal olefins is $RhH(CO)(PPh_3)_3$. Under mild conditions, 1-hexene gives n-hexane while cyclohexene and cis-4-methyl-2-pentene are not reduced.[98, 99] Hydrogenation and/or hydrogenolysis of aldehyde, hydroxy, nitrile, chloro, carboxylic acid, and ether groups does not occur. In addition there is no isomerization to internal olefins. Formation of stable complexes between the catalyst and coordinating dienes (e.g., 1,3-pentadiene) and acetylenes (1-hexyne) prevents hydrogenation. Sterically hindered terminal olefins (i.e., 2,2-disubstituted) are also unreactive. Table IX summarizes results obtained using the catalyst. (See also Table VIII, p. 34). The catalyst is conveniently prepared directly from rhodium trichloride trihydrate, isolation of the intermediate RhCl(CO)-$(PPh_3)_3$ not being necessary.[100a]

The trifluorophosphine analog of $RhH(CO)(PPh_3)_3$, $RhH(PF_3)(PPh_3)_3$. is also active for hydrogenation of terminal olefins, but isomerization to internal olefins is rapid.[100b]

[96] J. A. Osborn, G. Wilkinson, and J. F. Young, Chem. Commun., **1965**, 17.

[97] R. R. Schrock and J. A. Osborn, J. Amer. Chem. Soc., **93**, 3089 (1971).

[98] C. O'Connor, G. Yagupsky, D. Evans, and G. Wilkinson, Chem. Commun., **1968**, 420.

[99] C. O'Connor and G. Wilkinson, J. Chem. Soc., A, **1968**, 2665.

[100] (a) D. Evans, G. Yagupsky, and G. Wilkinson, J. Chem. Soc., A, **1968**, 2660; (b) J. F. Nixon and J. R. Swain, J. Organometal. Chem., **72**, C15 (1974); (c) D. E. Budd, D. G. Holah, A. N. Hughes, and B. C. Hui, Canad. J. Chem., **52**, 775 (1974); (d) D. G. Holah, I. M. Hoodless, A. N. Hughes, B. C. Hui, and D. Martin, ibid. **52**, 3758 (1974); (e) T. E. Paxson and M. F. Hawthorne, J. Amer. Chem. Soc., **96**, 4674 (1974); (f) G. F. Pregaglia, G. F. Ferrari, A. Andreetta, G. Capparella, F. Genoni, and R. Ugo, J. Organometal. Chem., **70**, 89 (1974); (g) H. Imai, T. Nishiguchi, and K. Fukuzumi, J. Org. Chem., **39**, 1622 (1974); (h) M. Gargano, P. Giannoccaro, and M. Rossi, J. Organometal. Chem., **84**, 389 (1975).

TABLE IX. Hydrogenations over
RhH(CO)(PPh$_3$)$_3$[99]
(Catalyst 1.25 × 10^{-3} M, substrate 0.6 M in benzene;
under 0.7 atm of hydrogen at 25°)

Hydrogenated	Not Hydrogenated
Ethylene	cis-2-Pentene
1-Hexene	cis-2-Heptene
1,5-Hexadiene	cis-4-Methyl-2-pentene
1-Decene	2-Methyl-1-pentene
1-Undecene	1,3-Pentadiene
Allyl alcohol	1-Hexyne
Allylbenzene	Cyclohexene
Allyl cyanide	Limonene
4-Vinylcyclohexene[a]	1-Chloro-1-propene
Styrene	2-Chloro-1-propene
	Allyl phenyl ether
	Acrylic acid
	Cinnamaldehyde

[a] The product is 4-ethylcyclohexene.

Selective hydrogenation of terminal olefins is also catalyzed by hydrido rhodium(I) derivatives of the cyclic phosphine 5-phenyl-5H-dibenzo-phosphole (DBP). Air-stable RhH(DBP)$_4$ allows rapid hydrogenation of terminal olefins (in benzene solution, 20°, 0.1 atm of hydrogen).[100c] Internal monoenes, conjugated dienes, and acetylenes are reduced far more slowly while cyclic monoenes, dienes, trans olefins, and cyano, nitro, and keto groups are not affected. Relative rates of reduction of 1-hexene over this and other rhodium catalysts are (relative rates in parenthesis): RhH(CO)(PPh$_3$)$_3$, (1); RhH(PPh$_3$)$_4$, (2.8); RhCl(PPh$_3$)$_3$, (3.5); RhH(DBP)$_4$, (24.4).

Dissociation of RhH(DBP)$_4$ in solution gives RhH(DBP)$_3$, which can be isolated and used as catalyst. Rates of hydrogenation over the tris(phosphole) complex are 10% greater than over RhH(DBP)$_4$.[100d]

Two isomeric hydridorhodium carborane complexes of formula RhH(C$_2$B$_9$H$_{12}$)(PPh$_3$)$_2$ catalyze hydrogenation of terminal olefins but lack the specificity shown by the above phosphole derivatives.[100e] Reduction of 1-hexene is accompanied by isomerization; the internal olefins produced are also hydrogenated.

Conjugated dienes are selectively hydrogenated to terminal monoolefins over RhH(PPh$_3$)$_4$ or [Rh(CO)$_2$(PPh$_3$)]$_2$·2 C$_6$H$_6$ in the presence of one equivalent of triethylphosphine (which gives increased solubility to the complexes).[100f] Hydrogenations of 1,3-butadiene and 1,3-pentadiene give high proportions of 1-butene and 1-pentene (60% and 80%, respectively) when the reactions are quenched at the appropriate stage. Isoprene does not give such clear cut results while 2,4-hexadiene does not react.

The high selectivity of the catalysts is ascribed to strong interaction of 1,3-diene with the metal, preventing coordination and hydrogenation of monoolefins until the diolefin concentration becomes very low. Reactions are carried out in cyclohexane at 50° to 100° under 15 atm of hydrogen.

Some cyclic diolefins are reduced to monoolefins over $[Rh(CO)_2(PPh_3)]_2 \cdot 2 C_6H_6$.[100f] 1,3-Cyclooctadiene gives cyclooctene as 99% of the product. $RhH(PPh_3)_4$ may not show the same specificity in this regard; it can be used to saturate cyclohexene.[100g]

Specific hydrogenation of 1,5-cyclooctadiene to cyclooctene using $[Ir(1,5\text{-cyclooctadiene})_2]^+PF_6^-$ at 30° and 1 atm of hydrogen is also worthy of mention.[97] Analogous conversions might perhaps be possible. Similarly 1,5-and 1,4-cyclooctadienes are hydrogenated to cyclooctene over $Ir_2H_2Cl_2(1,5\text{-cyclooctadiene})(PPh_3)_2$.[100h] 1,3-Cyclooctadiene does not react.

Chloroplatinic Acid—Tin(II) Chloride — and Related Systems

Preparation

Catalyst solutions are prepared by addition of chloroplatinic acid and tin(II) chloride dihydrate to the solvent, which usually contains methyl alcohol. Anhydrous tin(II) chloride gives identical results. For purposes of hydrogenation it is not necessary to isolate the resulting complex, but the anions $[Pt(SnCl_3)_5]^{3-}$ and $[PtCl_2(SnCl_3)_2]^{2-}$ can be isolated as phosphonium salts.[101, 102] Hydrogenated solutions yield salts of $[PtH(SnCl_3)_4]^{3-}$.

Related catalysts are prepared in a similar manner by addition of complexes $MX_2(QR_3)_2$ (where M = platinum or palladium, Q = phosphorus, arsenic, or antimony, X = halogen or similar, e.g., cyano, and R = alkyl or aryl) to a compound $M'X_2$ or $M'X_4$ (where M' = silicon, germanium, tin, or lead), most frequently tin(II) chloride, in a suitable solvent.[103–105] A diolefin, e.g., 1,5-cyclooctadiene, can replace the $(QR_3)_2$ in the platinum or palladium complex. Again, isolation of complexes is not necessary although $PtCl(SnCl_3)(PPh_3)_2$ is obtained from solutions of $PtCl_2(PPh_3)_2$ and tin(II) chloride.[101] The hydrides $PtH(SnCl_3)(PPh_3)_2$ and $[PtH(SnCl_3)_2(PPh_3)_2]^-$ are formed in hydrogenated solutions; the former has been

101 R. D. Cramer, E. L. Jenner, R. V. Lindsey, and U. G. Stolberg, J. Amer. Chem. Soc. 85, 1691 (1963).

102 R. V. Lindsey, G. W. Parshall, and U. G. Stolberg, J. Amer. Chem. Soc., 87, 658 (1965).

103 H. Itatani and J. C. Bailar, J. Amer. Oil Chem. Soc., 44, 147 (1967).

104 J. C. Bailar and H. Itatani, J. Amer. Chem. Soc., 89, 1592 (1967).

105 H. A. Tayim and J. C. Bailar, J. Amer. Chem. Soc., 89, 4330 (1967).

shown to be catalytically active.[102, 104-107] The complexes $MX_2(QR_3)_2$ are prepared by standard methods.[103, 104]

Platinum:tin ratios between 5 and 10 give highest rates of hydrogenation. Oxygen is excluded in all cases.

Mechanism

Platinum(II) complexes with general formula $[PtCl_x(SnCl_3)_{4-x}]^{2-}$ ($x = 0, 1, 2$) have been proposed as active intermediates derived from chloroplatinic acid.[101, 108, 109] Related catalysts give analogous species (Eq. 4).

$$PtCl_2(PPh_3)_2 + SnCl_2 \rightleftharpoons PtCl(SnCl_3)(PPh_3)_2 \qquad \text{(Eq. 4)}$$

Heterolytic activation of hydrogen follows.

$$PtCl(SnCl_3)(PPh_3)_2 + H_2 \rightleftharpoons PtH(SnCl_3)(PPh_3)_2 + H^+ + Cl^-$$

Activation of the catalysts by tin(II) chloride through formation of Pt-SnCl₃ linkages is attributed to the high π-acceptor ability of the trichlorotin ligand which decreases electron density at platinum with a twofold effect: it labilizes the hydridic hydrogen in complexes such as $PtH(SnCl_3)(PPh_3)_2$ and enhances coordination of olefin to the metal atom.[105, 110] The cyano group shows similar properties in $Pd(CN)_2(PPh_3)_2$ which is active without added tin(II) chloride.[103-105] Optimum rates are obtained using an apparent excess of tin(II) chloride.[104, 105, 109-114] The excess forces the equilibrium in Eq. 5 well to the right.

$$PtClL_x + SnCl_2 \rightleftharpoons Pt(SnCl_3)L_x \qquad \text{(Eq. 5)}$$

$$(L_x = \text{generalized ligands})$$

Hydrogenations proceed via stepwise hydrogen transfers from platinum to coordinated olefin; extensive isomerization of olefins is noted. With simple monoolefins reversible addition of the hydridoplatinum complex leads to isomerization, a second (slow) hydrogen activation gives hydrogenated product.

The isomerization of more bulky nonconjugated polyolefins to conjugated compounds and termination of hydrogenation at the monoene stage indicate that the presence or formation of a 1,3-diene unit is necessary for

106 J. C. Bailar and H. Itatani, *Inorg. Chem.*, **4**, 1618 (1965).

107 J. C. Bailar and H. Itatani, *J. Amer. Oil Chem. Soc.*, **43**, 337 (1966).

108 G. C. Bond and M. Hellier, *Chem. Ind.* (London) **1965**, 35.

109 G. C. Bond and M. Hellier, *J. Catal.*, **7**, 217 (1967).

110 H. A. Tayim and J. C. Bailar, *J. Amer. Chem. Soc.*, **89**, 3420 (1967).

111 H. van Bekkum, J. van Gogh, and G. van Minnen-Pathuis, *J. Catal.*, **7**, 292 (1967).

112 A. P. Khrushch, L. A. Tokina, and A. E. Shilov, *Kinet. Katal.*, **7**, 901 (1966) [*C.A.*, **66**, 37090t (1967)].

113 L. P. van't Hof and B. G. Linsen, *J. Catal.*, **7**, 295 (1967).

114 I. Yasumori and K. Hirabayshi, *Trans. Faraday Soc.*, **67**, 3283 (1971).

hydrogenation in these systems. The intermediate π-allylic complex subsequently generated is more stable than the σ-alkyl formed from an isolated double bond. Hydrogen transfer from platinum to 1,3-diene forms the π-allylic complex which retains the substrate at the metal center until a second hydrogen activation is completed to generate monoene. Addition of the latter hydrogen at either end of the π-allyl system gives the possibility of two products, as shown in Chart 2. Several

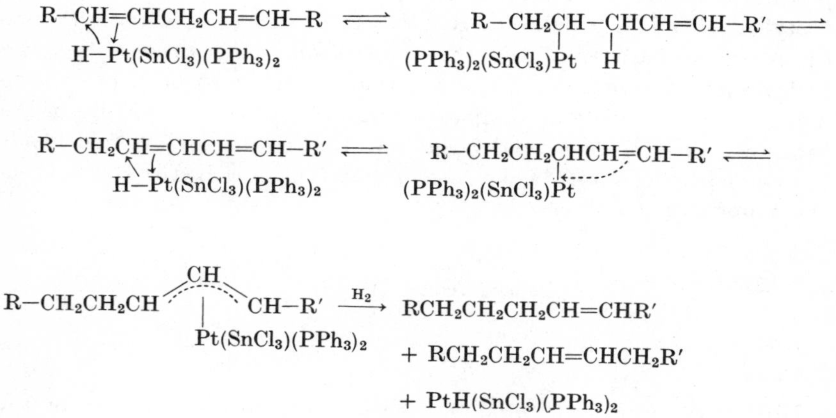

R—CH=CHCH₂CH=CH—R $\rightleftharpoons$ R—CH₂CH—CHCH=CH—R′ $\rightleftharpoons$
H—Pt(SnCl₃)(PPh₃)₂   (PPh₃)₂(SnCl₃)Pt H

R—CH₂CH=CHCH=CH—R′ $\rightleftharpoons$ R—CH₂CH₂CHCH=CH—R′ $\rightleftharpoons$
H—Pt(SnCl₃)(PPh₃)₂   (PPh₃)₂(SnCl₃)Pt

R—CH₂CH₂CH⟨CH⟩CH—R′ $\xrightarrow{H_2}$ RCH₂CH₂CH₂CH=CHR′
Pt(SnCl₃)(PPh₃)₂

+ RCH₂CH₂CH=CHCH₂R′

+ PtH(SnCl₃)(PPh₃)₂

CHART 2

hydridoplatinum-olefin intermediates, *e.g.*, PtH(SnCl₃)(octene))(PPh₃)₂, have been isolated from reaction mixtures.[105]

Hydrogen atoms added to the substrate do not come exclusively from hydrogen gas. Catalysis of hydrogen exchange between hydrogen gas and deuterated methanol has been noted, and considerable deuterium incorporation into products is observed using deuterated methanol as solvent.[112, 114, 115] Solvolysis of the platinum-substrate bond by alcohol probably also occurs; indeed, slow hydrogenation is observed in the absence of hydrogen in a few cases, the solvent acting as hydrogen donor.[103, 104, 116]

Scope and Limitations

Early work reports the reduction of ethylene platinous chloride, [PtCl(C₂H₄)₂]₂, by hydrogen to ethane and platinum metal.[3b] It is not clear whether metallic platinum is responsible for ethylene hydrogenation, although kinetic evidence points to genuine homogeneous hydrogen transfer.

[115] R. Cramer and R. V. Lindsey, *J. Amer. Chem. Soc.*, **88,** 3534 (1966).
[116] J. C. Bailar, H. Itatani, M. J. Crespi, and J. Geldard, *Adv. Chem. Ser.*, **62,** 103 (1966).

The catalyst derived from chloroplatinic acid and tin(II) chloride allows slow hydrogenation under mild conditions (1 atm, 20°) of ethylene, acetylene, 1-hexene, and cyclohexene.[101, 107, 111, 113] Variations in the structure of the catalyst precursor on activity, selectivity, and stereochemistry of products have been investigated, but all give qualitatively similar results in a practical sense. Highest rates are found for catalysts derived from $PtCl_2(AsPh_3)_2$, while those utilizing palladium in place of platinum require higher hydrogen pressures.[103, 105] Only olefinic and acetylenic bonds are reduced. Relative rates of the different reactions catalyzed decrease rapidly in the order olefin isomerization $\gg$ polyene hydrogenation $\gg$ monoene hydrogenation. Hydrogenation of polyene to monoene is thus easily accomplished and constitutes a major use of the catalysts. Both *cis-trans* and positional isomerization of olefinic bonds precede hydrogenation: in the case of monoolefins isomerization is often the major reaction observed.

(94%)   (6%)

Most work has concentrated on hydrogenation of polyolefins using the systems chloroplatinic acid-tin(II) chloride and $PtCl_2(PPh_3)_2$ or $PdCl_2(PPh_3)_2$-tin(II) chloride, with mixtures of isomeric monoenes generally resulting. The first catalyst shows higher activity toward hydrogenation of monoenes than the other two.[107] The ease of hydrogenation of polyenes decreases in the order open-chain > cyclic nonaromatic > cyclic semiaromatic $\gg$ aromatic. Conjugated and nonconjugated isomers are hydrogenated at equal rates because rapid conjugation precedes the slower hydrogenation. The strong driving force toward conjugation of bonds before hydrogenation is indicated by the product obtained from norbornadiene, rearrangement of the carbon skeleton occurring to allow conjugation.

The results of five hydrogenations are shown.[105, 111, 117] In many cases a detailed analysis of the isomeric products has not been carried out.

117 H. van Bekkum, F. van Rantwijk, G. van Minnen-Pathuis, J. D. Remijnse, and A. van Veen, *Rec. Trav. Chim. Pays-Bus*, **88**, 911 (1969).

$$\xrightarrow{\text{PtCl}_2(\text{PPh}_3)_2, \text{SnCl}_2, \text{H}_2} \quad + \text{ another isomer}$$

$$\xrightarrow[\text{(R = alkyl)}]{\text{H}_2\text{PtCl}_6, \text{SnCl}_2, \text{H}_2}$$

$$\xrightarrow{\text{H}_2\text{PtCl}_6, \text{SnCl}_2, \text{H}_2}$$

(trans : cis, 75 : 25)

$$\text{CH}_2\!\!=\!\!\text{CH(CH}_2)_4\text{CH}\!\!=\!\!\text{CH}_2 \xrightarrow{\text{PtCl}_2(\text{PPh}_3)_2, \text{SnCl}_2, \text{H}_2} \text{ mixture of octenes}$$

$$\text{CH}_2\!\!=\!\!\text{CHCH}_2\text{CH}\!\!=\!\!\text{CH(CH}_2)_3\text{CH}\!\!=\!\!\text{CH}_2 \xrightarrow{\text{PtCl}_2(\text{PPh}_3)_2, \text{SnCl}_2, \text{H}_2}$$
$$\text{mixture of decenes}$$

The hydrogenation of double bonds conjugated to (or able to become conjugated to) other unsaturated functional groups (*e.g.*, in α,β-unsaturated ketones) is noteworthy.[105] Similar application of the catalyst to reductions of α,β-unsaturated aldehydes would be of interest.

$$\text{CH}_3\text{COCH}\!\!=\!\!\text{C(CH}_3)_2 \xrightarrow{\text{PtCl}_2(\text{PPh}_3)_2, \text{SnCl}_2, \text{H}_2} \text{CH}_3\text{COCH}_2\text{CH(CH}_3)_2$$

$$\xrightarrow{\text{PtCl}_2(\text{PPh}_3)_2, \text{SnCl}_2, \text{H}_2}$$

$$\xrightarrow{\text{PtCl}_2(\text{PPh}_3)_2, \text{SnCl}_2, \text{H}_2}$$

Soybean oil and related fatty acid esters (chiefly linoleates and linol-enates) are converted to isomeric *trans* monoenes, with some dienes from methyl linolenate (Refs. 103, 104, 107, 113, 116, 118). The dienes contain widely separated double bonds—an indication of more facile conjugation of 1,4-dienes than of 1,5- or 1,6-isomers. Ester exchange is observed to a considerable extent between solvent alcohols and substrate esters, especially at temperatures above 60°.

The most obvious application of this class of catalyst is reduction to monoenes, rather than to the fully saturated systems. In view of the number of possible products inherent in the mechanism, the catalysts do not offer great promise for the synthesis of pure compounds except from some unsymmetrically substituted conjugated systems (including those containing heteroatoms).

Catalysts resembling those discussed above are obtained by dissolving platinum and palladium chlorides in dimethylformamide or dimethylacetamide. Hydrogenation of dicyclopentadiene, quinone, and 1,2- and 1,4-naphthaquinone are reported at 25–80° and 1 atm pressure, even in the presence of thiophene.[119] The products were not identified.

Experimental Conditions

Solvents most frequently employed are methanol and benzene-methanol (3:2 by volume). Addition of a small percentage of water to pure methanol does not affect results (chloroplatinic acid is conveniently added in aqueous solution). Higher alcohols can be used either alone or in mixtures with benzene. Those mentioned include ethyl, n-propyl, isopropyl, n-butyl, t-butyl, and n-pentyl alcohols. Toluene can be substituted for benzene.

Other classes of solvent are suitable but those with strong coordinating ability inhibit hydrogenation: decreasing series of rates shown by solvents are in the order dichloromethane $\sim$ chloroethane $>$ acetone $>$ tetrahydrofuran $\sim$ methyl alcohol $\gg$ pyridine.[105] Acetic acid and homologous carboxylic acids, 2-chloroethyl alcohol, diethyl and dipropyl ethers, nitrobenzene, 2-butanone, and 3-heptanone have also been employed.[111, 113] The use of ketones is complicated by promotion of aldol reactions by the catalysts.[105] Of the alternatives, dichloromethane seems most generally applicable, being a good solvent for catalysts, organic substrates, and reaction intermediates.

Catalyst concentration is generally 10^{-3} to 10^{-2} M with respect to platinum. Limits of concentrations reported are 10^{-8} and 0.5 M.

Temperatures range from 10 to 110°; catalyst decomposition which commences about 90° sets an upper limit. Hydrogen pressure varies from 1 to 70 atm. Though most reactions proceed at 1 atm, pressures around 40 atmospheres give higher rates.

[118] E. N. Frankel, E. A. Emken, H. Itatani, and J. C. Bailar, J. Org. Chem., **32**, 1447 (1967).

[119] P. N. Rylander, N. Himelstein, D. R. Steele, and J. Kreidl, Chem. Abstr., **57**, 15864e (1962).

Potassium Pentacyanocobaltate(II)—K₃[Co(CN)₅]

Preparation

Catalyst solution is prepared by mixing solutions of cobalt(II) chloride and potassium cyanide under nitrogen or hydrogen. The proportions of cobalt and cyanide used can have dramatic effects on the outcome of hydrogenations.

Solutions may be aqueous or nonaqueous. In the latter case the two salts are conveniently added in methanol to other solvents.

The solution thus prepared is activated before addition of substrate by subjecting it to 1 atm of hydrogen until absorption ceases, usually after 1 or 2 hours at 20°. This process generates the catalytic intermediate $[CoH(CN)_5]^{3-}$.

Potassium hydroxide or other bases are sometimes added to the system, the addition affecting the outcome of hydrogenations in a manner similar to addition of excess cyanide ion (*i.e.*, cyanide:cobalt ratios above 5).

Mechanism

Homolytic cleavage of hydrogen by the $[Co(CN)_5]^{3-}$ ion produces the catalytically active hydridocobalt complex. Some evidence suggests that

$$2 [Co(CN)_5]^{3-} + H_2 \rightleftharpoons 2 [CoH(CN)_5]^{3-}$$

it is a dimeric complex $[(CN)_5Co-H-Co(CN)_5]^{7-}$.[120]

Reaction with conjugated olefins follows to produce isolable enyl complexes (*e.g.*, with 1,3-butadiene).[121] Activated olefins (*e.g.*, α,β-unsaturated

$$[CoH(CN)_5]^{3-} + C_4H_6 \rightarrow [Co(C_4H_7)(CN)_5]^{3-}$$

acids) possibly react similarly.

Nuclear magnetic resonance studies of the butenyl and related allyl complexes mentioned above indicate the presence of σ- and π-bonded species depending on the cyanide ion concentration. This is shown for the allyl complex. In the butenyl case, analogous σ-π interconversions occur, mainly between the σ-2-butenyl and π-*syn*-butenyl complexes. They give

$$[CH_2{=}CH{-}CH_2{-}Co(CN)_5]^{3-} \underset{+CN^-}{\overset{-CN^-}{\rightleftharpoons}} \left[\begin{array}{c} H \\ C \\ H_2C^{\diagdown} \diagup CH_2 \\ | \\ Co(CN)_4 \end{array} \right]^{2-}$$

[120] N. Maki and Y. Ishiuchi, *Bull. Chem. Soc. Jap.*, **44**, 1721 (1971).

[121] (a) J. Kwiatek and J. K. Seyler, *J. Organometal. Chem.*, **3**, 421 (1965); (b) J. Kwiatek and J. K. Seyler, *Proc. 8th Int. Conf. Coord. Chem.*, Vienna, **1964**, 308.

rise to the major products noted at high and low cyanide ion concentrations, respectively.[122, 123] Variations in product ratios on addition of base are probably due to hydroxide ion forcing the equilibrium toward the σ-bonded intermediate.

$$[CH_3CH{=}CH{-}CH_2{-}Co(CN)_5]^{3-} \xrightleftharpoons[+CN^-]{-CN^-}$$

| 1-Butene | *trans*-2-Butene |

Allylic halides, acetates, and alcohols undergo hydrogenolysis through similar intermediates and, as would be expected from this mechanism, the butenylcyanocobalt complex prepared from α- or γ-crotyl bromide gives 1-butene with $[CoH(CN)_5]^{3-}$ in the presence of high cyanide ion concentrations and *trans*-2-butene at low concentrations.[121]

There are unexplained solvent effects on product ratios; they are most apparent in comparisons between aqueous and alcoholic solvents.

Scope and Limitations

The catalyst finds its most useful application to synthesis in the specific hydrogenation of conjugated dienes to monoenes. Activated mono-olefins and some nitrogen-containing unsaturated groups are also reduced.

Hydrogenation of 1,3-butadiene has received a great deal of attention.[123–131] The proportions of isomeric butenes produced depend on several factors including cyanide: cobalt ratio, base concentration, and solvent (see Table X). Considerable selectivity may be achieved by variation of these factors. Hydrogenations of isoprene and 1-phenylbutadiene show similar product distributions.[123, 132, 133]

[122] T. Funabiki and K. Tarama, *Chem. Commun.*, **1971**, 1177.
[123] T. Funabiki, M. Matsumoto, and K. Tarama, *Bull. Chem. Soc. Jap.*, **45**, 2723 (1972).
[124] M. S. Spencer and D. A. Dowden, U.S. Pat., 3,009,969 (1959) [*C.A.*, **56**, 8558d (1962)].
[125] J. Kwiatek, I. L. Madov, and J. K. Seyler, *Adv. Chem. Ser.*, **37**, 201 (1963).
[126] T. Suzuki and T. Kwan, *J. Chem. Soc. Jap.*, **86**, 713 (1965).
[127] T. Suzuki and T. Kwan, *Nippon Kagaku Zasshi*, **86**, 1198 (1965) [*C.A.*, **64**, 11070c (1966)]; *ibid.*, **86**, 713 (1965) [*C.A.*, **64**, 6473e (1966)].
[128] T. Suzuki and T. Kwan, *Bull. Chem. Soc. Jap.*, **41**, 1744 (1968).
[129] M. G. Burnett, P. J. Connolly, and C. Kemball, *J. Chem. Soc.*, A, **1968**, 991.
[130] T. Funabiki and K. Tarama, *Tetrahedron Lett.*, **1971**, 1111.
[131] T. Funabiki and K. Tarama, *Bull. Chem. Soc. Jap.*, **44**, 945 (1971).
[132] T. Suzuki and T. Kwan, *Nippon Kagaku Zasshi*, **86**, 1341 (1965) [*C.A.*, **65**, 12097d (1966)].
[133] T. Funabiki, M. Mohri, and K. Tarama, *J. Chem. Soc. (Dalton)*, **1973**, 1813.

TABLE X. HYDROGENATION OF 1,3-BUTADIENE OVER
$[Co(CN)_5]^{3-}$ [126,128,131]
(Under 1 atm hydrogen at 20° for 3 hours; [Co] = 0.2 M)

CN:Co Ratio	Solvent	Added Base	Products (mol %)		
			1-Butene	trans-2-Butene	cis-2-Butene
4.8	Water	Nil	16	81	2
5.7	,,	,,	94	5	1
4.8	,,	Potassium hydroxide	40	57	2
4.8	Glycerol/ methanol	Nil	10	84	5
5.7	,,	,,	50	9	41
5.7	,,	Ethylene diamine	32	10	58

1,3-Cyclohexadiene and cyclopentadiene are reduced to cyclohexene and cyclopentene, respectively.[125]

Sorbic acid (as its sodium salt) gives high yields of 2-hexenoic acid, conjugation in the product possibly leading to the high specificity.[134–137]

$$CH_3CH=CHCH=CHCO_2Na \xrightarrow{[Co(CN)_5]^{3-},\ H_2} CH_3CH_2CH_2CH=CHCO_2Na$$
$$(96\%)$$

Addition of deuterium with this catalyst is, however, not at all specific. Reduction of sorbate with deuterium gives high proportions of trideuterated hexenoic acid; d_0-, d_1- and d_2-butenes are obtained from butadiene, and 1-phenylbutadiene gives d_0 to d_4 products.[133]

Among activated monoolefins which can be hydrogenated are styrene, substituted styrenes, and α,β-unsaturated carboxylic acids and aldehydes.[125, 138–140] There are anomalous observations: acrylic acid and acrolein are not reduced, while their α- and β-substituted derivatives can give

[134] B. de Vries, Koninkl. Ned. Akad. Wetenschap., Proc. Ser. B, 63, 443 (1960) [C.A., 55, 9142i (1961)].

[135] A. F. Mabrouk, H. J. Dutton, and J. C. Cowan, J. Amer. Oil Chem. Soc., 41, 53 (1964).

[136] A. F. Mabrouk, E. Selke, W. K. Rohwedder, and H. J. Dutton, J. Amer. Oil Chem. Soc., 42, 432 (1965).

[137] T. Takagi, Nippon Kagaku Zasshi, 87, 600 (1966) [C.A., 65, 15217f (1966)].

[138] L. Simandi and F. Nagy, Acta Chem. Acad. Sci. Hung., 46, 137 (1965).

[139] W. Strohmeier and N. Iglauer, Z. Phys. Chem., 51, 50 (1966).

[140] M. Murakami, K. Suzuki, and J.-W. Kang, Nippon Kagaku Zasshi, 83, 1226 (1962) C.A.,59, 13868a (1963)].

$$C_6H_5\overset{\displaystyle CO_2H}{\underset{}{C}}=CH_2 \xrightarrow{[Co(CN)_5]^{3-},\ H_2} C_6H_5\overset{\displaystyle CO_2H}{\underset{}{C}}HCH_3$$

(100%)

$$C_6H_5CH=CHCH_2OH \xrightarrow{[Co(CN)_5]^{3-},\ H_2} C_6H_5CH_2CH_2CH_2OH$$

(85%)

high yields of saturated products.[125] Irreversible complex formation between cobalt and the unsubstituted compounds may be responsible.

$$\left.\begin{array}{c} CH_2=CHCHO \\[2ex] CH_2=CHCO_2H \end{array}\right\} \xrightarrow{[Co(CN)_5]^{3-},\ H_2,\ 20°} \text{No reaction}$$

$$CH_3CH=\overset{\displaystyle CH_3}{\underset{}{C}}CHO \xrightarrow{[Co(CN)_5]^{3-},\ H_2} CH_3CH_2\overset{\displaystyle CH_3}{\underset{}{C}}HCHO$$

(57%)

$$CH_2=\overset{\displaystyle CH_3}{\underset{}{C}}CO_2H \xrightarrow{[Co(CN)_5]^{3-},\ H_2} CH_3\overset{\displaystyle CH_3}{\underset{}{C}}HCO_2H$$

(97%)

Attempted hydrogenations of α,β-unsaturated acids and aldehydes at elevated temperatures give dimeric products.[125] Saturated aldehydes also undergo reductive dimerization.

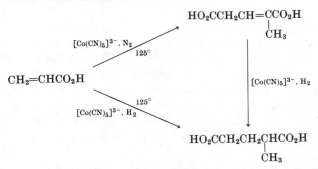

Hydrogenation of acetylenedicarboxylic acid gives, successively, fumaric acid and succinic acid.[140]

$$HO_2CC\equiv CCO_2H \xrightarrow[20°]{[Co(CN)_5]^{3-},\ H_2}$$

$$\underset{H}{\overset{HO_2C}{\diagup}}C=C\underset{CO_2H}{\overset{H}{\diagdown}} \xrightarrow[70°]{[Co(CN)_5]^{3-},\ H_2} HO_2CCH_2CH_2CO_2H$$

Ketoxime and nitro groups are reduced, although slowly, and various side reactions occur. Most notable is reductive dimerization of aryl nitro compounds to give azo and hydrazo derivatives.[125, 141] The most useful

$$m\text{-HOC}_6\text{H}_4\text{NO}_2 \xrightarrow{[\text{Co(CN)}_5]^{3-}, \text{H}_2} m\text{-C}_6\text{H}_4\text{NH}_2$$

$$
\begin{array}{c}
 \xrightarrow[\text{2 equiv aq CH}_3\text{CO}_2\text{Na}]{[\text{Co(CN)}_5]^{3-},\,\text{H}_2} \quad \text{C}_6\text{H}_5\text{N}=\text{NC}_6\text{H}_5 \\
(77\%)
\end{array}
$$

$$\text{C}_6\text{H}_5\text{NO}_2$$

$$
 \xrightarrow[{[\text{Co(CN)}_5]^{3-},\,\text{H}_2}]{\substack{1.2\ \text{equiv}\\ \text{aq CH}_3\text{CO}_2\text{Na}}} \quad \text{C}_6\text{H}_5\text{NHNHC}_6\text{H}_5
$$
$$(93\%)$$

application of the catalyst in this area is the hydrogenation of ketoximes and α-keto acid oximes, which lead to saturated amines.[142]

$$
\begin{array}{c}
\text{CH}_3\text{CCH}_3 \xrightarrow{[\text{Co(CN)}_5]^{3-},\,\text{H}_2} \text{CH}_3\text{CH(NH}_2)\text{CH}_3 \\
\underset{\text{NOH}}{\|} \qquad\qquad\qquad (40\%)
\end{array}
$$

$$
\begin{array}{c}
\text{C}_6\text{H}_5\text{CH}_2\text{CCO}_2\text{H} \xrightarrow{[\text{Co(CN)}_5]^{3-},\,\text{H}_2} \text{C}_6\text{H}_5\text{CH}_2\text{CH(NH}_2)\text{CO}_2\text{H} \\
\underset{\text{NOH}}{\|} \qquad\qquad\qquad\qquad (82\%)
\end{array}
$$

Related to this process is the reductive amination of α-keto acids in aqueous ammonia solution to good yields of α-amino acids.[142–144a] Reaction presumably occurs by a two-step process: formation of the α-iminoacid and then hydrogenation.

$$
\text{CH}_3\text{CH}_2\text{COCO}_2\text{H} \xrightarrow{[\text{Co(CN)}_5]^{3-},\,\text{H}_2,\,\text{NH}_3} \text{CH}_3\text{CH}_2\text{CH(NH}_2)\text{CO}_2\text{H}
$$
$$(85\%)$$

$$
\text{C}_6\text{H}_5\text{CH}_2\text{COCO}_2\text{H} \xrightarrow{[\text{Co(CN)}_5]^{3-},\,\text{H}_2,\,\text{NH}_3} \text{C}_6\text{H}_5\text{CH}_2\text{CH(NH}_2)\text{CO}_2\text{H}
$$
$$(95\%)$$

Catalysts with cyanide replaced by other ligands have not been extensively examined. Hydrogenations of butadiene and isoprene over $[\text{Co(CN)}_3\text{-amine}]^-$ (amine = ethylene diamine; 2,2'-bipyridine; 1,10-phenanthroline)

[141] M. Murakami, R. Kawai, and K. Suzuki, *J. Chem. Soc. Jap.*, **84**, 669 (1963).

[142] M. Murakami and J.-W. Kang, *Bull. Chem. Soc. Jap.*, **36**, 763 (1963).

[143] M. Murakami and J.-W. Kang, *Bull. Chem. Soc. Jap.*, **35**, 1243 (1962).

[144] (a) M. Murakami, K. Suzuki, M. Fujishige, and J.-W. Kang, *J. Chem. Soc. Jap.*, **85**, 235 (1964); (b) T. Funabiki, S. Kasaoka, M. Matsumoto, and K. Tarama, *J. Chem. Soc. (Dalton)*, **1974**, 2043.

are more rapid than those over $[Co(CN)_5]^{3-}$ but selectivity is not significantly improved.[144b] Similarly, dicyanobipyridinecobalt, $Co(CN)_2$(bipyridine), catalyzes hydrogenations more rapidly than $[Co(CN)_5]^{3-}$; for example, reduction of 1,3-cyclohexadiene to cyclohexene is three times faster.[145] The chloropentamminecobaltate, $[CoCl(NH_3)_5]Cl_2$, in the presence of potassium cyanide, is effective in reductive aminations.[140]

Experimental Conditions

Initial investigations of the catalyst were conducted in aqueous or aqueous alcoholic media. Later use of solvent systems such as methanol with ethylene glycol or glycerol has extended the range of application in organic synthesis and may change the nature of the product.

Cobalt concentrations about 0.2 M have generally been used with hydrogen pressures of 0.2–1 atm and temperatures of 20–30° (occasionally up to 90° for difficult reductions). The influence of higher hydrogen pressures is not reported.

Octacarbonyldicobalt—$Co_2(CO)_8$

This complex has been used for many years as a catalyst for hydroformylation of olefins, the commercially important "oxo" process.

$$RCH{=}CHR \xrightarrow{\text{Co}_2(\text{CO})_8,\ \text{H}_2,\ \text{CO}} RCH_2CH(R)CHO$$

Hydrogenation of the product aldehyde to the corresponding alcohol is often used as an integral part of the process; but in some instances hydrogenation of the initial olefin is a competing reaction. Hydrogenation of aldehydes, olefins, condensed aromatic compounds, and others can become the predominant reaction under appropriate conditions.

Preparation and Properties

A convenient preparation of octacarbonyldicobalt from cobalt(II) carbonate with hydrogen and carbon monoxide (1:1) at 200 atm pressure and 160° has been described.[146, 147] The complex forms orange crystals (mp 51°, with decomposition) soluble in benzene and light petroleum. The solid spontaneously evolves carbon monoxide, leaving pyrophoric dodecacarbonyltetracobalt, $Co_4(CO)_{12}$. Storage in an atmosphere of carbon monoxide reverses the process and limits decomposition.

The hydridocarbonyl complex $CoH(CO)_4$ is formed on treatment of $Co_2(CO)_8$ with hydrogen and carbon monoxide at high pressure. It can be

[145] G. M. Schwab and G. Mandre, *J. Catal.*, **12**, 103 (1968).
[146] I. Wender, H. Greenfield, and M. Orchin, *J. Amer. Chem. Soc.*, **73**, 2656 (1951).
[147] I. Wender, H. W. Sternberg, S. Metlin, and M. Orchin, *Inorg. Synth.*, **5**, 190 (1957).

more conveniently synthesized by disproportionation of $Co_2(CO)_8$ (in hydrocarbon solution) with dimethylformamide (DMF) or pyridine, followed by acidification.[148, 149] The aqueous dimethylformamide phase can be re-

$$3 Co_2(CO)_8 + 12 DMF \rightarrow 2 [Co(DMF)_6][Co(CO)_4]_2 + 8 CO$$

$$[Co(DMF)_6][Co(CO)_4]_2 + 2 H^+ \rightarrow 2 CoH(CO)_4 + 6 DMF + Co^{2+}$$

moved; a hydrocarbon (toluene or pentane) solution of $CoH(CO)_4$ is left. The pure complex can be distilled in a stream of carbon monoxide. The hydrido complex is an extremely toxic, volatile, light-yellow liquid (mp $-26°$) which decomposes to $Co_2(CO)_8$ above $-33°$ in the absence of carbon monoxide. It is stable in an atmosphere of carbon monoxide and can be stored at $-78°$.

Alternative preparations involve acidification of salts obtained from (1) an alkaline suspension of cobalt(II) cyanide with carbon monoxide, and (2) cobalt(II) carbonate with pyridine, carbon monoxide, and hydrogen.[149, 150]

Mechanism

Precise details of the mechanism of the hydroformylation and accompanying hydrogenation reactions are not clear, owing in part to the extreme conditions under which the reactions take place. Some insight has been gained by studies of reactions of $CoH(CO)_4$ with substrates.[151]

It is accepted that the following equilibria are established when solutions of the octacarbonyl are subjected to pressures of hydrogen and carbon monoxide.[152, 153a]

$$Co_2(CO)_8 + H_2 \rightleftharpoons 2 CoH(CO)_4$$

$$CoH(CO)_4 \rightleftharpoons CoH(CO)_3 + CO$$

Inhibition of hydroformylation and hydrogenation by high carbon monoxide pressures indicate participation of $CoH(CO)_3$ in the catalytic process. The role of $CoH(CO)_3$ in stoichiometric hydroformylations has been examined.[153b] This coordinatively unsaturated complex is able to bind the substrate before its insertion into the cobalt-hydrogen (or cobalt-carbon monoxide) bond. It seems probable that aldehydes and olefins

[148] L. Kirch and M. Orchin, J. Amer. Chem. Soc., **80**, 4428 (1958).
[149] H. W. Sternberg, I. Wender, and M. Orchin, Inorg. Synth., **5**, 192 (1957).
[150] P. Gilmont and A. A. Blanchard, Inorg. Synth., **2**, 238 (1946).
[151] R. F. Heck, Adv. Chem. Ser., **49**, 181 (1965).
[152] R. F. Heck and D. S. Breslow, J. Amer. Chem. Soc., **83**, 4023 (1961).
[153] (a) F. Ungvary, J. Organometal. Chem., **36**, 363 (1972); (b) A. C. Clark, J. F. Terapane, and M. Orchin, J. Org. Chem., **39**, 2405 (1974).

insert into the cobalt-hydrogen bond to give unstable alkoxy or alkyl complexes.[148, 152, 154, 155] The instability of alkylcarbonylcobalt complexes has been demonstrated.[156] In support of this hypothesis, reactions between

$$RCHO + CoH(CO)_3 \rightleftharpoons RCH_2O-Co(CO)_3 \text{ or } R\overset{\overset{\displaystyle OH}{|}}{C}H-Co(CO)_3$$
$$RCH{=}CH_2 + CoH(CO)_3 \rightleftharpoons RCH_2CH_2-Co(CO)_3$$

$CoH(CO)_4$ and conjugated diolefins lead to isolable π-allyl intermediates.[157–159] 1,3-Butadiene, for example, gives a mixture of *syn* and *anti* π-methylallyl complexes. Reversible addition of cobalt and hydrogen

$$CoH(CO)_4 + C_4H_6 \longrightarrow \underset{Co(CO)_3}{/\!\!\diagdown\!\!/\!\!\diagdown\!\!/CH_3} + \underset{\underset{Co(CO)_3}{|}}{/\!\!\diagdown\!\!/\!\!\diagdown\, CH_3} + CO$$

to the substrate is indicated by occurrence of several isomerizations.[160, 161]

A second hydrogen activation, or hydrogen transfer from $CoH(CO)_4$, completes the hydrogenation reaction. Hydroformylation requires inser-

$$RCH_2CH_2-CoH_2(CO)_3 \longrightarrow RCH_2CH_3 + CoH(CO)_3$$

$$RCH_2CH_2-Co(CO)_3 \overset{H_2}{\overbrace{}}$$

$$\underset{CoH(CO)_4}{\searrow}$$

$$RCH_2CH_3 + Co_2(CO)_7 \overset{CO}{\longrightarrow} Co_2(CO)_8$$

tion of carbon monoxide into the substrate-cobalt bond before the second hydrogen transfer.

$$RCH_2CH_2-Co(CO)_3 \overset{CO}{\longrightarrow} RCH_2CH_2Co(CO)_4$$
$$\longrightarrow RCH_2CH_2\overset{\overset{\displaystyle O}{||}}{C}-Co(CO)_3$$
$$\overset{H_2}{\longrightarrow} RCH_2CH_2CHO + CoH(CO)_3$$

[154] L. Marko, *Proc. Chem. Soc.*, **1962,** 67.
[155] G. L. Aldridge and H. B. Jonassen, *J. Amer. Chem. Soc.*, **85,** 886 (1963).
[156] W. Hieber, O. Vohler, and G. Braun, *Z. Naturforsch.*, *B*, **13,** 192 (1958).
[157] D. W. Moore, H. B. Jonassen, T. B. Joyner, and J. A. Bertrand, *Chem. Ind.* (London), **1960,** 1304.
[158] J. A. Bertrand, H. B. Jonassen, and D. W. Moore, *Inorg. Chem.*, **2,** 601 (1963).
[159] W. Rupilius and M. Orchin, *J. Org. Chem.*, **36,** 3604 (1971).
[160] P. Taylor and M. Orchin, *J. Organometal. Chem.*, **26,** 389 (1971).
[161] P. D. Taylor and M. Orchin, *J. Org. Chem.*, **37,** 3913 (1972).

In both hydrogenation and hydroformylation a substantial pressure of carbon monoxide is necessary to prevent dissociation of the catalyst beyond the tricarbonyl stage, leading to deposition of metallic cobalt.

Scope and Limitations

Catalytic hydrogenations over $Co_2(CO)_8$ (using hydrogen and carbon monoxide) or reductions with stoichiometric quantities of $CoH(CO)_4$ (under carbon monoxide alone) are possible. Both methods give similar results. In view of the high pressures and temperatures necessary to effect reduction in the presence of $Co_2(CO)_8$ (typically 20–30 atm, 150°), use of preformed $CoH(CO)_4$ (under 1 atm at room temperature) may be preferable.

An obvious possible side reaction is the hydroformylation of olefinic linkages, especially if they are not highly substituted (see below). Applications are correspondingly restricted. Halogenated compounds can deactivate the catalyst by formation of cobalt halides but do not always do so.[162a]

The most useful applications are hydrogenations of aldehydes and ketones, polycyclic aromatic compounds, sulfur-containing compounds, and Schiff bases. Reductive amination of aldehydes and ketones is also catalyzed by $Co_2(CO)_8$ and related complexes[162b]. The reaction is thought to occur in a two-step process analogous to reductive aminations over $[Co(CN)_5]^{3-}$ (see p. 51).

Monoolefins are hydrogenated only if one of the carbon atoms bears two alkyl substituents, less highly substituted olefins being preferentially hydroformylated.[163] Thus propene and cyclohexene undergo almost exclusive hydroformylation, while isobutylene gives only 36% of hydroformylation product and 53% of saturated hydrocarbon.

Conjugated olefinic bonds are reduced more readily. Dialdehydes are never observed as products from 1,3-dienes, rapid hydrogenation of one double bond being the initial reaction.[164] The remaining double bond may then be hydrogenated or hydroformylated. It is possible that hydrogenation of unconjugated dienes is preceded by isomerization to conjugated

$$CH_2=C-C=CH_2 \xrightarrow{Co_2(CO)_8,\ H_2,\ CO} CH_3CH-CHCH_2CHO$$

with CH_3 and CH_3 substituents on the left carbons, and CH_3 and CH_3 substituents on the product.

(Major product)

162 (a) H. Adkins and G. Krsek, *J. Amer. Chem. Soc.*, **71**, 3051 (1949); (b) L. Marko and J. Bakos, *J. Organometal. Chem.*, **81**, 411 (1974).

163 (a) L. Marko, *Khim. Tekhnol. Top. Masel*, **5**, 19 (1960) [*C.A.*, **55**, 2075e (1961)]; (b) M. Freund, L. Marko, and J. Laki, *Acta Chim. Acad. Sci. Hung.*, **31**, 77 (1962); (c) L. Marko, *Chem. Ind.* (London), **1962**, 260.

164 H. Adkins and J. L. R. Williams, *J. Org. Chem.*, **17**, 980 (1952).

systems. Unsaturated fats, for example, are hydrogenated to monoenes, the last double bond being neither hydroformylated nor reduced.[165]

An extension of more practical significance is the ready hydrogenation of α,β-unsaturated carbonyl compounds.[162a, 166] Crotonaldehyde, acrolein, methyl vinyl ketone, mesityl oxide, and ethyl cinnamate all undergo hydrogenation. Yields range from 50 to 90%. Other examples include the furyl derivatives. Olefins conjugated to benzene rings (*e.g.*, *cis*-stilbene) are

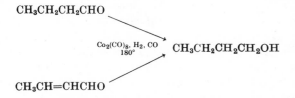

similarly hydrogenated.[167] A notable exception to this general rule is methyl acrylate which is hydroformylated only.[168]

Acetylenes usually give low yields of hydroformylation products from the corresponding olefins, although diphenylacetylene is reduced to diphenylethane.[167]

At higher temperatures α,β-unsaturated aldehydes and ketones, and aliphatic aldehydes and ketones, are reduced to saturated alcohols.[154, 155, 169]

$$CH_3CH_2CH_2CHO$$
$$\searrow$$
$$\xrightarrow[180°]{Co_2(CO)_8, H_2, CO} CH_3CH_2CH_2CH_2OH$$
$$\nearrow$$
$$CH_3CH=CHCHO$$

Dimerization of aldehydes has been noted as a side reaction under these conditions but does not usually occur to a large extent.[170]

Aromatic aldehydes and ketones are also hydrogenated, but hydrogenolysis of the alcohols produced gives hydrocarbons as major products.[146]

$$C_6H_5COCH_3 \xrightarrow{Co_2(CO)_8, H_2, CO} C_6H_5CH_2CH_3$$
$$(67\%)$$

[165] (a) I. Ogata and A. Misono, *Yukagaku,* **14,** 16 (1965) [*C.A.,* **63,** 17828d (1965)]; (b) E. N. Frankel, E. P. Jones, V. L. Davison, E. Emken, and H. J. Dutton, *J. Amer. Oil Chem. Soc.,* **42,** 130 (1965).

[166] R. Ercoli and R. F. Torregrosa, *Chim. Ind.* (Milan), **40,** 552 (1958) [*C.A.,* **53,** 3186g (1959)].

[167] H. Greenfield, J. H. Wotiz, and I. Wender, *J. Org. Chem.,* **22,** 542 (1957).

[168] H. Uchida and K. Bando, *Bull. Chem. Soc. Jap.,* **29,** 953 (1956).

[169] (a) I. Wender, R. Levine, and M. Orchin, *J. Amer. Chem. Soc.,* **72,** 4375 (1950); (b) I. Wender, M. Orchin, and H. H. Storch, *ibid.,* **72,** 4842 (1950).

[170] H. Uchida and A. Matsuda, *Bull. Chem. Soc. Jap.,* **36,** 1351 (1963).

Aryl and furyl carbinols are similarly hydrogenolyzed to hydrocarbons.[146, 166, 171]

$$1\text{-}C_{10}H_7CH_2OH \xrightarrow{\text{Co}_2(\text{CO})_8,\ \text{H}_2,\ \text{CO}} 1\text{-}C_{10}H_7CH_3$$

(80%)

Cyclic anhydrides of dibasic acids yield aldehyde acids; acyclic anhydride of monobasic acids yield the aldehyde and the acid.[172]

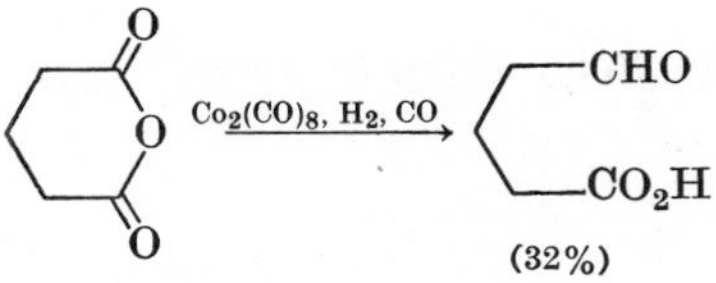

(32%)

Many polycyclic aromatic systems can be partially hydrogenated to products of a single structure, e.g., anthracene to 9,10-dihydroanthracene (99%) and naphthacene to 5,12-dihydronaphthacene (70%).[161, 173] The phenanthrene nucleus is particularly stable to hydrogenation under these conditions, as shown by the products from perylene and pyrene. Isolated benzene rings are not affected.

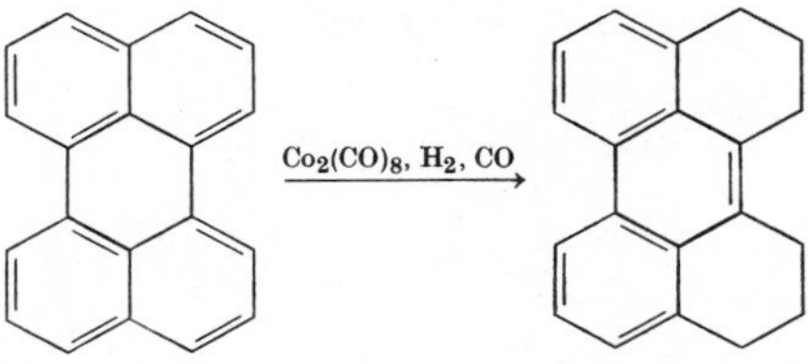

(72%)

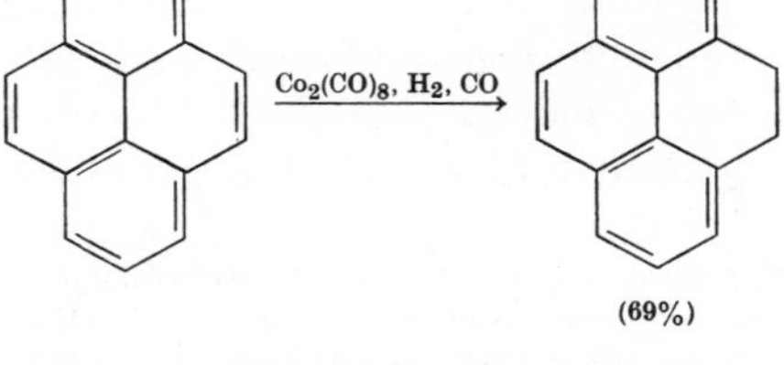

(69%)

[171] W. Dawydoff, Chem. Technol. (Berlin), 11, 431 (1959).
[172] H. Wakamatsu, J. Furukawa, and N. Yamakami, Bull. Chem. Soc. Jap., 44, 288 (1971).
[173] S. Friedman, S. Metlin, A. Svedi, and I. Wender, J. Org. Chem., 24, 1278 (1959).

In contrast to many heterogeneous catalysts, $Co_2(CO)_8$ is not poisoned by the presence of some sulfur compounds.[174] Thiophene inhibits catalysis to a very small extent; advantage is taken of this fact in the saturation of thiophene and its derivatives.[175] Hydrogenolysis of α-oxygen substitu-

$$\overset{}{\underset{S}{\fbox{}}}R \xrightarrow{\text{Co}_2(\text{CO})_8,\ \text{H}_2,\ \text{CO}} \overset{}{\underset{S}{\fbox{}}}R$$

$$\begin{array}{cccc} R = H, & CH_3, & C_2H_5 \\ \text{Yield} & 66\% & 77\% & 82\% \end{array}$$

ents occurs in the thiophene series, as noted above for benzene and furan derivatives.

$$\overset{}{\underset{S}{\fbox{}}}COCH_3 \xrightarrow{\text{Co}_2(\text{CO})_8,\ \text{H}_2,\ \text{CO}} \overset{}{\underset{S}{\fbox{}}}CH_2CH_3 \longrightarrow \overset{}{\underset{S}{\fbox{}}}CH_2CH_3$$

Schiff bases are hydrogenated to the corresponding amines; carbon monoxide insertion to form amides is not observed.[176] In contrast, azo-

$$p\text{-}RC_6H_4N{=}CHC_6H_5 \xrightarrow{\text{Co}_2(\text{CO})_8,\ \text{H}_2,\ \text{CO}} p\text{-}RC_6H_4NHCH_2C_6H_5$$

$$(R = H,\ CH_3,\ CH_3O,\ Cl,\ NO_2) \qquad 80\%$$

benzene and nitrobenzene are both hydrogenated to aniline, but some carbonylation to diphenylurea also occurs.

$$\begin{array}{c} C_6H_5NO_2 \\ \searrow \\ \qquad \xrightarrow{\text{Co}_2(\text{CO})_8,\ \text{H}_2,\ \text{CO}} C_6H_5NH_2 + C_6H_5NHCONHC_6H_5 \\ \qquad \qquad (70\text{-}95\%) \qquad\qquad (5\text{-}30\%) \\ \nearrow \\ C_6H_5N{=}NC_6H_5 \end{array}$$

Related Catalysts

Several complexes related to $Co_2(CO)_8$ and $CoH(CO)_4$, notably tertiary phosphine-substituted analogs, have been examined for activity in hydrogenation. The presence of stabilizing ligands other than the carbonyl group should remove the need for an atmosphere containing carbon monoxide and so eliminate competing carbonyl insertion reactions.

Effective catalysts are indeed obtained by addition of phosphines to solutions of $Co_2(CO)_8$, or by synthesis of complexes such as $[Co(CO)_3(PR_3)]_2$

174 (a) L. Marko, Proc. Sympos. Coord. Chem., Tihany, Hungary, **1964**, 271; (b) J. Laky, P. Szabo, and L. Marko, Acta Chim. Acad. Sci. Hung., **46**, 247 (1965).

175 H. Greenfield, S. Metlin, M. Orchin, and I. Wender, J. Org. Chem., **23**, 1054 (1958).

176 (a) S. Murahashi and S. Horiie, Bull. Chem. Soc. Jap., **33**, 78 (1960); (b) A. Nakamura and N. Hagihara, Osaka Univ. Mem., **15**, 195 (1958); (c) S. Murahashi, S. Horiie, and T. Jo, Nippon Kagaku Zasshi, **79**, 68 (1958) [C.A., **54**, 5558d (1960) .

(R = phenyl, n-butyl, cyclohexyl). These do not decompose when used under hydrogen alone and have been applied to the selective hydrogenation of 1,5,9-cyclododecatriene, giving cyclododecene under low hydrogen pressures.[177] However, high substrate:catalyst ratios cause decreased selectivity and catalyst decomposition.[178] Mixtures of products are obtained from hydrogenations of other dienes (e.g., 1,5-cyclooctadiene and 1,5-hexadiene).

Analogs of $CoH(CO)_4$ are active for catalytic hydrogenations of olefins, conjugated diolefins, acetylenes, and aldehydes, and for the isomerization of olefins.[179a–c] These include the series $CoH(CO)_x(PR_3)_{4-x}$ (where $x =$ 1, 2; R = phenyl, n-butyl). However, they require high hydrogen pressures for hydrogenation of monoolefins, and isomerization is more rapid than hydrogenation. Selectivity is not high in hydrogenations of conjugated diolefins.

The tri(n-butyl)phosphine-derived catalysts are potentially useful for selective reductions of alkynes to alkenes. Partial hydrogenation proceeds under mild conditions where olefin hydrogenation and isomerization do not interfere. Thus 1-pentyne gives 1-pentene in 95% yield (4 hours at 40° and 30 atm); 2-pentyne gives cis-2-pentene in high yield. Aldehydes are

$$C_2H_5C{\equiv}CCH_3 \xrightarrow{\text{CoH(CO)[P(\textit{n}-C}_4\text{H}_9)_3]_3, H_2}} \begin{array}{c} H \\ \diagdown \\ C_2H_5 \end{array} C{=}C \begin{array}{c} H \\ \diagup \\ CH_3 \end{array}$$

(85%)

also hydrogenated over these tri(n-butyl)phosphine derivatives. Formation of $CoH(CO)_3[P(C_4H_9 - n)_3]$, analogous to $CoH(CO)_4$, in the system $Co_2(CO)_8/P(C_4H_9 - n)_3/H_2/CO$ indicates that catalysis probably follows a pathway similar to that postulated for $Co_2(CO)_8$.[179d] An unexplained increase in rate is observed on addition of excess phosphine (up to a phosphine:cobalt ratio of 5).

The related cluster compounds $[Co(CO)_2(PR_3)]_3$ (R = phenyl, n-butyl) show catalytic properties similar to those of the monomeric complexes.[180]

177 (a) A. Misono and I. Ogata, Bull. Chem. Soc. Jap., 40, 2718 (1967); (b) I. Ogata and A. Misono, Discuss. Faraday Soc., 46, 72 (1968).

178 D. R. Fahey, J. Org. Chem., 38, 80 (1973).

179 (a) M. Hidai, T. Kuse, T. Hikita, Y. Uchida, and A. Misono, Tetrahedron Lett., 1970, 1715; (b) G. F. Pregaglia, A. Andreetta, G. F. Ferrari, and R. Ugo, J. Organometal. Chem., 30, 387 (1971); (c) G. F. Ferrari, A. Andreetta, G. F. Pregaglia, and R. Ugo, ibid., 43, 213 (1972); (d) M. van Boven, N. Alemdaroglu, and J. M. L. Penninger, ibid., 84, 65 (1975).

180 (a) G. F. Pregaglia, A. Andreetta, G. F. Ferrari, and R. Ugo, Chem. Commun., 1969, 590; (b) G. F. Pregaglia, A. Andreetta, G. F. Ferrari, G. Montrasi, and R. Ugo, J. Organometal. Chem., 33, 73 (1971).

Experimental Conditions

Solvents used for catalytic hydrogenations over $Co_2(CO)_8$ or reductions with $CoH(CO)_4$ include benzene, toluene, n-pentane, n-hexane, n-decane, cyclohexane, diethyl ether, and di-n-hexyl ether. Vigorous conditions are required for $Co_2(CO)_8$: temperatures in the range 100–200° with pressures between 100 and 300 atm. Decomposition of the catalyst occurs above 200°. Mixtures of hydrogen and carbon monoxide contain from 30 to 70% of hydrogen.

In contrast, very mild conditions are used with $CoH(CO)_4$: 1 atm of carbon monoxide at 25°.

The phosphine-substituted catalysts are employed in a similar range of solvents and require intermediate reaction conditions: 50–150° under 15–50 atm of hydrogen.

Soluble Ziegler Catalysts

A large number of the Ziegler-type polymerization catalysts [group (IV)–(VIII) transition metal complexes with trialkylaluminum or similar organometallic compounds] form heterogeneous systems. However, some are soluble in hydrocarbon solvents, and applications as homogeneous hydrogenation catalysts have been investigated.

Preparation of catalysts involves addition of the alkylaluminum, Grignard, or organolithium reagent to a hydrocarbon solution of the transition metal complex. The solutions produced are air- and moisture-sensitive. Their preparation is carried out in an atmosphere of nitrogen, and then the nitrogen is exchanged with hydrogen. (*Caution: Handling trialkylaluminum compounds requires special precautions.*)

Mechanism

Reactions of alkylaluminum compounds with transition metal complexes follow complicated pathways involving the formation of metal alkyls which often react further to give metal hydrides or reduced species.[181] The mechanism of hydrogenation has not been studied extensively but is

$$M(ligand) + Al(alkyl)_3 \rightleftharpoons M(alkyl) + Al(alkyl)_2(ligand)$$

$$M(alkyl) \rightleftharpoons M\text{—}H + olefin$$

$$M\text{—}H + substrate\ olefin \rightleftharpoons M(alkyl) \xrightarrow{H_2} M\text{—}H + alkane$$

(M = transition metal)

[181] T. Mole, *Organometal. Reactions*, **1**, 41 (1970).

taken to involve the accompanying sequence of steps.[182, 183] Residual ligands present prevent complete reduction of the catalyst to the free metal.

Scope and Limitations

The majority of reductions studied involve saturation of monoolefins and full or partial hydrogenation of conjugated diolefins. Ketone, aldehyde, nitrile, nitro, azo, and ester groups are not reduced.[182] One noteworthy application is the hydrogenation of isolated benzene rings. With the exception of a recent report on the use of $Co(C_3H_5)[P(OCH_3)_3]_3$ (see p. 65), Ziegler catalysts are the only homogeneous systems capable of catalyzing this reaction. In addition, a catalyst closely related to the Ziegler systems shows high specificity in the semihydrogenation of terminal alkynes to alkenes.

Catalysts derived from nickel, cobalt, iron, chromium, and copper 2-ethylhexanoates with triethylaluminum (transition metal:aluminum ratio of 3 or 4:1) allow hydrogenation of the aromatic nucleus.[184] The most active (the nickel and cobalt species) are more efficient than Raney nickel and other supported nickel catalysts, although nitrobenzene and p-nitrophenol, which are readily reduced by the heterogeneous systems, are inert to the Ziegler catalysts. Other aromatics reduced include phenol, dimethyl phthalate, and naphthalene (to tetralin, 84%, and decalin, 13%).

$$\underset{\text{CH}_3}{\overset{\text{CH}_3}{\bigcirc}} \xrightarrow{\text{Ni}(O_2CC_7H_{15})_2,\ Al(C_2H_5)_3,\ H_2} \underset{\text{CH}_3}{\overset{\text{CH}_3}{\bigcirc}} \text{(65%)} + \underset{\text{CH}_3}{\overset{\text{CH}_3}{\bigcirc}} \text{(35%)}$$

$$\underset{N}{\bigcirc} \xrightarrow{\text{Ni}(O_2CC_7H_{15})_2,\ Al(C_2H_5)_3,\ H_2} \underset{\overset{\text{N}}{\text{H}}}{\bigcirc} \text{(98%)}$$

$$\underset{\text{CO}_2\text{CH}_3}{\overset{\text{CO}_2\text{CH}_3}{\bigcirc}} \xrightarrow{\text{Ni}(O_2CC_7H_{15})_2,\ Al(C_2H_5)_3,\ H_2} \underset{\text{CO}_2\text{CH}_3}{\overset{\text{CO}_2\text{CH}_3}{\bigcirc}} \text{(100%)}$$

[182] M. F. Sloan, A. S. Matlack, and D. S. Breslow, J. Amer. Chem. Soc., 85, 4014 (1963).
[183] F. Ungvary, B. Babos, and L. Marko, J. Organomental. Chem., 8, 321 (1967).
[184] S. J. Lapporte and W. R. Schuett, J. Org. Chem., 28, 1947 (1963).

Triethylaluminum with acetylacetonates of vanadium, chromium, manganese, iron, cobalt, nickel, and copper, or with π-cyclopentadienyl complexes of titanium and iron, forms active catalysts for hydrogenation of benzene.[185] Optimum ratios of trialkylaluminum:transition metal are given for these systems but applications to other aromatics are not mentioned.

Dicarbonylbis-(π-cyclopentadienyl)titanium, $Ti(C_5H_5)_2(CO)_2$, which probably reacts via an intermediate similar to reduction products from $Ti(C_5H_5)_2Cl_2$ with trialkylaluminums,[186, 187] is a catalyst for the hydrogenation of terminal acetylenes and activated olefins.[188] If the acetylene is not activated by conjugation, hydrogenation ceases with production of the corresponding olefin. Activated acetylenes and olefins give fully saturated products; internal acetylenes are not affected. Acetylene itself is not

$$CH_3(CH_2)_nC\equiv CH \xrightarrow[(n\ =\ 2,\ 4)]{Ti(C_5H_5)_2(CO)_2,\ H_2} CH_3(CH_2)_nCH=CH_2$$
$$(90\%)$$

$$CH_3CH_2C\equiv C(CH_2)_2CH_3 \xrightarrow{Ti(C_5H_5)_2(CO)_2,\ H_2} \text{No reaction}$$

$$C_6H_5C\equiv CH \xrightarrow{Ti(C_5H_5)_2(CO)_2,\ H_2} C_6H_5CH_2CH_3$$
$$(95\%)$$

hydrogenated efficiently; production of polymeric products is the major reaction. Trimerization of substituted acetylenes to benzene derivatives occurs over some Ziegler catalysts.[182, 189]

Complexes formed by reduction of $Ti(C_5H_5)_2Cl_2$ with sodium, magnesium, calcium, sodium naphthalide, or butyllithium allow hydrogenation of a variety of acetylenes and olefins.[187, 190, 191]

[185] V. G. Lipovich, F. K. Schmidt, and I. V. Kalechits, *Kinet. Katal.*, 8, 939 (1967) [*C.A.*, 68, 59185w (1968)].

[186] J. E Bercaw, R. H. Marvich, L. G. Bell, and H. H. Brintzinger, *J. Amer. Chem. Soc.*, 94, 1219 (1972).

[187] (a) R. H. Grubbs, C. Gibbons, L. C. Kroll, W. D. Bonds, and C. H. Brubaker, *J. Amer. Chem. Soc.*, 95, 2373 (1973); (b) E. E. van Tamelen, W. Cretney, N. Keaentschi, and J. S. Miller, *Chem. Commun.*, 1972, 481.

[188] K. Sonogashira and N. Hagihara, *Bull. Chem. Soc. Jap.*, 39, 1178 (1966).

[189] D. V. Sokol'skii, G. N. Sharifkanova, and N. F. Noskova, *Dokl. Akad. Nauk SSSR*, 194, 599 (1970) [*C.A.*, 74, 12410z (1971)].

[190] K. Shikata, K. Nishino, and K. Azuma, *Kogyo Kagaku Zasshi*, 68, 490 (1965) [*C.A.*, 63, 7111a (1965)].

[191] M. Shimoi, M. Ichikawa, and K. Tamara, *Abstr. 20th Symp. Organometal. Chem.*, Kyoto, Japan 1972.

Olefins hydrogenated over Ziegler catalysts include acyclic and cyclic olefins (Refs. 182, 183, 187a, 190–198), conjugated and nonconjugated diolefins[187a, 193, 199–201] (including soybean oil methyl esters) and unsaturated polymers (cis-1,4-polybutadiene and butadiene-styrene copolymer).[192, 193, 202] Ease of hydrogenation follows the order normally observed over homogeneous catalysts (monosubstituted > disubstituted > trisubstituted olefins; cis > trans; etc.).

Transition metal complexes employed are acetylacetonates (Refs. 182, 185, 189, 192, 193, 195, 196, 198, 199, 201–203), alkoxides,[182] carboxylates,[183, 184, 202a, 204] and π-cyclopentadienyl (Refs. 182, 185, 187a, 190, 191, 194, 195, 197, 200) derivatives of titanium, zirconium, vanadium, chromium, molybdenum, manganese, iron, ruthenium, cobalt, nickel, palladium, and copper. They are most commonly used in conjunction with triethylaluminum or tri-isobutylaluminum (Refs. 182, 184, 185, 189, 192, 193, 195–203), but n-butyllithium and other alkyllithium compounds (Refs. 182, 194, 200, 202a, 204), ethylmagnesium chloride, phenylmagnesium chloride, and other Grignard reagents[183, 194, 200] are also suitable in many cases.

Butadiene and isoprene are selectively reduced to monoolefins over some catalysts, and to fully saturated products over others. Only slight differences in the catalytic systems are involved.[200, 201] (See p. 64.)

Soybean oil methyl esters (linoleate and linolenate) give mainly trans-monoene products, conjugation of double bonds followed by rapid hydrogenation of diene to monoene being the presumed pathway.[199] Cis-trans

[192] B. I. Tokhomirov, I. A. Klopotova, and A. I. Yakubchik, Vysokomol. Soedin., Ser. B, 9, 427 (1967) [C.A., 67, 82418n (1967)].

[193] B. I. Tikhomirov, I. A. Klopotova, and A. I. Yakubchik, Vestn. Leningrad. Univ., 22, Fiz. Khim., 147 (1967) [C.A., 68, 59020p (1968)].

[194] K. Shikata, K. Nishino, K. Azama, and Y. Takegami, Kogyo Kagaku Zasshi, 68, 358 (1965) [C.A., 65, 10452b (1966)].

[195] I. V. Kalechits and F. K. Schmidt, Kinet. Katal., 7, 614 (1966) [C.A., 65, 16817b (1966)].

[196] I. V. Kalechits, V. G. Lipovich, and F. K. Schmidt, Neftekhim., 6, 813 (1966) [C.A., 66, 94632v (1967); Katal. Reakts. Zhidk. Faze, Tr. Vses. Konf., 2nd, Alma-Ata, Kaz. SSR, 1966, 425 [C.A., 69, 76204q (1968)].

[197] I. V. Kalechits, V. G. Lipovich, and F. K. Schmidt, Kinet. Katal., 9, 24 (1968) [C.A., 69, 2424q (1968)].

[198] D. V. Sokol'skii, N. F. Noskova, and M. I. Popandopulo, Tr. Inst. Khim. Nauk, Akad. Nauk Kaz. SSR, 26, 106 (1969) [C.A., 72, 78193w (1970)].

[199] Y. Tajima and E. Kunioka, J. Amer. Oil Chem. Soc., 45, 478 (1968).

[200] Y. Tajima and E. Kunioka, J. Org. Chem., 33, 1689 (1968).

[201] Y. Tajima and E. Kunioka, J. Catal., 11, 83 (1968).

[202] (a) C. J. Falk, Makromol. Chem., 160, 291 (1972); (b) A. I. Yakubchik, B. I. Tikhomirov, I. A. Klopotova, and L. N. Mikhailova, Dokl. Akad. Nauk SSSR, 161, 1365 (1965) [C.A., 63, 4412g (1965)].

[203] D. V. Sokol'skii, G. N. Sharifkanova, N. F. Noskova, A. D. Dembitskii, and M. I. Goryaev, Zh. Org. Khim., 7, 1556 (1971) [C.A., 76, 3191t (1972)].

[204] J. Falk, J. Org. Chem., 36, 1445 (1971).

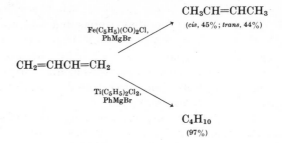

$$CH_3CH=CHCH_3$$
$$(cis, 45\%; trans, 44\%)$$

Fe(C$_5$H$_5$)(CO)$_2$Cl,
PhMgBr

$$CH_2=CHCH=CH_2$$

Ti(C$_5$H$_5$)$_2$Cl$_2$,
PhMgBr

$$C_4H_{10}$$
$$(97\%)$$

and positional isomerization of monoolefins also occurs.[195, 197, 203] Hydrogenation of unsaturated fats occurs over nickel, copper, cobalt, and iron acetylacetonates even in the absence of trialkylaluminum compounds.[205]

The molar ratio of transition metal complex:organometallic compound has considerable effect on the efficiency of the catalytic systems. Values are typically within the range 1:3 to 1:12.

One catalyst is active under exceptionally mild conditions: Ti(C$_5$H$_5$)$_2$Cl$_2$ reduced with magnesium or calcium allows hydrogenation of cyclohexene at $-20°$ and 1 atm.[191]

A number of related catalytic systems which have been designated "heavy metal hydrides" appear to be very similar if not identical to Ziegler catalysts. They are generated by reaction of iron, nickel, cobalt, and other transition metal halides with aluminum hydrides or Grignard reagents and are active for the hydrogenation of various mono- and di-olefins.[206a–h]

Experimental Conditions

Hydrogenations over these catalysts are usually carried out in hydrocarbon solvents, which include n-alkanes, cyclohexane, decalin, benzene, toluene, and xylene. Ethers or mixed hydrocarbon-ether systems are used when a Grignard reagent or an alkali metal is added to the transition metal complex.

[205] E. A. Emken, E. N. Frankel, and R. O. Butterfield, *J. Amer. Oil Chem. Soc.*, **43**, 14 (1966).
[206] (a) Y. Takegami and T. Fujimaki, *Kogko Kagaku Zasshi*, **64**, 287 (1961)[*C.A.*, **57**, 4271f (1962)]; (b) Y. Takegami, T. Ueno, K. Shinoki, and T. Sakata, *ibid.*, **67**, 316 (1964) [*C.A.*, **61**, 6618c (1964)]; (c) Y. Takegami, T. Ueno, and T. Fujii, *ibid.*, **67**, 1009(1964) [*C.A.*, **61**, 13931f (1964)]; (d) Y. Takegami, T. Ueno, and T. Fujii, *Bull. Chem. Soc. Jap.*, **38**, 1279 (1965); (e) Y. Takegami, T. Ueno, and T. Sakata, *Kogku Kagaku Zasshi*, **68**, 2373 (1965)[*C.A.*, **65**, 16884c (1966)]; (f) Y. Takegami, T. Ueno, and K. Kawajiri, *ibid.*, **66**, 1068 (1963) [*C.A.*, **62**, 7661b (1965)]; (g) Y. Takegami, T. Ueno, and T. Fujii, *ibid.*, **69**, 1467 (1966) [*C.A.*, **66**, 22605r (1967)]; (h) R. Stern and L. Sajus, *Tetrahedron Lett.*, **1968**, 6313; (i) E. L. Muetterties and F. J. Hirsekorn, *J. Amer. Chem. Soc.*, **96**, 4063 (1974); (j) F. J. Hirsekorn, M. C. Takowski, and E. L. Muetterties, *ibid.*, **97**, 237 (1975).

Temperatures and hydrogen pressures vary widely: 15–200° and 1–150 atm. For the hydrogenation of aromatics and partial hydrogenation of acetylenes the conditions are, respectively, 150–200°, 70 atm, and 50–65°, 50 atm. Hydrogenations of simple olefins proceed under 1–4 atm at 20–50°.

Other Catalysts

This section briefly reviews other systems which catalyze homogeneous hydrogenation. Included are catalysts not mentioned in the foregoing sections for one of two reasons: (1) while appearing potentially useful, they have been only superficially investigated, or (2) they do not appear to promise useful applications in organic synthesis. Those in the former category are discussed first. Several of them, notably complexes of ruthenium and chromium, show high selectivity in certain situations and deserve more thorough studies. However, much more work is often required before firm conclusions about practical utility can be reached.

Hydrogenation of benzene under mild conditions (25°, 1 atm of hydrogen) is reported using $Co(C_3H_5)[P(OCH_3)_3]_3$ as catalyst.[206i] Substituted benzenes are also successfully reduced.

R = H, Alkyl, OCH₃

The catalyst is also active for hydrogenation of olefins, including cyclohexadiene and cyclohexene, but competition experiments show it to be selective for benzene over cyclohexene.[206j] It is suggested that an important intermediate in hydrogenation involves a benzene ring bound to cobalt through four of the six aromatic π electrons.

By analogy with $RhCl(PPh_3)_3$, $RuCl_2(PPh_3)_3$ was expected to catalyze reduction of olefins. It does so, but only in the presence of bases such as ethyl alcohol or triethylamine, which promote formation of the hydride $RuHCl(PPh_3)_3$ in the presence of hydrogen.[207a–c] The presumed catalytic

207 (a) P. S. Hallmann, B. R. McGarry and G. Wilkinson, *J. Chem. Soc., A*, **1968**, 3143; (b) P. S. Hallmann, D. Evans, J. A. Osborn, and G. Wilkinson, *Chem. Commun.*, **1967**, 305; (c) I. Jardine and F. J. McQuillin, *Tetrahedron Lett.*, **1968**, 5189; (d) B. R. James, L. D. Markham, and D. K. W. Wang, *Chem. Commun.*, **1974**, 439.

intermediate RuHCl(PPh$_3$)$_2$ has been generated in solution recently, but its use in hydrogenation is not reported.[207d] The catalyst allows specific hydrogenation of 1-alkenes; internal double bonds in chains are not affected. Rates of isomerization of olefins (both positional and *cis-trans* conversions) are very low, and disproportionation of cyclohexadienes does not occur in the presence of the complex.[208] Although they are not hydrogenated, internal olefins undergo exchange with deuterium. Diolefins are reduced to monoolefins; in many cases only one major product results.[209] Selective hydrogenations of acetylenes to the corresponding olefins and of 1,4-androstadien-3,17-dione to the 4-en-3,17-dione in high yield are reported.[210, 211] Reduction of cyclohexanone to the alcohol at slightly higher temperatures (84–140°, compared to 40° for olefin hydrogenation) indicates that the presence of carbonyl groups require cautious use of elevated temperatures.[212a] Olefin isomerization is also enhanced at higher temperatures (80°).[212b] The analogous triphenylarsine derivative, RuCl$_2$(AsPh$_3$)$_3$, and related nitrosyl complexes, RuH(NO)(tertiary phosphine)$_3$, are also active for hydrogenation of olefins.[213, 214a] The crystal structure of RuH(CO$_2$CH$_3$)(PPh$_3$)$_3$ has been discussed in terms of its catalytic properties.[214b]

Selective hydrogenation of polyenes to monoenes is also effectively catalyzed by RuCl$_2$(CO)$_2$(PPh$_3$)$_2$ in the presence of excess triphenyl-

RuCl$_2$(CO)$_2$(PPh$_3$)$_2$
$\xrightarrow{\text{PPh}_3, \text{ H}_2}$

(97%)

phosphine.[178, 215] Chlorocarbonylphosphine complexes of ruthenium appear, in general, to be poor catalysts for hydrogenation of monoolefins.[216]

Hydrogenated solutions of Ti(1-methylallyl)(C$_5$H$_5$)$_2$ catalyze rapid hydrogenation of unhindered olefins under mild conditions, but complications due to polymerization of acetylenes and to formation of stable

[208] J. E. Lyons, *J. Catal.*, **28**, 500 (1973).

[209] E. F. Litvin, L. Kh. Freidlin, and K. G. Karimov, *Neftekhim.*, **12**, 318 (1972) [*C.A.*, **77**, 100613k (1972)].

[210] E. F. Litvin, A. Kh. Freidlin, and K. K. Karimov, *Izv. Akad. Nauk SSSR, Ser. Khim.*, **1972**, 1853 [*C.A.*, **77**, 151382s (1972)].

[211] (a) S. Nishimura, T. Ichino, A. Akimoto, and K. Tsuneda, *Bull. Chem. Soc. Jap.*, **46**, 279 (1973); (b) S. Nishimura and K. Tsuneda, *ibid.*, **42**, 852 (1969).

[212] (a) L. Kh. Freidlin, V. Z. Sharf, V. N. Krutii, and S. I. Shcherbakova, *Zh. Org. Khim.*, **8**, 979 (1972) [*C.A.*, **77**, 61310n (1972)]; (b) D. Bingham, D. E. Webster, and P. B. Wells, *J. Chem. Soc.* (*Dalton*), **1974**, 1519.

[213] S. T. Wilson and J. A. Osborn, *J. Amer. Chem. Soc.*, **93**, 3068 (1971).

[214] (a) M. M. Taqui Khan, R. K. Andal, and P. T. Manoharan, *Chem. Commun.*, **1971**, 561; (b) A. C. Skapski and F. A. Stephens, *J. Chem. Soc.* (*Dalton*), **1974**, 390.

[215] D. F. Fahey, *J. Org. Chem.*, **38**, 3343 (1973).

[216] B. R. James, L. D. Markham, B. C. Hui, and G. L. Rempel, *J. Chem. Soc.* (*Dalton*), **1973**, 2247.

allylic complexes with some dienes (accompanied by loss of catalytic activity) may limit general application of this system.[217a, b] Rapid hydrogenation of acetylenes and terminal and internal olefins is catalyzed by $Ti(CO)(C_5H_5)_2(PhC\equiv CPh)$ at ambient temperature and pressure.[217c]

Tricarbonyl(methyl benzoate)chromium, $Cr(C_6H_5CO_2CH_3)(CO)_3$, catalyzes specific 1,4-hydrogen addition to *trans,trans*-1,3-diolefins able to assume the *s-cis* configuration, and to cyclic 1,3-dienes that are held in this configuration. *Cis* monoolefins are the principal products. Thus methyl

s-trans   s-cis

sorbate is reduced to methyl 3-hexenoate, deuteration studies demonstrating exclusive 1,4-hydrogen addition.[218] 1,3-Cyclohexadiene is reduced

easily to cyclohexene; 1,3,5-cycloheptatriene gives 1,3-cycloheptadiene and cycloheptene in successive steps.[219, 220]

The products obtained from numerous acyclic dienes, including unsaturated fats, point to specific 1,4-hydrogen addition and low rates of *cis-trans* isomerization over the catalyst.[220–223] (This is one of the very few systems which produce mainly *cis* monoenes in hydrogenations of unsaturated fats.) There is evidence that conjugation of 1,4-to 1,3-dienes can occur relatively rapidly, especially in cyclic compounds.[219, 220] The

[217] (a) H. A. Martin and R. O. DeJongh, *Chem. Commun.*, **1969**, 1366; (b) H. A. Martin and R. O. DeJongh, *Rec. Trav. Chim. Pays-Bas*, **90**, 713 (1971); (c) G. Fachinetti and C. Floriani, *Chem. Commun.*, **1974**, 66.

[218] (a) M. Cais, E. N. Frankel, and A. Rejoan, *Tetrahedron Lett.*, **1968**, 1919; (b) E. N. Frankel, E. Selke, and C. A. Glass, *J. Amer. Chem. Soc.*, **90**, 2446 (1968).

[219] E. N. Frankel, *J. Org. Chem.*, **37**, 1549 (1972).

[220] E. N. Frankel and R. O. Butterfield, *J. Org. Chem.*, **34**, 3930 (1969).

[221] E. N. Frankel, E. Selke, and C. A. Glass, *J. Org. Chem.*, **34**, 3936 (1969).

[222] E. N. Frankel and F. L. Little, *J. Amer. Oil Chem. Soc.*, **46**, 256 (1969).

[223] E. N. Frankel and F. L. Thomas, *J .Amer. Oil Chem. Soc.*, **49**, 70 (1972).

conjugation is, of course, followed by hydrogenation. Isomerizations of 1,4-cyclohexadiene and 1,4-hexadiene occur in the absence of hydrogen.[224] Similar selectivity toward hydrogenation of conjugated dienes and trienes is shown by $[Cr(C_5H_5)(CO)_3]_2$.[225] The catalyst is generated *in situ* from chromocene and carbon monoxide. The reactions could be useful for the stereoselective synthesis of trisubstituted olefins. Analogous

molybdenum and tungsten hydrido complexes, $MH(C_5H_5)(CO)_3$ (M = molybdenum, tungsten), allow noncatalytic reduction by the same pathway as does the chromium catalyst.[226] Catalytic hydrogenations of cyclic 1,3-and 1,4-dienes to monoenes with yields between 50 and 90 % are achieved over $MoH_2(C_5H_5)_2$.[227] α,β-Unsaturated carbonyl compounds are also reduced. Hexacarbonylchromium, $Cr(CO)_6$, and $Cr(norbornadiene)(CO)_4$ are photolytically activated to catalyze 1,4-hydrogen addition to 1,3-dienes able to attain the *s-cis* conformation.[228, 229a, b] Reaction occurs under very mild conditions to give monoene products with high specificity.

[224] E. N. Frankel, *J. Catal.*, **24**, 358 (1972).

[225] A. Miyake and H. Kondo, *Angew. Chem., Int. Ed. Engl.*, **7**, 631 (1968).

[226] A. Miyake and H. Kondo, *Angew. Chem., Int. Ed. Engl.*, **7**, 880 (1968).

[227] (a) A. Nakamura and S. Otsuka, *Tetrahedron Lett.*, **1973**, 4529; (b) A. Nakamura and S. Otsuka, *J. Amer. Chem. Soc.*, **95**, 7262 (1963).

[228] J. Nasielski, P. Kirsch, and L. Wilputte-Steinert, *J. Organometal. Chem.*, **27**, C13 (1971).

[229] (a) M. Wrighton and M. A. Schroeder, *J. Amer. Chem. Soc.*, **95**, 5764 (1973); (b) G. Platbrood and L. Wilputte-Steinert, *J. Organometal. Chem.*, **70**, 407 (1974); (c) G. Platbrood and L. Wilputte-Steinert, *ibid.*, **70**, 393 (1974); (d) G. Platbrood and L. Wilputte-Steinert, *ibid.*, **85**, 199 (1975); (e) G. Platbrood and L. Wilputte-Steinert, *Tetrahedron Lett.*, **1974**, 2507; (f) M. A. Schroeder and M. S. Wrighton, *J. Organometal. Chem.*, **74**, C29 (1974).

Hexacarbonyl-molybdenum and -tungsten are also effective catalysts under similar conditions, but they promote isomerization of dienes and monoenes. Consequently the reactions are not so specific as those carried out over the chromium catalyst.

The mechanism of hydrogenation over these catalysts is not entirely clear. It has been proposed that photolytic cleavage of a chromium-norbornadiene π bond to form a pentacoordinate intermediate gives an activated complex which then participates in thermal catalytic cycles.[229b–d]

Such a mechanism explains the observed continuation of hydrogenation after initial irradiation is discontinued, and the high quantum yields obtained.

Selectivity for hydrogenation of *trans,trans*-2,4-hexadiene over the *cis,trans* and *cis,cis* isomers is induced by addition of acetone (0.36 M) to the reaction mixture.[229e] The overall rate is increased four-fold by the addition but at the same time hydrogenation of *cis,trans*- and *cis,cis*-2,4-hexadienes is hindered.

Similar 1,4-hydrogenation of 1,3-dienes is catalyzed by $Cr(CO)_3(CH_3CN)_3$ in the absence of ultraviolet irradiation and under far milder conditions than required for the $Cr(arene)(CO)_3$ complexes ($40°/1.5$ atm of hydrogen opposed to $160°/30$ atm).[229f] This catalyst thus offers synthetic advantages over those described above. The tungsten analog, $W(CO)_3(CH_3CN)_3$, is also an active catalyst.

Phosphite complexes of cobalt, $CoCl[P(OC_2H_5)_3]_n$ (where $n = 3$ or 4), catalyze more rapid hydrogenation of acetylenes than of olefins, giving the possibility of selective alkyne hydrogenation.[230a] Elevated temperatures ($75°$) and hydrogen pressures are required. The corresponding hydrido complex, $CoH[P(OCH_3)_3]_4$, is reported to show very slow ligand exchange and very low activity as a catalyst for the hydrogenation of 1-hexene.[230b]

Bis(dimethylglyoximato)(pyridine)cobalt(II) is a useful catalyst for reduction of nitro and azo groups, and of olefins and carbonyl groups conjugated to electron-withdrawing systems.[231]

[230] (a) M. E. Vol'pin and I. S. Kolomnikov, *Katal. Reakts. Zhidk. Faze, Tr. Vses. Konf., 2nd, Alma-Ata, Kaz. SSR*, **1966**, 429. [*C.A.*, **69**, 46340p (1968)]; (b) E. L. Muetterties and F. J. Hirsekorn, *J. Amer. Chem. Soc.*, **96**, 7920 (1974).

[231] (a) Y. Ohgo, S. Takeuchi, and J. Yoshimura, *Bull. Chem. Soc. Jap.*, **44**, 283 (1971); (b) S. Takeuchi, Y. Ohgo, and J. Yoshimura, *ibid.*, **47**, 463 (1974); (c) M. N. Ricroch and A. Gaudemer, *J. Organometal. Chem.*, **67**, 119 (1974).

$$p\text{-}O_2NC_6H_4CH{=}CHCO_2CH_3 \xrightarrow{\;Co(DMG)_2,\,H_2\;} p\text{-}H_2NC_6H_4CH{=}CHCO_2CH_3$$

(78%)

$$C_6H_5COCOC_6H_5 \xrightarrow{\;Co(DMG)_2,\,H_2\;} C_6H_5CHOHCOC_6H_5$$

(99%)

Reduction of dimethyl sulfoxide to dimethyl sulfide is catalyzed by rhodium(III); both the trichloride trihydrate and the trichlorotris(diethyl sulfide) complex are active.[232] Ruthenium tribromide similarly catalyzes reduction of dimethyl sulfoxide but is itself slowly reduced to inactive $RuBr_2(dimethyl\ sulfoxide)_4$.[233] Rhodium and iridium chlorides, as well as the trichlorotris(dimethyl sulfoxide) and trichlorotris(organic sulfide) rhodium complexes in dimethyl sulfoxide solution catalyze hydrogenation of olefins, activated olefins, cyclohexanone (to cyclohexanol), and phenylacetylene.[89a, 234, 235]

Catalytic hydrogenation of olefins is briefly mentioned in the literature as a property of a wide variety of transition metal complexes. The majority of catalysts referred to below require quite mild conditions of temperature and pressure (80° or lower, 1 atm). Isomerization is often a competing reaction.

Olefins are hydrogenated over $RhCl_3(pyridine)_3$,[236] the pentamethylcyclopentadienyl complexes $MHCl_3[C_5(CH_3)_5]_2$ (M = rhodium, iridium),[237] $RhHCl_2(tertiary\ phosphine)_2$, and $RhH_2Cl(tertiary\ phosphine)_2$.[238, 239] The last two catalysts also allow saturation of acetylenes. The N-formylpiperidine complex, $RhCl_3(C_5H_{10}NCHO)_3$, catalyzes hydrogenation of olefins including steroidal double bonds, and of nitrobenzene to aniline.[240] The chloro-bridged polymeric complex $[RhCl_2(2\text{-methylallyl})]_n$ is active in the presence of phosphines, sulfides, and amines.[241]

[232] B. R. James, F. T. T. Ng, and G. L. Rempel, *Can. J. Chem.*, **47**, 4521 (1969).

[233] B. R. James, E. Ochiai, and G. L. Rempel, *Inorg. Nucl. Chem. Lett.*, **7**, 781 (1971).

[234] L. Kh. Freidlin, Yu A. Kopyttsev, N. M. Nazarova, and T. I. Varava, *Izv. Akad. Nauk SSSR, Ser. Khim.*, **1972**, 1420 [*C.A.*, **77**, 125793g (1972)].

[235] B. R. James and F. T. T. Ng, *J. Chem. Soc. (Dalton)*, **1972**, 1321.

[236] R. D. Gillard, J. A. Osborn, P. B. Stockwell, and G. Wilkinson, *Proc. Chem. Soc.*, **1964**, 284.

[237] C. White, D. S. Gill, J. W. Kang, H. B. Lee, and P. M. Maitlis, *Chem. Commun.*, **1971**, 734.

[238] (a) C. Masters, W. S. McDonald, G. Raper, and B. L. Shaw, *Chem. Commun.*, **1971**, 210; (b) C. Masters and B. L. Shaw, *J. Chem. Soc., A*, **1971**, 3679.

[239] D. G. Holah, A. N. Hughes, and B. C. Hui, *Can. J. Chem.*, **50**, 3714 (1972).

[240] I. Jardine and F. J. McQuillin, *Tetrahedron Lett.*, **1972**, 173.

[241] F. Pruchnik, *Inorg. Nucl. Chem. Lett.*, **9**, 1229 (1973).

Cationic complexes of rhodium(I) or iridium(I) with 1,5-cyclooctadiene, acetonitrile, and tertiary phosphine ligands are active for hydrogenation of a variety of olefins.[55, 242] Rhodium salts of tyrosine and other organic acids,[243] and [Ir(cyclooctene)$_2$Cl]$_2$, are active in dimethylacetamide solution but the last-named catalyst deposits metallic iridium in the absence of olefin (*i.e.*, on completion of hydrogenation).[244] Olefinic bonds activated by highly polar substituents (*e.g.*, that in maleic acid) are hydrogenated in dimethylacetamide solution over rhodium trichloride trihydrate,[245] RhCl$_3$(dimethyl sulfide),[23, 246] and [Rh(cyclooctene)$_2$Cl]$_2$ with lithium chloride.[247]

Coordinated bonds are reduced in certain complexes. Hydrogen is absorbed by solutions of Rh[diphenyl-(*o*-vinylphenyl)phosphine]$_2$$^+$, giving solvated Rh[diphenyl-(*o*-ethylphenyl)phosphine]$_2$$^+$;[248] [Rh(norbornadiene)$_2$]$^+$ gives a hydrodimer of norbornadiene.[97, 249]

The ruthenium complexes [Ru(PPh$_3$)$_2$]$^{2+}$,[250] RuH$_2$(PPh$_3$)$_4$,[251] and RuH(NO)(tertiary phosphine)$_3$[213] catalyze hydrogenation of olefins, while olefins activated by polar substituents are saturated over ruthenium(I), (II), and (III) chlorides containing complexed dimethylacetamide.[252, 253a, b] In dimethylacetamide solution ruthenium(II) or (III) is reduced to ruthenium(I) which, in equilibrium with a dihydridoruthenium(III) complex, is the active catalyst.[253c] Aqueous acidic solutions of ruthenium(II) also catalyze hydrogenation of activated olefinic bonds.[254] Hydrogen is added stereospecifically *cis* to fumaric acid, although isotopic labelling studies show it to originate predominantly from the solvent, not from gaseous hydrogen. The related complex ion [RuCl$_4$(bipyridine)]$^{2-}$ shows

[242] M. Green, T. A. Kuc, and S. H. Taylor, *Chem. Commun.*, **1970**, 1553.
[243] V. A. Avilov, Y. G. Borod'ko, V. B. Panov, M. L. Khidekel, and P. S. Shekric, *Kinet. Katal.*, **9**, 698 (1968) [*C.A.*, **69**, 63198r (1968)].
[244] C. Y. Chan and B. R. James, *Inorg. Nucl. Chem. Lett.*, **9**, 135 (1972).
[245] (a) B. R. James and G. L. Rempel, *Discuss. Faraday Soc.*, **46**, 48 (1968); (b) B. R. James and G. L. Rempel, *Can. J. Chem.*, **44**, 233 (1966).
[246] B. R. James and F. T. T. Ng, *J. Chem. Soc.* (*Dalton*), **1972**, 355.
[247] B. R. James and F. T. T. Ng, *Chem. Commun.*, **1970**, 908.
[248] P. R. Brookes, *J. Organometal. Chem.*, **43**, 415 (1972).
[249] R. J. Roth and T. J. Katz, *Tetrahedron Lett.*, **1972**, 2503.
[250] R. W. Mitchell, A. Spencer, and G. Wilkinson, *J. Chem. Soc.* (*Dalton*), **1973**, 846.
[251] S. Komiya, A. Yamamoto, and S. Ikeda, *J. Organometal. Chem.*, **42**, C65 (1972).
[252] B. Hui and B. R. James, *Chem. Commun.*, **1969**, 198.
[253] (a) B. R. James, R. S. McMillan, and E. Ochiai, *Inorg. Nucl. Chem. Lett.*, **8**, 239 (1972); (b) B. C. Hui and B. R. James, *Canad. J. Chem.*, **52**, 3760 (1974); (c) B. C. Hui and B. R. James, *ibid.*, **52**, 348 (1974).
[254] (a) J. Halpern, J. F. Harrod, and B. R. James, *J. Amer. Chem. Soc.*, **83**, 753 (1961); (b) J. Halpern, J. F. Harrod, and B. R. James, *ibid.*, **88**, 5130 (1966); (c) J. Halpern, *Proc. Symp. Coord. Chem.*, Tihany, Hungary, **1964**, 45.

similar activity.[255] Acrylonitrile gives propionitrile (45%) and dihydro-
dimers over $RuCl_2(acrylonitrile)_3$.[256, 257]

Several complexes of palladium(0) and (II), including tertiary phosphine
and dimethyl sulfoxide derivatives, catalyze hydrogenation of monoenes
and dienes.[258, 259] Specific reduction of dienes to monoenes occurs in
several cases.

Butadiene is reduced by $[CoH_2(bipyridine)(tertiary\ phosphine)_2]^+$ and
by the analogous 1,10-phenanthroline complex.[260] The trihydrido complex
$CoH_3(PPh_3)_3$ allows hydrogenation of olefins.[179a, 261, 262] The dinitrogen
cobalt complex $CoHN_2(PPh_3)_3$ catalyzes hydrogenation of monoolefins.[263]
A similar complex of iron, $FeH_2N_2[P(C_2H_5)Ph_2]_3$, catalyzes reduction of
ethylene to ethane.[264]

Hydrogenation of olefins with only a slight degree of accompanying
isomerization is catalyzed by $Mn_2(CO)_{10}$.[265] Hydroformylation takes place
when carbon monoxide is added to the system but, in contrast to $Co_2(CO)_8$,
its presence is not necessary to prevent catalyst decomposition.

The hexacyanodinickel anion, $[Ni_2(CN)_6]^{4-}$, catalyzes hydrogenation of
acetylene to ethylene and of 1,3-butadiene to mixtures of butenes.[266, 267]
The products from the diene show isomeric distributions similar to those
obtained from hydrogenations over $[Co(CN)_5]^{3-}$.

Bis(triphenylphosphine)nickel halides catalyze hydrogenation of poly-
unsaturated fatty esters; the products are mainly monoenes.[268] Similarly,
hydrogenation of unsaturated fats over nickel(II) chloride-sodium borohydride in dimethylformamide gives monoenes. Little isomerization of
double bonds is observed with this system.[269, 270] Hydrogenation of

[255] B. C. Hui and B. R. James, *Inorg. Nucl. Chem. Lett.*, **6**, 367 (1970).
[256] A. Misono, Y. Uchida, M. Hidai, and H. Kanai, *Chem. Commun.*, **1967**, 357.
[257] J. D. McClure, R. Owyang, and L. H. Slaugh, *J. Organometal. Chem.*, **12**, 8 (1968).
[258] (a) E. W. Stern and P. K. Maples, *J. Catal.*, **27**, 120 (1970); (b) E. W. Stern and
P. K. Maples, *ibid.*, **27**, 134 (1972).
[259] (a) L. Kh. Freidlin, N. M. Nazarova, and Yu. A. Kopyttsev, *Izv. Akad. Nauk SSSR*,
Ser. Khim., **1972**, 201 [*C.A.*, **77**, 4634x (1972)]; (b) N. M. Nazarova, L. Kh. Freidlin, Yu. A.
Kopyttsev, and T. I. Varava, *ibid.*, **1972**, 1422 [*C.A.*, **77**, 100943 (1972)].
[260] A. Camus, C. Cocevar, and G. Mestroni, *J. Organometal. Chem.*, **39**, 355 (1972).
[261] A. Misono, Y. Uchida, T. Saito, and K. M. Song, *Chem. Commun.*, **1967**, 419.
[262] J. L. Hendrikse and J. W. E. Coenen, *J. Catal.*, **30**, 72 (1973).
[263] (a) S. Tyrlic, *J. Organometal. Chem.*, **50**, C46 (1973); (b) E. Balogh-Hergovich, G. Speier,
and L. Marko, *ibid.*, **66**, 303 (1974).
[264] V. D. Bianco, S. Doronzo, and M. Aresta, *J. Organometal. Chem.*, **42**, C63 (1972).
[265] T. A. Weil, S. Metlin, and I. Wender, *J. Organometal. Chem.*, **49**, 227 (1973).
[266] (a) M. G. Burnett, *Chem. Commun.*, **1965**, 507; (b) D. Bingham and M. G. Burnett,
J. Chem. Soc., A, **1971**, 1782.
[267] M. S. Spencer, U.S. Pat. 2,966,534 [*C.A.*, **55**, 8288 (1961)].
[268] H. Itatani and J. C. Bailar, *J. Amer. Chem. Soc.*, **89**, 1600 (1967).
[269] P. Abley and F. J. McQuillin, *J. Catal.*, **24**, 536 (1972).
[270] A. G. Hinze and D. J. Frost, *J. Catal.*, **24**, 541 (1972).

unsaturated fats over iron carbonyls has received considerable attention.[165a, 271–273] The iron carbonyls themselves, $Fe(CO)_5$, $Fe_2(CO)_9$, and $Fe_3(CO)_{12}$, or tricarbonyldieneiron complexes (formed by reaction of an iron carbonyl with unsaturated fat) are active. Products containing mainly *trans* monoenes are obtained: the results are not of great interest to the synthetic organic chemist, although possibly of industrial importance.

Stearates and other fat-soluble salts of nickel(II), cobalt(II), iron(III), manganese(II), chromium(III), and copper(II) catalyze hydrogenations. Reductions of cyclohexene under mild conditions have been used to study mechanisms.[274] Industrial interest centers on partial hydrogenation of fatty acids, including conversion to alcohols.[275] The industrial processes frequently involve mixtures of salts, *e.g.*, copper(II) and cadmium(II), under drastic conditions where it is far from clear that the catalysts remain homogeneous.[276] However, Russian work demonstrates that there are conditions involving genuinely homogeneous hydrogenation.[274, 275, 277] The range of olefins hydrogenated (cyclohexene, cyclopentene, 2-pentene, and oleic acid) is too limited to indicate the structural scope of the processes.

A combination of two reactions: (1) addition of boron hydrides to olefins, giving alkylboranes, and (2) reaction of hydrogen with trialkylboranes to produce dialkylborane hydrides plus alkanes gives a procedure for catalytic hydrogenation of olefins.[278, 279] Trialkylboranes and N-trialkylborazanes are effective catalysts although only at elevated temperatures (200°). Cyclohexene is reduced to cyclohexane and the method has also been applied to hydrogenation of polymers, *e.g.*, *cis*-1,4-polybutadiene.[279–281a]

271 (a) I. Ogata and A. Misono, *Nippon Kagaku Zasshi*, **85**, 748 (1964); *ibid.*, **85**, 753 [*C.A.*, **62**, 1203lc, e (1965)]; (b) T. Hashimoto and H. Shiina, *Yukagaku*, **8**, 259 (1959) [*C.A.*, **54**, 25898i (1960)].

272 M. Cais and N. Maoz, *J. Chem. Soc.*, A, **1971**, 1811.

273 (a) E. N. Frankel, H. M. Peters, E. P. Jones, and H. J. Dutton, *J. Amer. Oil Chem. Soc.*, **41**, 186 (1964); (b) E. N. Frankel, E. A. Emken, H. M. Peters, V. L. Davison, and R. O. Butterfield, *J. Org. Chem.*, **29**, 3292 (1964); (c) E. N. Frankel, E. P. Jones, and C. A. Glass, *J. Amer. Oil Chem. Soc.*, **41**, 392 (1964); (d) E. N. Frankel, E. A. Emken, and D. L. Davison, *J. Org. Chem.*, **30**, 2739 (1965).

274 V. A. Tulupov, *Russ. J. Phys. Chem.*, **39**, 1251 (1965).

275 A. J. Pantula and K. T. Achaya, *J. Amer. Oil Chem. Soc.*, **41**, 511 (1964).

276 B. Stouthamer and J. C. Vlugter, *J. Amer. Oil Chem. Soc.*, **42**, 646 (1965).

277 (a) V. A. Tulupov, *Russ. J. Phys. Chem.*, **41**, 456 (1967); (b) V. A. Tulupov, *Proc. Symp. Coord. Chem.*, Tihany, Hungary, **1964**, 57; (c) A. I. Tulupova and V. A. Tulupov, *Russ. J. Phys. Chem.*, **37**, 1449 (1963).

278 (a) H. C. Brown, *Tetrahedron*, **12**, 117 (1961); (b) R. Köster, *Angew. Chem.*, **68**, 383 (1956); (c) R. Köster, G. Bruno, and P. Binger, *Ann.*, **644**, 1 (1961).

279 E. J. DeWitt, F. L. Ramp, and L. E. Trapasso, *J. Amer. Chem. Soc.*, **83**, 4672 (1961).

280 A. I. Yakubchik, B. I. Tikhomirov, and N. I. Shapranova, *Zh. Prikl. Khim.*, **41**, 377 (1968) [*C.A.*, **68**, 96611c (1968)].

281 (a) F. L. Ramp, E. J. DeWitt, and L. E. Trapasso, *J. Org. Chem.*, **27**, 4268 (1961); (b) G. Filardo, M. Galluzzo, B. Giannici, and R. Ercoli, *J. Chem. Soc.* (*Dalton*), **1974**, 1787; (c) J. F. Knifton, *J. Catalysis*, **33**, 289 (1974).

Functional groups that interfere with borane-catalyzed bond isomerization also inhibit hydrogenation, e.g., alcohol, ketone, ester, secondary amine, chlorinated groups. The method would appear to have limited use.

Hydrogenation of aldehydes and ketones is catalyzed by copper(I) and/or copper(II) species produced at the anode during electrolysis under hydrogen pressure (100 to 150 atm).[281b] Acetophenone, benzophenone, cyclohexanone, 2-ethylhexanal, 2-ethyl-2-hexenal, and 2-butanone are reduced by this electrocatalytic method. The necessary use of a high-pressure electrolytic cell would limit the practical utility of the procedure.

Nitroalkanes are hydrogenated to oximes over copper(I), copper(II), and silver(I) salts in alkylpolyamide solvents.[281c] Elevated temperatures and hydrogen pressures are required (105°, 35 atm).

ASYMMETRIC HYDROGENATION

Enzymatic hydrogenations are not only stereospecific, but they also generate optically pure isomers. Soluble catalysts offer an opportunity to imitate such processes. Attempts at modification of heterogeneous catalysts to produce asymmetric environments for hydrogenation of olefinic bonds have usually given low optical yields (often less than 20%). (Optical yield is defined as the excess of one enantiomorph over the racemic mixture.) The highest resolutions (up to 70%) are reported in hydrogenations over palladium deposited on silk fibroin.[282] Modification of Raney nickel with various amino acids and tartaric acid derivatives gives up to 56% optical yield in hydrogenation of a ketonic substrate.[283]

The obvious approach with homogeneous systems is to synthesize asymmetric catalyst molecules by incorporation of optically active ligands and find those which give the highest optical yields. (Resolution of complexes in which the optical activity lies at the metal would appear to be an alternative, and perhaps preferable, procedure, but racemization of such complexes by dissociation of ligands during the catalytic cycle would be almost certain to occur.) Results so far are very encouraging, but there is an indication that catalysts may have to be tailored to individual substrates in order to obtain optimum yields.

Catalysts, probably of the Wilkinson type, are generated by hydrogenation of rhodium(III) precursors $RhCl_3L_3^*$ (L^* is an optically active tertiary phosphine). Using the complex derived from (−)-methylpropylphenylphosphine, α-phenylacrylic acid (atropic acid) is hydrogenated to optically

[282] J. D. Morrison and H. S. Mosher, *Asymmetric Organic Reactions*, Prentice-Hall, Englewood Cliffs, New Jersey. 1972, pp. 292, 297.
[283] S. Tatsumi, *Bull. Chem. Soc. Jap.*, **41**, 408 (1968), and references therein.

active hydratropic acid in 22% optical yield.[284a, 285] Hydrogenation of itaconic acid results in lower optical yield.

$$CH_2=\overset{R}{\underset{|}{C}}CO_2H \xrightarrow[\underset{CH_2CO_2H}{\underset{C_6H_5}{R}}]{RhCl_3L_3^*,\ H_2} \begin{array}{c} Optical\ Yield \\ 22\% \\ 4\% \end{array} CH_3\overset{R}{\underset{*}{C}}HCO_2H$$

The Wilkinson catalyst generated *in situ* by reaction of (+)-methyl-propylphenylphosphine with [Rh(1,5-hexadiene)Cl]$_2$ gives 8 and 4% optical yields from α-ethylstyrene and α-methoxystyrene, respectively.[286] A model based on conformational and steric considerations explains the products observed. The same system gives lower yields when applied to other substrates.[287] Hydrogenation of α-acylaminoacrylic acids using *o*-anisylcyclohexylmethylphosphine and other optically active phosphines gives particularly high optical yields (Table XI).[288] Since the D or L α-amino acid derivative is often easily separated from the DL mixture by crystallization, this reaction constitutes a practical synthesis of optical isomers that does not involve a classical resolution step.

Chiral phosphines in which the asymmetric center lies not at phosphorus but in an alkyl side chain can also be employed. One approach using (2-methylbutyl)diphenylphosphine, PPh$_2$[CH$_2$$\overset{*}{C}$H(CH$_3$)CH$_2CH_3$], gives low optical yields (1%, hydrogenation of α-phenylacrylic acid) in analogs of RhCl(PPh$_3$)$_3$ and somewhat higher yields (14%, hydrogenation of itaconic acid) in analogs of RhH(CO)(PPh$_3$)$_3$.[284] Good results are obtained, however, using neomenthyldiphenylphosphine in conjunction with rhodium(I) complexes, hydrogenation of atropic acid and cinnamic acid derivatives giving optical yields of 28–61%.[289] Hydrogenation of noncar-

$$\begin{array}{c} CH_3 \\ \diagdown \\ C_6H_5 \end{array}\!C=C\!\begin{array}{c} CO_2H \\ \diagup \\ H \end{array} \longrightarrow C_6H_5-\overset{CH_3}{\underset{*}{C}}H-CH_2CO_2H$$

(Optical yield 61%)

boxylic acid substrates gives lower yields: α-ethylstyrene gives 7% optical yield (compared to 8% when the asymmetry is at phosphorus). An obvious

[284] (a) W. S. Knowles and M. J. Sabacky, *Chem. Commun.*, **1968**, 1445; (b) W. R. Cullen, A. Fenster, and B. R. James, *Inorg. Nucl. Chem. Lett.*, **10**, 167 (1974).

[285] Anon., *Chem. Eng. News*, **48**, (29), 41 (1970).

[286] L. Horner, H. Siegel, and H. Buthe, *Angew. Chem., Int. Ed. Engl.*, **7**, 942 (1968).

[287] W. S. Knowles, M. J. Sabacky, and B. D. Vineyard, *Ann. N.Y. Acad. Sci.*, **172**, 232 (1970).

[288] W. S. Knowles, M. J. Sabacky, and B. D. Vineyard, *Chem. Commun.*, **1972**, 10.

[289] J. A. Morrison, R. E. Burnett, A. M. Aguiar, C. J. Morrow, and C. Phillips, *J. Amer. Chem. Soc.*, **93**, 1301 (1971).

TABLE XI. ASYMMETRIC HYDROGENATION OF α-ACYLAMINOACRYLIC ACIDS[288]

(Catalyst: [Rh(1,5-hexadiene)Cl]$_2$ + 2 equiv of chiral phosphine)

Chiral Phosphine,

$$R^2\!-\!\overset{*}{P}\!-\!R^3$$
(with R^1)

Substrate,

$$R^4CH\!=\!\overset{NHCOR^5}{\underset{|}{C}}\!-\!CO_2H$$

Product,

$$R^4CH_2\!-\!\overset{NHCOR^5}{\underset{*}{\underset{|}{CH}}}\!-\!CO_2H$$

R^1	R^2	R^3	R^4	R^5	Optical yield (%)
o-Anisyl	CH$_3$	C$_6$H$_5$	3-CH$_3$O-4-HOC$_6$H$_5$	C$_6$H$_5$	61
CH$_3$	C$_6$H$_5$	n-C$_3$H$_7$	"	"	31
"	"	i-C$_3$H$_7$	"	"	31
o-Anisyl	Cyclohexyl	CH$_3$	"	"	92–95
"	"	"	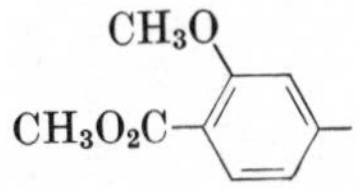	CH$_3$	86–93
"	"	"	C$_6$H$_5$	"	89
"	"	"	"	C$_6$H$_5$	89
"	"	"	H	CH$_3$	63

practical advantage in using phosphines such as the neomenthyl derivative is their preparation from readily available optically active materials.

Yet another approach to the problem is the use of chiral diphosphines.[290a, b] Once again an advantage lies in preparation from naturally occurring optically active compounds, e.g., (+)-ethyl tartrate. Catalysts

formed from **2** [2,3-O-isopropylidene-2,3-dihydroxy-1,4-bis(diphenylphosphino)butane, abbreviated diop] and [Rh(cyclooctene)$_2$Cl]$_2$ (diop: rhodium = 1:1) give good optical yields in hydrogenation of acrylic acid derivatives (Table XII). The free carboxylic acids generally give the best

TABLE XII. ASYMMETRIC HYDROGENATION OF ACRYLIC
ACID DERIVATIVES[290a,b]
(Catalyst: [Rh(cyclooctene)$_2$Cl]$_2$ + 1 equiv of **2**)

Substrate, $\begin{array}{c}R^1 \\ H\end{array}C=C\begin{array}{c}R^2 \\ COR^3\end{array}$			Product, $\begin{array}{c}COR^3 \\ H-\!\!\!\!-R^2 \\ CH_2R^1\end{array}$
R^1	R^2	R^3	Optical yield (%)
H	C_6H_5	OH	63
H	$NHCOCH_3$	OH	73
H	$NHCOCH_2C_6H_5$	OH	68
C_6H_5	$NHCOCH_3$	OH	72
HOC_6H_4	,,	OH	80
	,,	OH	79
C_6H_5	,,	OCH_3	55
C_6H_5	,,	NH_2	71
HOC_6H_4	$NHCOC_6H_5$	OH	62
i-C_3H_7	,,	OH	22

[290] (a) T. P. Dang and H. B. Kagan, Chem. Commun., **1971**, 481; (b) H. B. Kagan and T. P. Dang, J. Amer. Chem. Soc., **94**, 6429 (1972); (c) A. Levi, G. Modena, and G. Scorrano, Chem. Commun., **1975**, 6; (d) T. Hayashi, K. Yamamoto, and M. Kumada, Tetrahedron Lett., **1974**, 4405.

optical yields, although amides and methyl esters can give good results. As expected, no hydrogenation occurs when the olefinic bond is highly hindered; an example is $(CH_3)_2C=C(NHCOC_6H_5)CO_2H$.

Other catalyst systems using diop as the asymmetric ligand also give good optical yields. Hydrogenation of α-acetamidoacrylic acid over $RhH[(+)\text{-diop}]_2$ gives N-acetyl-(S)-alanine of 60% optical purity.[284b] Use of $\{Rh(norbornadiene)[(+)\text{-diop}]\}^+ClO_4^-$ as catalyst for hydrogenation of α-acetamido-(Z)-cinnamic acid gives N-acetyl-(R)-phenylalanine in 78 to 85% optical yield (depending on solvent).[290c]

Yet another approach to the synthesis of chiral phosphines involves stereoselective lithiation of (S)- or (R)-α-ferrocenylethyldimethylamine (an easily resolved derivative of ferrocene), with the introduction of one or two phosphino groups to give mono- or di-phosphines.[290d] These ferro-

$$R = CH_3, Ph$$

cenylphosphines have been applied in asymmetric hydrosilylation of ketones, but may also be useful for hydrogenation reactions.

Many of the hydrogenations described above are carried out in the presence of a base, usually triethylamine, which leads to increased optical yields and increased rates of hydrogenation.[284a, 285, 287–290a,b] This does not universally occur, however; addition of triethylamine has little influence in some hydrogenations.[290a,b]

An extension of the McQuillin catalyst, $[RhCl(pyridine)_2(dimethyl-formamide)(BH_4)]Cl$, where coordinated dimethylformamide is replaced by optically active amides, gives optical yields up to 60% in hydrogenation of methyl β-methylcinnamate.[291a] The enantiomer of methyl 3-phenyl

$$\underset{\overset{|}{C_6H_5-C=CHCO_2CH_3}}{CH_3} \longrightarrow \underset{\overset{|}{C_6H_5-CH-CH_2CO_2CH_3}}{CH_3}$$

butanoate produced by each of a series of amides is related to the stereochemistry of the individual amide and its coordination to rhodium. The majority of amides can be prepared from readily available optically active substances (lactic acid, camphor, and glucosamine). The optical activity induced does not depend on whether the solvent is the amide itself or the amide in dilute solution in diethylene glycol monoethyl

[291] (a) P. Abley and F. J. McQuillin, J. Chem. Soc., C, 1971, 844; (b) P. H. Boyle and M. T. Keating, Chem. Commun., 1974, 375.

ether. Thus the product results from a truly asymmetric catalytic process and not merely from asymmetric solvation. Use of the amide in dilute solution has obvious practical advantages.

Similar use of the McQuillin catalyst in $(+)$ or $(-)$-N-(1-phenylethyl)-formamide allows asymmetric hydrogenation of the 5,6 carbon-nitrogen double bond of folic acid.[291b]

$$R = MHC_6H_4CONH.CH(CO_2H)CH_2CH_2CO_2H$$

Attempts to carry out asymmetric hydrogenations using cyanocobalt catalysts in conjunction with optically active amines have met with only limited success.[292, 293] Potassium atropate hydrogenated over $K_4[(CN)_4Co$-$(\mu$-CH$_3$NĊH(CH$_3$)CH$_2$NHCH$_3)$Co(CN)$_4]$ gives hydratropic acid in 7% optical yield.

Asymmetric homogeneous hydrogenation of ketones has also been attempted. Using [Rh(norbornadiene)(tertiary phosphine)$_2$]$^+$ClO$_4^-$ containing $(+)$-benzylmethylphenylphosphine, hydrogenations of acetone and 2-butanone give optical yields of 7 and 2% respectively.[294] [Rh(norbornadiene)(diop)]$^+$ClO$_4^-$ gives a similarly low optical yield in the hydrogenation of acetophenone, but its use in saturation of the carbon-nitrogen double bond of acetophenone benzylimine results in improved optical yields (16 to 22% depending on solvent).[290c] Hydrogenation of benzil to $(+)$-benzoin (up to 61% optical yield) is achieved using bis(dimethylglyoximato)quininecobalt(II) as catalyst.[295] A crystallographic study of a

$$C_6H_5COCOC_6H_5 \rightarrow C_6H_5\overset{*}{C}HOHCOC_6H_5$$

related complex does not provide evidence of direct interaction between the asymmetric portion of the catalyst and the substrate.[296]

None of the catalysts so far obtained is ideal, but the results promise the availability of a useful range of catalysts in the near future.

[292] Y. Ohgo, S. Takeuchi, and J. Yoshimura, Bull. Chem. Soc. Jap., **43**, 505 (1970).

[293] Y. Ohgo, K. Kobayashi, S. Takeuchi, and J. Yoshimura, Bull. Chem. Soc., Jap., **45**, 933 (1972).

[294] P. Bonvicini, A. Levi, G. Modena, and G. Scorrano, Chem. Commun., **1972**, 1188.

[295] Y. Ohgo, S. Takeuchi, and J. Yoshimura, Bull. Chem. Soc. Jap., **44**, 583 (1971).

[296] Y. Ohashi, Y. Sasada, Y. Tashiro, Y. Ohgo, S. Takeuchi, and J. Yoshimura, Bull. Chem. Soc. Jap., **46**, 2589 (1973).

EXPERIMENTAL PROCEDURES

The following experimental procedures include, where appropriate, preparations of catalysts and one or more examples of their use in hydrogenation. Except for the more thoroughly investigated systems, explicit experimental details are lacking in much of the literature. Usually, however, procedures and apparatus are those commonly employed for hydrogenations over heterogeneous catalysts, with two notable modifications: (1) many of the homogeneous catalysts are unstable toward oxygen (especially when in solution) and precautions must be taken to deoxygenate solvents and exclude air; (2) the catalysts cannot be separated by filtration on completion of hydrogenation. Their removal often involves chromatography (the organometallic complexes are usually decomposed and strongly adsorbed on silica gel or alumina columns) or distillation of products (the complexes are generally nonvolatile).

General experimental conditions (solvents, temperature, pressure, catalyst concentration) for each catalyst are given in the preceding discussions of individual catalysts.

Chlorotris(triphenylphosphine)rhodium—RhCl(PPh$_3$)$_3$

Preparation.[8] To a solution of freshly recrystallized triphenylphosphine (12 g, 6 M excess) in hot ethyl alcohol (350 ml) was added a solution of rhodium trichloride trihydrate (2 g) in hot ethyl alcohol (70 ml) and the solution was heated under reflux for 30 minutes. The hot solution was filtered, cooled, and the burgundy-red crystals of the complex were washed with degassed ether (50 ml) and dried under reduced pressure. The yield was 6.25 g (88 % based on rhodium); mp 157–158°.

If more concentrated solutions were used (200 ml or less of ethyl alcohol), orange crystals of the complex were obtained after heating under reflux for 5 minutes; they were often mixed with a small quantity of the above-mentioned red complex. On continued heating under reflux gradual conversion of the orange crystals to the deep-red form occurred. The infrared spectra of the two forms of the complex from 4000–600 cm^{-1} showed no differences. Their chemical properties appeared to be identical, and in particular there was no difference in catalytic behavior when they were dissolved.

The preparation of the complex could also be carried out in aqueous acetone. Thus 0.1 g of rhodium trichloride trihydrate in 5 ml of water was added to a hot solution of 0.6 g of triphenylphosphine in 25 ml of acetone and heated under reflux. Orange crystals of the complex were deposited after a few minutes.

The excess triphenylphosphine used in the preparations was recovered by addition of water to the filtrates until precipitation began; on standing for 2–3 days, triphenylphosphine crystallized. Recrystallization from ethyl alcohol and ethyl alcohol-benzene gave pure material.

Hydrogenation Procedure and General Observations.[8] The catalyst (10^{-4} mol, used in a total volume of 80 ml) was weighed into a small glass bucket suspended from a side arm of the reaction flask. Rotation of the side arm enabled the bucket to drop into the solution. The stirrer was a Teflon-coated magnet driven by the external motor so that it operated at the gas-liquid interface, *i.e.*, on the side wall of the round flask as shown in Figure 2. This gave very efficient stirring in of hydrogen.

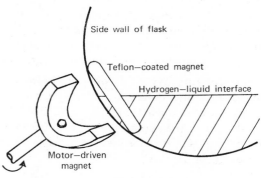

Side wall of flask

Teflon—coated magnet

Hydrogen—liquid interface

Motor—driven magnet

FIG. 2.

The solvent was first degassed by careful evacuation and the system was flushed three times with hydrogen. The catalyst was added by means of the bucket and the solution stirred until the solid dissolved, giving a pale-yellow solution. Stirring was stopped and the freshly distilled (or otherwise deoxygenated) substrate was added; the color of the solution became deep brown, but absorption of hydrogen did not occur until stirring was recommenced.

Additional precautions taken when using this catalyst, including purification of gases and solvents, have been enumerated.[19b]

4,5,6,7-Tetrahydroindane (Selective Hydrogenation of 4,7-Dihydroindane).[45]

(*a*) 4,7-Dihydroindane was prepared by Birch reduction using a simplification of the procedure of Giovannini and Wegmuller.[297] Indane (6 g) in 5 ml of methyl alcohol was added with vigorous stirring to 150 ml of liquid

---

[297] E. Giovannini and H. Wegmuller, *Helv. Chim. Acta*, **41**, 933 (1958).

ammonia and treated with sufficient sodium to keep the solution blue for
1 hour. After addition of 10 ml of ethyl alcohol and evaporation of the
ammonia, 150 ml of water was added and the mixture extracted with
pentane (2 × 50 ml). Distillation of the product using a nitrogen bleed
gave 4,7-dihydroindane as a colorless liquid, bp 64–64.5°/15 mm (4.02 g,
66%).

(b) The dihydroindane (1 g) absorbed 1 M proportion of hydrogen in
2 hours when treated with 100 mg of RhCl(PPh₃)₃ in 20 ml of benzene
under 1 atm at room temperature. After filtration through Florisil in
benzene, 4,5,6,7-tetrahydroindane was distilled as a colorless oil (718
mg, 72%).

The mass spectrum showed m/e 122; gas-liquid chromatography on
Carbowax 1540 showed a purity of 91% with 3% of indane (present also
in the dihydroindane). The nature of the product was deduced from the
amount of hydrogen absorbed and from the complete absence of vinylic
proton resonances in the nmr spectrum. No other satisfactory method for
the preparation of this compound is reported in the literature.

4-Methyl-4-trichloromethyl-2-cyclohexen-1-one (Lack of Hydrogenolysis of Carbon-Chlorine Bonds).[45] Hydrogenation of 4-methyl-4-
trichloromethyl-2,5-cyclohexadien-1-one (1 g) using RhCl(PPh₃)₃ (400 mg)
in benzene (40 ml) at room temperature and 1 atm for 24 hours, followed
by filtration through Florisil (elution with ether) and two recrystallizations
of the product from light petroleum (bp 40–60°) gave 4-methyl-4-trichloro-
methyl-2-cyclohexen-1-one as colorless plates (640 mg), mp 58–62°.
Nmr spectroscopy showed the presence of a small proportion of starting
material which was not removed by seven more recrystallizations (the
melting point also remained unchanged) or by chromatography on alumina.

The very much slower absorption of further hydrogen to give the cyclohexanone derivative was possibly due to the stronger deactivating effect
of the carbonyl group in the monoene.

13,14-Dihydroeremophilone (Selective Hydrogenation of Eremophilone).[30] A solution of 0.102 g of eremophilone and 0.07 g of RhCl(PPh₃)₃
in 15 ml of benzene was stirred under an atmosphere of hydrogen for
8 hours. The solution was then passed through an alumina column (3 g)
and, on evaporation of the solvent and distillation of the residue, afforded
pure 13,14-dihydroeremophilone (0.097 g, 94%), bp 100°/1 mm.

4-Cholesten-3-one (Selective Hydrogenation of Steroid 1,4-
Dien-3-one Systems).[298, 299] 1,4-Cholestadien-3-one (1.5 g) together

[298] D. E. M. Lawson, B. Pelc, P. A. Bell, P. W. Wilson, and E. Kodicek, *Biochem. J.*,
121, 673 (1971).
[299] B. Pelc and E. Kodicek, *J. Chem. Soc.*, C, **1971**, 3415.

with 0.7 g of RhCl(PPh$_3$)$_3$ in 75 ml of benzene was hydrogenated at atmospheric pressure. After 4–5 hours uptake of hydrogen stopped. The solution was evaporated to dryness and the residue refluxed with 50 ml of light petroleum (bp 60–80°) to decompose the catalyst. The resulting solution was filtered while hot and the residue washed with three portions of hot light petroleum (20 ml each) and one portion of diethyl ether. Filtrate and washings were combined and evaporated, giving 1.5 g of 4-cholesten-3-one. The product showed only one spot when chromatographed on thin-layer chromatography plates of silica gel with chloroform, identical with authentic 4-cholesten-3-one (R^f 0.3).

[1,2-^{3}H$_2$]-4-Cholesten-3-one (Selective Tritiation of Steroid 1,4-Dien-3-one Systems).[298, 299] Repetition of the procedure above with 1,4-cholestadien-3-one (2 g), RhCl(PPh$_3$)$_3$ (0.5 g), and tritium gas (58 Ci/mmol, 350 Ci) gave a crude product in which 67% of the radioactivity was present in 4-cholesten-3-one. Chromatography of this material on 200 g of Florisil and elution with light petroleum(bp 60–80°)–benzene (3:1, v/v) gave successively cholestan-3-one (50 Ci), a mixture of cholestan-3-one and 4-cholesten-3-one (100 Ci, containing 56% of 4-cholesten-3-one) and 4-cholesten-3-one (111 Ci). The impure fractions were rechromatographed on a second Florisil column where further separation was achieved.

Analogous procedures gave 4,22-ergostadien-3-one and [1,2-^{3}H$_2$]-4,22-ergostadien-3-one from 1,4,22-ergostatrien-3-one.

Cholestan-3-one Dimethyl Ketal (Simultaneous Hydrogenation and Ketalization of 4-Cholesten-3-one).[50] 4-Cholesten-3-one (200 mg) and 500 mg of RhCl(PPh$_3$)$_3$ were weighed into a hydrogenation flask and the flask was purged with hydrogen. Methyl alcohol (100 ml) was added slowly and the solution stirred under hydrogen at 1 atm pressure for 4 days. The reaction mixture was evaporated to dryness, the residue taken up in a little chloroform and run on to a preparative thin-layer chromatography plate (silica gel GF$_{254}$; bands were located using ultraviolet light and compounds were eluted with dichloromethane). The plate was developed using cyclohexane/ethyl acetate (96:4). Unreacted ketone (63 mg) was recovered, and reaction products totaling 117 mg were obtained. Further chromatography of the latter fraction using cyclohexane/ethyl acetate (98:2) gave cholestan-3-one (9 mg, 4%), mp 124–127°, and cholestan-3-one dimethyl ketal (90 mg, 40%), mp 77–78°.

Chlorotris(trisubstituted Phosphine)rhodium(Preparation in situ).
A. *Chlorobis(cyclooctene)rhodium*.[300] Rhodium trichloride trihydrate (2g)

---

[300] A. van der Ent and A. L. Onderdelinden, *Inorg. Synth.*, **14**, 93 (1973).

was dissolved in an oxygen-free mixture of 40 ml of isopropyl alcohol and 10 ml of water. Cyclooctene (6 ml) was added and the solution stirred for about 15 minutes under nitrogen. The flask was then closed and allowed to stand at room temperature for 5 days. The resulting reddish-brown crystals were collected on a filter, washed with ethyl alcohol, dried under vacuum, and stored under nitrogen at $-5°$. The yield was 2 g (74%).

B. *The Catalyst*.[35] (See p. 8 for equation.) The reaction flask containing chlorobis(cyclooctene)rhodium (3.9 mg, 1.1×10^{-2} mmol) was purged with hydrogen, then a degassed solution of a phosphine ligand ($2.3 - 3.2 \times 10^{-2}$ mmol) in 3 ml of benzene and 1 ml of ethyl alcohol was added via hypodermic syringe. The system was first shaken gently under hydrogen for 5 minutes then strongly agitated for an additional 5 minutes.

The substrate was added and the rest of the procedure was identical to that used with an externally prepared catalyst. The lower ratio of phosphine to rhodium gave higher rates of hydrogenation.

N-Acetylamino Acids (Preparation of Catalyst Containing Chiral Diphosphine and Asymmetric Hydrogenation of Substituted Acrylic Acids).[290b]

$$[\text{RhCl}(\text{C}_8\text{H}_{14})_2]_n + \begin{array}{c} \text{H} \\ \text{CH}_3 \quad \text{O} \quad \text{CH}_2\text{PPh}_2 \\ \diagup\!\!\!\!\diagdown \qquad \diagdown \\ \text{CH}_3 \quad \text{O} \quad \text{CH}_2\text{PPh}_2 \\ \text{H} \end{array} \longrightarrow \begin{array}{c} \text{P} \\ \diagup \quad \diagdown \\ \quad \quad \text{RhCl(solvent)} + 2\ \text{C}_8\text{H}_{14} \\ \diagdown \quad \diagup \\ \text{P} \end{array}$$

(a) To a benzene solution of chlorobis(cyclooctene)rhodium (3×10^{-3} M) under argon was added the diphosphine (1 equiv of diphosphine per rhodium). The solution was stirred for 15 min and was introduced into the hydrogenation flask by means of a syringe, avoiding any contact with air.

(b) The order of addition of reactants into the flask was substrate, hydrogen, ethyl alcohol, catalyst solution. The ratio of ethyl alcohol to benzene was 2:1 to 4:1. On completion of hydrogenation at atmospheric pressure and room temperature the solution was evaporated to dryness and the following procedures were used to isolate the products mentioned.

A. For N-acetylalanine and N-acetyltyrosine the residue was dissolved in water and separated from the insoluble catalyst by filtration. Evaporation to dryness afforded the product.

B. For N-acetylphenylalaninamide, N-acetylphenylalanine methyl ester, and 1-(N-acetylamino)-1-phenylpropane the product was isolated by thin-layer chromatography on silica gel. The eluents were acetone-methyl alcohol for the first compound and ethyl acetate-hexane for the other two.

C. For other N-acetylamino acids the residue was dissolved in 0.5 *M* sodium hydroxide and separated from the insoluble catalyst by filtration. The filtrate was acidified with dilute hydrochloric acid, extracted with ether, and washed with a little water. The ethereal phase was dried over sodium sulfate and evaporated to dryness.

Rhodium Borohydride Complex

Preparation.[57] Rhodium trichloride was refluxed with pyridine in ethyl alcohol solution. Evaporation under reduced pressure gave yellow crystals of RhCl$_3$(pyridine)$_3$.

Sodium borohydride (1 equiv) was added to a saturated solution of RhCl$_3$(pyridine)$_3$ in dimethylformamide. Addition of diethyl ether precipitated [RhCl(C$_5$H$_5$N)$_2$(HCONMe$_2$)(BH$_4$)]Cl which was obtained as airstable, dark-red crystals by recrystallization from chloroform.

In situ Preparation and Hydrogenation Procedure.[57, 59] Finely ground sodium borohydride (1 equiv) was added to a warm solution of RhCl$_3$(pyridine)$_3$ in dimethylformamide (10^{-3} to 10^{-2} *M* in rhodium) under hydrogen. The substrate was introduced through a side arm. Hydrogenation products were isolated by dilution with water followed by extraction and further purification by chromatography when necessary.

For *in situ* preparation using other amides (*e.g.*, optically active amides), a solution of RhCl$_3$(pyridine)$_3$–sodium borohydride (4.5 × 10^{-3} *M*) was prepared either in the neat amide, or in a 5% solution of the amide in diethylene glycol monoethyl ether or diethylene glycol monoethyl ether-water (10:1). The mixture was shaken under hydrogen and hydrogenation was carried out by the procedure above.

Piperidine and 1,2,3,4-Tetrahydroquinoline.[57] Pyridine (7.5 × 10^{-2} *M*) was hydrogenated in dimethylformamide with RhCl$_3$(pyridine)$_3$-sodium borohydride (7.5 × 10^{-3} *M*) at room temperature and 1 atm. The rate of reaction accelerated markedly with hydrogen uptake. The product, piperidine, was isolated by chromatography on alumina and characterized as the hydrochloride, mp 237°, and benzenesulfonyl derivative, mp 92°.

The same procedure with quinoline furnishes the 1,2,3,4-tetrahydro derivative (hydrochloride, mp 178°, and platinichloride, mp 203°).

Hydrogenation of Nitro Compounds (Preparation of Cyclohexylamine, Aniline, p-Aminobenzoic Acid, and p-Toluidine).[62a] Nitrocyclohexane (204 mg) with 200 mg of the catalyst in 10 ml of dimethylformamide absorbed 100 ml of hydrogen in 17 hours to give cyclohexylamine, identified by thin-layer chromatography and conversion to N-cyclohexylbenzamide, mp 148°.

Nitrobenzene (171 mg) similarly absorbed 94 ml of hydrogen in 12 hours to give aniline, identified by nmr spectroscopy and as acetanilide, mp 112°.

4-Nitrobenzoic acid (190 mg) similarly absorbed 67 ml of hydrogen in 6 hr to give 4-aminobenzoic acid, identified by nmr spectroscopy and as 4-acetamidobenzoic acid, mp 253°.

4-Nitrotoluene gave p-toluidine identified by nmr spectroscopy and as 4-acetamidotoluene, mp 145°.

Diphenylcarbinol (Benzhydrol).[62a] Benzophenone (210 mg) with 60 mg of RhCl$_3$(pyridine)$_3$-sodium borohydride in 10 ml of dimethylformamide absorbed 56 ml of hydrogen in 4 days to give diphenylmethyl alcohol and a trace of benzophenone. (In the absence of hydrogen under the same conditions, no diphenylcarbinol was detected.)

Acetophenone similarly gives 1-phenylethyl alcohol.

Henbest Catalyst—Chloroiridic Acid-Trimethyl Phosphite

3-Sterols (General Procedure for Reduction of Steroid 3-ketones).[90] The ketone (1 part, 0.1–10 g), chloroiridic acid (1/20 part), trimethyl phosphite (2 parts, v/w), and 90% aqueous isopropyl alcohol (25 parts, v/w) were heated together under reflux. Drops of the solution were withdrawn at intervals varying from 8 to 24 hours and spotted directly on a thin-layer chromatography plate (silica gel) together with a reference spot of the starting material in isopropyl alcohol solution. Development of the plate with suitable solvents (usually benzene-ethyl acetate) and exposure to iodine vapor gave distinct brown spots corresponding to ketone, axial C$_3$ alcohol and, usually, also a trace of equatorial C$_3$ alcohol. Nonsteroidal products derived from the reagents remained near the base of the plate as white spots barely affected by iodine. Reaction was stopped when little or none of the oxo-steroid was detected, usually after 48–72 hours. Greatly prolonged reaction times led to the gradual appearance of faster-moving spots than those of the original ketones, probably resulting from formation of steroid isopropyl ethers or elimination products.

The reaction mixture was finally cooled, poured into water, and the products were extracted by use of ether-benzene or other suitable solvents. The organic phase was washed with water and dilute aqueous sodium bicarbonate, and the solvents were removed. The axial alcohols frequently crystallized easily from such crude products, or they could be separated from contaminants by chromatography if necessary.

Potassium Pentacyanocobaltate(II)—$K_3[Co(CN)_5]$

2-Hexenoic Acid (Selective Hydrogenation of Sorbic Acid).[142] Sorbic acid (1 g, 0.0089 mol), sodium hydroxide (0.36 g, 0.009 mol), potassium cyanide (0.685 g) and cobalt(II) chloride (0.5 g) in 25 ml of distilled water were placed in a 100 ml stainless steel autoclave and hydrogenated at 70° under 50 atm of hydrogen for 8 hours. The cooled reaction mixture was acidified with 2 M hydrochloric acid and repeatedly extracted with ether. After drying the combined extracts, the ether was removed, and 2-hexenoic acid was distilled at 101–102°/12 mm to give 0.61 g (60 %).

Phenylalanine A. (Hydrogenation of the Oxime of Phenylpyruvic Acid.)[142] To a solution of the oxime of phenylpyruvic acid (0.37 g, 0.0021 mol) and sodium hydroxide (0.082 g, 0.0025 mol) in 25 ml of distilled water was added potassium cyanide (0.685 g, 0.0105 mol). The mixture was poured into an autoclave, fine crystals of cobalt(II) chloride (0.5 g, 0.0021 mol) were added, and the autoclave was immediately sealed. The reaction went to completion at 70° under a hydrogen pressure of 50 atm in 8 hours. (The reaction mixture was homogeneous. When 2 ml were treated with 2,4-dinitrophenylhydrazine in hydrochloric acid only a trace of hydrazone derivative was detected.)

The solution was evaporated to dryness under reduced pressure. The residue was dissolved in the minimum amount of water and filtered. The filtrate was run on to a column of Ion-Exchange Amberite I.R.120 (washed with 2 M ammonium hydroxide, water, 2 M hydrochloric acid and water) and eluted with water until the eluent was free of halide ions. The amino acid was then eluted with 2 M ammonium hydroxide and the excess ammonia was evaporated to dryness under reduced pressure to give phenylalanine (0.28 g, 82 %). After recrystallization from water the product melted at 267–270° (dec). The identity of the phenylalanine obtained was confirmed by paper chromatographic analysis and by mixed melting-point determination.

B. (Reductive Amination of Phenylpyruvic Acid).[142] A solution of phenylpyruvic acid (1 g, 0.0061 mol) in 6% ammonium hydroxide, potassium cyanide (0.685 g, 0.0105 mol), and cobalt(II) chloride (0.5 g, 0.0021 mol) were hydrogenated according to (a) above. The reaction was complete after 8 hours at 40° under 50 atm of hydrogen. The cooled solution was treated with hydrogen sulfide to remove cobalt ion, evaporated to dryness under reduced pressure, and worked up as detailed above to give phenylalanine (0.856 g, 85.6 %).

Replacing ammonium hydroxide and cobalt(II) chloride with sodium hydroxide and chloropentamminecobalt(III) chloride gave similar reductive amination (70°, 50 atm, 5 hours); yield 94 %.

Octacarbonyldicobalt—$Co_2(CO)_8$

Full preparative details for this complex are given in References 146–150. It is toxic, air-sensitive, and unstable in the absence of an atmosphere of carbon monoxide. Suitable precautions must therefore be taken.

5,12-Dihydronaphthacene (Partial Hydrogenation of Naphthacene).[173] A solution of naphthacene (2.8 g, 0.012 mol) and 2 g of the catalyst in 85 ml of benzene was placed in a 200-ml stainless steel autoclave. Synthesis gas (hydrogen:carbon monoxide, 1:1) was added to a pressure of 200 atm, the autoclave was heated with rocking to 140° within 90 minutes and held at this temperature for 5 hours. The autoclave was allowed to cool overnight and gases were vented to the atmosphere.

The benzene was removed by evaporation and replaced by toluene. The solution was refluxed for 24 hours, during which time the catalyst was completely decomposed. The solution was filtered and the toluene removed under reduced pressure. The residue was dissolved in light petroleum (bp 60–80°) and chromatographed on activated alumina using light petroleum, benzene, chloroform, and ethyl alcohol as eluents.

Two principal fractions were isolated, one identified as unreacted naphthacene (30%), the other as 5,12-dihydronaphthacene (70%), mp 209–210°.

Dicarbonylbis-(π-cyclopentadienyl)titanium—$Ti(C_5H_5)_2(CO)_2$

Preparation and Properties.[301] To a solution of n-butyllithium (0.15 mol) in 80 ml of diethyl ether at $-10°$ was added to $Ti(C_5H_5)_2Cl_2$ (16.9 g, 0.07 mol). The mixture was stirred for 1 hr under nitrogen while warming to room temperature. It was added to a 1-l rocking autoclave under nitrogen, pressurized to 240 atm with carbon monoxide, and heated to 150° for 8 hours. After cooling, the contents were removed under nitrogen and the autoclave was washed out with deaerated benzene.

The solvent was removed from the resulting dark red-brown solution by distillation under reduced pressure, the residue was treated with 100 ml of hot oxygen-free hexane, filtered, and the residue extracted with additional hot hexane (30 ml). The combined hexane extracts were cooled

301 J. G. Murray, J. Amer. Chem. Soc., 83, 1287 (1961).

overnight at −78°. The red-brown needles were removed by filtration, washed thoroughly with cold hexane, and dried at room temperature under reduced pressure.

Cyclopentadienylsodium in benzene or tetrahydrofuran could replace n-butyllithium. The yield was about 18 % based on $Ti(C_5H_5)_2Cl_2$. Preparation directly from titanium tetrachloride and cyclopentadienylsodium by an analogous method was also successful.

The complex melted above 90° with decomposition, even under nitrogen. It reacted extremely readily with oxygen and was pyrophoric in air, but stable for months in sealed vials under nitrogen. It sublimed with appreciable decomposition at 90°/1 mm. It was readily soluble in common organic solvents but decomposed by chlorinated solvents.

1-Pentene (Partial Hydrogenation of 1-Pentyne).[188] Since the catalyst was sensitive to air and moisture, all operations were carried out under nitrogen or argon.

A 100-ml autoclave was charged with 5 g of 1-pentyne and 20 ml of benzene containing 0.5 g of $Ti(C_5H_5)_2(CO)_2$, closed, and hydrogen was admitted to a pressure of 50 atm. Upon heating and shaking, a rapid hydrogen uptake started at 50° and the hydrogenation was finished within 10 minutes. After cooling, analysis of the reaction mixture by gas chromatography indicated 1-pentene in 95 % yield. 1-Pentene (4.5 g, 90 %) was isolated by fractional distillation.

Dichlorotris(triphenylphosphine)ruthenium—$RuCl_2(PPh_3)_3$[302]

$$RuCl_3 \cdot 3\ H_2O + PPh_3 \rightarrow RuCl_2\ (PPh_3)_3\ (+ \text{ other products})$$

Ruthenium trichloride trihydrate (1 g, 3.8 mmol) was dissolved in 250 ml of methyl alcohol and the solution refluxed under nitrogen for 5 minutes. After cooling, triphenylphosphine (6 g, 22.9 mmol) was added and the solution again refluxed under nitrogen for 3 hours. The complex precipitated from the hot solution as shiny black crystals; on cooling, they were filtered under nitrogen, washed several times with degassed ether, and dried under vacuum. The yield was about 2.7 g (74 % based on ruthenium), mp 132–134°.

The complex was moderately soluble in warm chloroform, acetone, benzene and ethyl acetate giving yellow-brown solutions which were air-sensitive, becoming green.

4-Androsten-3,17-dione (Selective Hydrogenation of 1,4-Androstadien-3,17-dione).[211] 1,4-Androstadien-3,17-dione (500 mg) was hydrogenated

302 P. S. Hallman, T. A. Stephenson, and G. Wilkinson, *Inorg. Synth.*, **12**, 237 (1970).

with $RuCl_2(PPh_3)_3$ (50 mg, 0.052 mmol) and triethylamine (5.3 mg, 0.052 mmol) in 10 ml of benzene at 40° under a hydrogen pressure of 130 atm for 8 hours. The benzene solution was passed through alumina, then eluted with benzene-ether. Evaporation of the solvent gave a solid residue (498 mg). Recrystallization from acetone-hexane gave colorless needles of 4-androsten-3,17-dione (447 mg, 89%), mp 169.5–170°. The purity by gas chromatography was 98%.

TABULAR SURVEY

Tables XIII to XXXI summarize homogeneous hydrogenations reported in the literature to the end of December 1974. Table XIII lists catalysts most suited to particular applications; references are to page numbers in this chapter. The remaining tables are divided according to type of substrate and list experimental conditions, products, and literature references.

The organization of Tables XIV to XXXI is in accord with the following points.

1. Several substrates fall into the categories of more than one table. To avoid duplication of entries these appear as set out below,

(i) Cinnamic acid and its derivatives are in Table XXII, not Table XXIII.

(ii) 4-Vinylcyclohexene is in Table XIX, not Table XIV or XVII.

(iii) Table XXVI does not include aromatic compounds when hydrogenation is in a side chain only; these cases appear in other tables as appropriate.

(iv) Table XXVIII similarly includes nitro compounds only when reduction of the nitro group is involved.

2. Within each table, compounds are arranged in order of increasing number of carbon atoms and then in approximate order of increasing complexity (e.g., saturated before unsaturated, straight-chain before branched-chain). Derivatives of unsaturated carboxylic acids are listed under the carbon atom content of the parent acid. General classes of compounds are listed at the beginning of the appropriate section.

3. Kinetic studies are indicated by (K) following the product; deuteration and tritiation experiments are shown by (D_2) and (T_2), respectively, with the pressure.

4. Products are quoted when they have been identified or can be reasonably inferred. A dash (—) under conditions or product indicates that they are not given in the literature. Yields are frequently not given in the literature and products are often deducible only from hydrogen uptake. The inferences are that conversions are normally high but may in some cases be incomplete under the experimental conditions. Yields are entered with the products when available.

Common abbreviations used throughout are:

R = alkyl

X = halide

Ph = phenyl

Ac = acetyl

n-Pr = n-propyl

i-Pr = isopropyl

n-Bu = n-butyl

i-Bu = isobutyl

DBP = 5-phenyl-5H-dibenzophosphole

diop = 2,3-O-isopropylidene-2,3-dihydroxy-1,4-bis(diphenylphosphino)butane

COD = cyclooctadiene

DMA = N,N-dimethylacetamide

DMF = N,N-dimethylformamide

DMSO = dimethyl sulfoxide

THF = tetrahydrofuran

ether = diethyl ether

diglyme = diethylene glycol diethyl ether

TABLE XIII. PREFERRED CATALYSTS FOR SPECIFIC HYDROGENATIONS—COMPARISON WITH HETEROGENEOUS CATALYSTS

Reaction	Catalyst	Text Pages	Action of Heterogeneous Catalysts
Selective hydrogenation of terminal olefins	$RhH(CO)(PPh_3)_3$	39–40	Less selective reaction leading to complex mixtures
	$RuCl_2(PPh_3)_3$	65–66	''
	$RhH(DBP)_4$	40	''
Selective hydrogenation of mono- and di-substituted olefins	$RhCl(PPh_3)_3$	13–22	''
Hydrogenation of olefins without hydrogenolysis of other groups	$RhCl[PPh_2(piperidyl)]_3$	24	''
	$RhCl(PPh_3)_3$	21	Frequent cleavage of carbon-heteroatom bonds (minimized by altering catalyst and conditions)
Specific deuteration of olefins	$RhCl(PPh_3)_3$	15–17	Scrambling of hydrogens in substrates, particularly at allylic position. Frequent lack of stereo-specificity
Hydrogenation of 1,4-dihydroaromatics without disproportionation	$RhCl(PPh_3)_3$	19–22	Frequent rapid disproportionation to aromatic and monoolefin
	$IrCl(CO)(PPh_3)_2$	33	
Hydrogenation of terminal acetylenes to olefins (without affecting internal acetylenes)	$Ti(C_5H_5)_2(CO)_2$	62	Mixtures often result, including complete reduction
Hydrogenation of terminal and internal acetylenes to olefins (internal to cis-olefins)	$CoH(CO)(n\text{-}Bu_3P)_3$	59	''
Hydrogenation of internal acetylenes to cis-olefins	$[Rh(diene)(tertiary\ phosphine)_2\ or\ 3]^+$	39	Mixtures often result, including complete reduction
Hydrogenation of di- and higher olefins to monoolefins	$RuCl_2(PPh_3)_3$	65–66	Frequent low selectivity
	$[Co(CN)_5]^{3-}$	48–49	''
	$RuCl_2(CO)_2(PPh_3)_2$	66	''

Reaction	Catalyst	References	Remarks
Hydrogenation of 1,3-dienes to terminal olefins	[Cr(C₅H₅)(CO)₃]₂	68	hydrogenation (Some disproportionation with cyclohexadienes)
	Cr(CO)₆/hν	68	,,
	Cr(CO)₃(CH₃CN)₃	69	,,
	RhH(PPh₃)₄	40–41	Mixtures on partial hydrogenation; frequently complete saturation
Asymmetric hydrogenation of olefins	[Rh(CO)₂(PPh₃)]₂·2 C₆H₆	40–41	Low optical yields only
	Various catalysts	74–79	
Hydrogenation of the benzene nucleus	Ziegler catalysts	61–62	Heterogeneous catalysts generally more efficient
Partial hydrogenation of polycyclic aromatics	Co(C₃H₅)[P(OCH₃)₃]₃	65	,,
	Co₂(CO)₈	57	Frequently more efficient; products may differ
Hydrogenation of the thiophene nucleus	Co₂(CO)₈	58	Poisoning of most heterogeneous catalysts
Hydrogenation of Schiff bases	RhCl₃(pyridine)₃/NaBH₄	29	Heterogeneous catalysts also efficient
	Co₂(CO)₈	58	,,
	[Co(CN)₅]³⁻	51	,,
Hydrogenation of ketoximes to saturated amines	RhCl₃(pyridine)₃/NaBH₄	29	Similar results over heterogeneous catalysts
Reduction of nitro to amino groups	Co(dimethylglyoximato)₂	69–70	,,
Reductive amination of α-keto acids to saturated amino acids	[Co(CN)₅]³⁻	51	,,
Hydrogenation of aliphatic aldehydes and ketones	Co₂(CO)₈	56	Heterogeneous catalysts generally more active
Hydrogenation and specific deuteration of ketones	[Rh,IrH₂(tertiary phosphine)₂(solvent)₂]⁺	38–39	Scrambling of hydrogens in substrates
Reduction of cyclohexanones to axial alcohols	H₂IrCl₆/P(OCH₃)₃	35–37	Less stereospecific reduction
Hydrogenolysis of acid anhydrides to acid and aldehyde	Co₂(CO)₈	57	When applicable heterogeneous catalysts give alcohols

Note: References 303–344 are on pp. 185–186.

93

TABLE XIV. HYDROCARBONS—TERMINAL OLEFINS

Substrate	Catalyst	Solvent	Temperature (°)/ Pressure (atm)	Product (yield %)	Refs.
"1-Alkenes"	$[Ru(PPh_3)_2]^{2+}$	CH_3OH/H^+	40/0.5	Alkanes	250
"Olefins"	$RhH_2Cl(t-Bu_3P)_2$,	20/1		Alkanes	238
	$RhHCl_2(t-Bu-n-Pr_2P)_2$				
	$RhCl_3/C_6H_3Ph_3$	DMA	..	..	303
	$CoCl[P(OC_2H_5)_3]_{3\&4}$	C_2H_5OH	>75/"High"	..	230a
C_2 Ethylene	$RhCl(PPh_3)_3$	C_6H_6	22/1	Ethane (K)	31a
	$IrCl(CO)(PPh_3)_2$	DMA	50/1	" (K)	71
	$IrH(CO)(PPh_3)_2$,	DMA	50/1	" (K)	85, 86a
	$IrH_3(CO)(PPh_3)_2$				
	Rh, $IrX(CO)(PPh_3)_3$	C_6H_6, $C_6H_5CH_3$	40-60/1(D_2)	" (K)	68b, 78, 82
	$RhH(DBP)_4$	C_6H_6	20/0.1	" (K)	100c
	$RhCl_3$	DMA	40/1	" (K)	245a
	$RhCl_3[S(C_2H_5)_2]_3$	DMA	50, 80/1	" (K)	23, 246
	Organic acids, Rh salts	DMF	—/—	"	243
	$H_2PtCl_6/SnCl_2$	CH_3OH	10-20/1(D_2)	" (K)	101, 112, 114
	$Pd_2(Ph_2PCH_2PPh_2)_3$,	$C_6H_5CH_3$, $C_6H_5CH_3/$	Room temp/	..	258
	$PdCl_2(Ph_2PCH_2PPh_2)$	C_6H_5CN	7(D_2)		
	$RuH_2(PPh_3)_4$	C_6H_6	—/—	..	251
	$TiCl_2(C_5H_5)_2/Mg$, Ca	THF	−20 to 20/1	..	191
	$FeH_2N_2[PPh_2(C_2H_5)]_3$	C_6H_6, $C_6H_5CH_3$, THF	30/1	..	264

		Catalyst	Solvent	Conditions	Product (K)	References
C₃	Propene	RhCl(PPh₃)₃	C₆H₆	22, 25/1	Propane (K)	8, 31a
		Rh, IrX(CO)(PPh₃)₂	C₆H₆, C₆H₅CH₃	40–60/1	"	68b, 82
		Pd₂[Ph₂P(CH₂)ₙPPh₂]₃, n = 1, 2, 3	C₆H₅CH₃	Room temp/7	"	258a
		Co₂(CO)₈	—	200/ 300(H₂ + CO)	Butanol (75), propane (0.2)	163c
C₄	1-Butene	[Co(CO)₂(n-Bu₃P)]₃	Heptane	66/15	Propane	180a, b
		TiCl₂(C₅H₅)₂/Mg, Ca	THF	–20 to 20/1	"	191
		RhCl(PPh₃)₃	C₆H₆	25/1	Butane	8
		Ir carbonyls	—	—/—	"	304
		IrCl(CO)(PPh₃)₂	C₆H₆; C₆H₅CH₃	50, 60/1	", butenes	69, 78
		Pd₂(Ph₂PCH₂PPh₂)₃	C₆H₅CH₃	Room temp, 110/7	", 2-butenes	258a
	(CH₃)₂C=CH₂	Co₂(CO)₈	—	200/ 300(H₂ + CO)	Pentanol (35), (CH₃)₃CH (53)	163c
C₅	1-Pentene	RhCl (tertiary phosphine)₃	C₆H₆/C₂H₅OH	24, 30/1	Pentane	9, 36
		Rh(O₂CCH₃)(PPh₃)₃	C₆H₆	25/1	" (K)	19b
		[RhCl(pyridine)₂(DMF)(BH₄)]⁺	DMF	Room temp/1	" (K)	57
		RhCl₃(DMSO)₃	—	—/—	", pentenes	234
		RuHCl(PPh₃)₃	C₆H₆, C₂H₅OH	Room temp, 50/1	" (K)	207a,b, c
		Pt(SnCl₃)₂Cl₂	H₂O/CH₃OH	25/1	", pentenes	305
		CoH(CO)ₓ(n-Bu₃P)₄₋ₓ, x = 1, 2	Heptane	45, 115/30	", " (K)	179b
	2-Methyl-1-butene	[Co(CO)₂(n-Bu₃P)]₃	"	66/15	", "	180a, b
		Co, Ni, Fe, Ti salts/ LiAlH(OR)₃	"/THF	20/1	"	206h
	2-Methylbutane	Co, Ni, Fe, Ti salts/ LiAlH(OR)₃	"/".	20/1	2-Methylbutane	206h

95

Note: References 303–344 are on pp. 185–186.

TABLE XIV. HYDROCARBONS—TERMINAL OLEFINS (*Continued*)

Substrate	Catalyst	Solvent	Temperature (°)/Pressure (atm)	Product (yield %)	Refs.
3-Methyl-1-butene	RhCl$_3$(DMSO)$_3$	—	—/—	2-Methylbutane, 2-methyl-butenes	234
	Pd$_2$(Ph$_2$PCH$_2$PPh$_2$)$_3$	C$_6$H$_5$CH$_3$	Room temp/7(D$_2$)	Mixture of deuterated 2-methyl-butanes	258b
	RhCl(PPh$_3$)$_3$	C$_6$H$_6$	22–25/1	—C$_2$H$_5$ (85), pentanes (15)	39a
C$_6$ 1-Hexene	RhCl(PPh$_3$)$_3$	C$_6$H$_6$, C$_6$H$_5$CH$_3$	22–30/1(D$_2$)	Hexane (K)	8, 26, 31a, 36
	RhCl(PPh$_3$)$_3$, polymer supported	C$_6$H$_6$, C$_6$H$_6$/C$_2$H$_5$OH	25/1	"	5a, 5d
	IrCl(PPh$_3$)$_3$	C$_6$H$_6$	25/1	", 2-hexenes (K)	54
	RhCl(tertiary phosphine)$_3$	C$_6$H$_6$	30/1	"	36
	RhCl(tertiary phosphine)$_{2–3}$	C$_6$H$_6$, C$_6$H$_6$/C$_2$H$_5$OH	20–30/1	" (K)	56c
	RhCl(phosphole)$_3$	C$_6$H$_6$	20/1	"	239
	Rh(NO)(tertiary phosphine)$_3$	CH$_2$Cl$_2$	25/1, 4	"	14, 56a
	Rh(O$_2$CR)(PPh$_3$)$_3$	C$_6$H$_6$	25, 40/1	"	19b
	RhH(CO)(PPh$_3$)$_3$	C$_6$H$_6$	15–30/0.2–0.8	" (K)	98, 99
	RhH(DBP)$_{3,4}$	C$_6$H$_6$	20/0.1	" (K)	100c, d
	RhH(C$_2$B$_9$H$_{11}$)(PPh$_3$)$_4$	C$_6$H$_6$	35/1	"	100e
	IrCl(CO)(PPh$_3$)$_2$	C$_6$H$_5$CH$_3$	30–110/30	"	81
	RhCl$_3$; RhCl$_3$(pyridine)$_3$	C$_2$H$_5$OH	Room temp/1	"	236

RhHCl$_2$(triphenylphosphole)$_3$	C$_6$H$_6$/(C$_2$H$_5$)$_3$N	20/1	..	239
RhHCl$_2$(t-Bu-n-Pr$_2$P)$_2$	i-PrOH/i-PrONa	20/1	..	238
IrHX$_2$(PPh$_3$)$_3$, IrHCl$_2$(CO)(PPh$_3$)$_3$, IrH$_2$Cl(CO)(PPh$_3$)$_3$	C$_6$H$_5$CH$_3$	30–110/30	..	81
[RhH$_2$(PPh$_3$)(solvent)$_2$]$^+$	THF	25/1	..	94
[Rh, Ir(1,5-COD)(CH$_3$CN)$_2$]$^+$	Acetone, THF, AcOH	—/—	..	242
[Ir(cyclooctene)$_2$Cl]$_2$	DMA	Room temp/1	..	244
[RhCl(pyridine)$_2$(DMF)(BH$_4$)]$^+$	DMF	Room temp/1	.. (K)	57
RuCl$_2$(PPh$_3$)$_3$	C$_6$H$_6$/C$_2$H$_5$OH	20/1	.. (K)	210
RuHCl(PPh$_3$)$_3$	C$_6$H$_6$	50/1	.. (K)	207a,b
RuH(NO)(tertiary phosphine)$_3$	C$_6$H$_6$	20/1	.., hexenes	213
[Ru(PPh$_3$)$_2$]$^{2+}$	CH$_3$OH/H$^+$	40/0.5	.. (K)	250
H$_2$PtCl$_6$/SnCl$_2$	AcOH	20/1	.., hexenes	113
TiCl$_2$(C$_5$H$_5$)$_2$/Mg, Ca, n-BuLi, Na naphthalide, AlR$_3$	THF, hexane, C$_6$H$_5$CH$_3$	−20 to 30/1–2	.., ..	187a, 191, 196b, 197
TiCl$_2$(C$_5$H$_5$)$_2$/BuLi, Na naphthalide (polymer supported)	THF, hexane	Room temp/1	..	187a
Cr, Fe, Co, Ni, Mn, V acetylacetonates/AlR$_3$	—	30/2	.. (K)	196a, b
Co 2-ethylhexanoate/ (alkyl)Li, (aryl)Li	Cyclohexane	50/3	..	204
CoHN$_2$(PPh$_3$)$_3$/Na naphthalide	C$_6$H$_6$/THF	20/1	..	263a
NiCl$_2$/NaBH$_4$	DMF	25/1	..	270

Note: References 303–344 are on pp. 185–186.

TABLE XIV. HYDROCARBONS—TERMINAL OLEFINS (Continued)

Substrate	Catalyst	Solvent	Temperature (°)/Pressure (atm)	Product (yield %)	Refs.
2-Methyl-1-pentene	RhCl(PPh$_3$)$_3$	C$_6$H$_6$	22/1	2-Methylpentane (K)	26a, 31a
	RuHCl(PPh$_3$)$_3$	C$_6$H$_6$	50/1	" (K)	207a
4-Methyl-1-pentene	Rh$_2$, Ir$_2$HCl$_3$[C$_5$(CH$_3$)$_5$]$_2$	i-PrOH	24/1	2-Methylpentane	237
t-BuCH=CH$_2$	RhCl(PPh$_3$)$_3$	C$_6$H$_6$	22/1	2,2-Dimethyl-butane	31a
	RhCl(PPh$_3$)$_3$	C$_6$H$_6$	22–25/1	—C$_2$H$_5$ (86), 2-methyl-pentane (11)	39a
	RhCl(PPh$_3$)$_3$	C$_6$H$_6$	22–25/1	—C$_3$H$_7$-i (97), hexanes (3)	39a
C$_7$ 1-Heptene	RhCl(PPh$_3$)$_3$	C$_6$H$_6$, C$_6$H$_6$/C$_2$H$_5$OH, C$_6$H$_5$CH$_3$	22, 25/1(D$_2$)	Heptane (K)	8, 26b, 31a, 41a, 56b, 306
	RhCl(PPh$_3$)$_3$; polymer supported	—	25, 65/24, 35	"	5b
	IrCl(PPh$_3$)$_3$	C$_6$H$_5$CH$_3$	25/1	" (K)	56b, 306
	Rh, IrCl(CO)(PPh$_3$)$_3$	"	80/1	" (K)	74, 77

98

Substrate	Catalyst	Solvent	Conditions	Product	Refs.
	Rh, IrX(CO)L₂, X = Cl, Br, I, L = SCN, PPh₃, P(OPh)₃, P(C₆H₁₁)₃	"	80/1	" (K)	72, 73; 80d–i
	Rh, IrH(CO)(PPh₃)₃	"	25, 30/1	" (K)	56b, 74, 83
	RhH(DBP)₄	C_6H_6	20/0.1	:	100c
	Rh, Ir(NO)(PPh₃)₃	$C_6H_5CH_3$	25/1	" (K)	56b
	[RhCl(pyridine)₂(DMF)(BH₄)]⁺	DMF	Room temp/1	" (K)	57
	RuCl₂(PPh₃)₃	C_6H_6/C_2H_5OH	25/1(D₂)	" (K)	307
	RuHCl(PPh₃)₃	"	Room temp, 50/1	" (K)	207a, b, c
	TiCl₂(C₅H₅)₂/Al(C₂H₅)₃	$C_6H_5CH_3$	30/2	", heptenes	195, 197
	Metal acetylacetonates/Al(C₂H₅)₃	—	30/2	", "	195
	Fe, Co, Ni stearates/Grignard reagent	ether/hydrocarbon	Room temp/1	:	183
Methylenecyclohexane	RhCl(PPh₃)₃		—/—	Methylcyclohexane	36
	H₂PtCl₆/SnCl₂	i-PrOH	25/1	", 1-methylcyclohexene	111
C₈ 1-Octene	RhCl(PPh₃)₃	C_6H_6, C_6H_6/various co-solvents	22/1	Octane (K)	31a
	RhCl(PPh₃)₃ in presence of thiophene, sulfides	C_6H_6	Room temp./1	:	308
	[RhCl(pyridine)₂(DMF)(BH₄)]⁺	DMF	Room temp/1	" (K)	57, 58
	RhCl₃(N-formylpiperidine)₃	DMF	—/—	:	240
	RhH(PF₃)(PPh₃)₃	C_6H_6	25/1	:	100b
	RhH(DBP)₄	C_6H_6	20/0.1	:	100c
	RuHCl(PPh₃)₃	C_2H_5OH, $C_6H_5CH_3$	Room temp, 50/1	" (K)	207a, c

Note: References 303–344 are on pp. 185–186.

TABLE XIV. HYDROCARBONS—TERMINAL OLEFINS (Continued)

Substrate	Catalyst	Solvent	Temperature (°)/Pressure (atm)	Product (yield %)	Refs.
C_8 1-Octene (contd.)	Ziegler catalysts	$C_6H_5CH_3$, C_7H_{16}	25–40/3.5	Octane	182
	$Ti(CO)(C_5H_5)_2(PhC \equiv CPh)$	C_7H_{16}	Room temp/1	"	217c
	$Mn_2(CO)_{10}$	C_6H_6, dioxane, methylcyclohexane	160/20	"	265
2-Ethyl-1-hexene	$n\text{-}Bu_3B$, $i\text{-}Bu_3B$	None	220/67, 130	"	279, 281a
	RhCl(chiral diphosphine), polymer supported	C_6H_6	Room temp/1	2-Ethylhexane	344
$t\text{-}C_4H_9CH_2C\text{-}(CH_3)=CH_2$ (Terminal olefin portion of diisobutylene)	$RhCl(PPh_3)_3$	C_6H_6, C_2H_5OH	22/1	$t\text{-}C_4H_9CH_2CH_2C_3H_7\text{-}i$ (K)	6b, 31a
	$Co_2(CO)_8$	—	200/ 300(H_2 + CO)	C_8H_{18} (53), $C_8H_{17}OH$ (25)	163
	$RhCl(PPh_3)_3$	C_6H_6	22–25/1	$-C_3H_7\text{-}i$ (99)	39a
Vinylcyclohexane	$RhCl(PPh_3)_3$	C_6H_6	22/1	Ethylcyclohexane	31a
	$RhCl(PPh_3)_3$	C_6H_6/C_2H_5OH	25/1(D_2)	(cis and trans)	40a
C_9 1-Nonene	$RhCl(PPh_3)_3$	C_6H_6	22/1	Nonane (K)	31a
Allylbenzene	$PdCl_2(DMSO)_2$	DMSO	—/—	n-Propylbenzene	259b

	Substrate	Catalyst	Solvent	Conditions	Product	Refs.
C_{10}	1-Decene	$RhH(CO)(PPh_3)_3$	C_6H_6	25/0.7	,, (K)	99
		$Rh_2HCl_3[C_5(CH_3)_5]_2$	i-PrOH	24/100	,,	237
		$RhCl(PPh_3)_3$	C_6H_6	22/1(D_2)	Decane (K)	31a, 32b
		$RhH(CO)(PPh_3)_3$	C_6H_6	25/0.7	,, (K)	99
		$RhH(DBP)_{3,4}$	C_6H_6	20/0.1	,,	100c, d
		$RuHCl(PPh_3)_3$	C_6H_6	50/1	,, (K)	207a, b
		$TiCl_2(C_5H_5)_2$/Na, Li naphthalide	THF	Room temp/1	,,	187b
C_{11}	1-Undecene	$RhCl(PPh_3)_3$	C_6H_6	22/1(D_2)	Undecane (K)	31a, 32b
		$RhH(CO)(PPh_3)_3$	C_6H_6	25/0.7	,, (K)	99
		$RuHCl(PPh_3)_3$	C_2H_5OH	Room temp/1	,,	207c
		$RhCl(PPh_3)_3$, $RhH(CO)(PPh_3)_3$	C_6H_6	—/0.1-100	(cis and trans)	41a, 309
C_{12}	1-Dodecene	$RhCl(PPh_3)_3$	C_6H_6, C_6H_6/C_2H_5OH	Room temp/1(D_2)	Dodecane (K)	26a, 31a, 32b, 51
C_{13}	1-Tridecene	$RhCl(PPh_3)_3$	C_6H_6	Room temp/1(D_2)	Tridecane	32b
C_{14}	1-Tetradecene	$RhCl(PPh_3)_3$	C_6H_6	Room temp/1(D_2)	Tetradecane	32b
C_{16}	1-Hexadecene	$RhH(DBP)_4$	C_6H_6	20/0.1	Hexadecane	100c
C_{18}	"Octadecenes"	$RhCl(PPh_3)_3$, polymer supported	C_6H_6	25/1	Octadecane	5a
C_{22}	1-Docosene	$RhH(DBP)_4$	C_6H_6	20/0.1	Docosane	100c

Note: References 303–344 are on pp. 185–186.

101

TABLE XV. HYDROCARBONS—INTERNAL ACYCLIC OLEFINS

Substrate	Catalyst	Solvent	Temperature (°)/ Pressure (atm)	Product (Yield %)	Refs.
"Olefins"	$RhHCl_2(t\text{-}Bu\text{-}i\text{-}Pr_2P)_2$, $RhH_2Cl(t\text{-}Bu_3P)_2$	—	20/1	Alkanes	238a
	$RhCl_3/C_6H_3Ph_3$	DMA	—/—	''	303
	$CoCl[P(OC_2H_5)_3]_{3\&4}$	C_2H_5OH	>75/''High''	''	230a
C_4 2-Butene	Ir carbonyls	—	—/—	n-Butane	304
C_5 2-Pentene	Co, Ni, Fe, Ti salts/ LiAlH(OR)$_3$	THF/C_7H_{16}	20/1	n-Pentane	206h
cis-2-Pentene	$RhCl(PPh_3)_3$	C_6H_6, C_6H_6/C_2H_5OH	25, 30/1	'', trans-2-pentene (K)	9, 26a
	$RhCl_3(DMSO)_3$	—	—/—	n-Pentane, pentenes	234
	$[Co(CO)_2(n\text{-}Bu_3P]_3$	C_7H_{16}	66/15	''	180a, b
	Co 2-ethylhexanoate/ n-BuLi	Cyclohexane	50/3	''	204
	$CoHN_2(PPh_3)_3/$ Na naphthalide	C_6H_6/THF	20/1	—	263a
trans-2-Pentene	$RhCl(PPh_3)_3$	C_6H_6/C_2H_5OH	30/1	n-Pentane (K)	9
	$RhCl_3(DMSO)_3$	—	—/—	'', pentenes	234
	Cr acetylacetonate/i-Bu$_3$Al	$C_6H_5CH_3$, C_7H_{16}	30/3.5	''	182
	Co 2-ethylhexanoate/n-BuLi	Cyclohexane	50/3	''	204
	$[Co(CO)_2(n\text{-}Bu_3P]_3$	C_7H_{16}	66/15	''	180a, b

102

Substrate	Catalyst	Solvent	Temp/pressure	Product	Ref.
2-Methyl-2-butene	Cr acetylacetonate/i-Bu$_3$Al	C$_6$H$_5$CH$_3$, C$_7$H$_{16}$, decalin	20, 30/3.5	2-Methylbutane	182, 193
C$_6$					
2-Hexene	RuHCl(PPh$_3$)$_3$	C$_6$H$_6$	50/1	Slow reaction (K)	207b
	TiCl$_2$(C$_5$H$_5$)$_2$/Al(C$_2$H$_5$)$_3$	C$_6$H$_5$CH$_3$	30/2	Hexane, hexenes	197
cis-2-Hexene	RhCl(PPh$_3$)$_3$	C$_6$H$_6$	25/1	,, (K)	8
	[RhH$_2$(PPh$_3$)$_2$(solvent)$_2$]$^+$	THF	25/1	,,	94
$trans$-2-Hexene	RhCl(PPh$_3$)$_3$	C$_6$H$_6$	25/1	,,	8
	Rh(NO)(PPh$_3$)$_3$	CH$_2$Cl$_2$	25/1	,,	56a
	[RhH$_2$(PPh$_3$)$_2$(solvent)$_2$]$^+$	THF	25/1	,,	94
$trans$-3-Hexene	RhCl(PPh$_3$)$_3$	C$_6$H$_6$	Room temp/1	,, (K)	26a
2-Methyl-2-pentene	Co 2-ethylhexanoate/n-BuLi	Cyclohexane	50/3	2-Methylpentane	204
3-Methyl-2-pentene	Cr, Co, Fe, Ni acetylacetonates/Al(C$_2$H$_5$)$_3$	—	30/2	3-Methylpentane	196a
cis-4-Methyl-2-pentene	RhCl(PPh$_3$)$_3$	C$_6$H$_6$	Room temp/1	2-Methylpentane (K)	26a
$trans$-4-Methyl-2-pentene	Rh$_2$, Ir$_2$HCl$_3$[C$_5$(CH$_3$)$_5$]$_2$	i-PrOH	24/1	,,	237
	RhCl(PPh$_3$)$_3$	C$_6$H$_6$	Room temp/1	,, (K)	26a
2,3-Dimethyl-2-butene (tetramethylethylene)	Rh$_2$, Ir$_2$HCl$_3$[C$_5$(CH$_3$)$_5$]$_2$	i-PrOH	24/1	,,	237
	RhCl(PPh$_3$)$_3$	C$_6$H$_6$	Room temp/1	2,3-Dimethylbutane (K)	26a
	Cr, Co, Fe, Ni acetylacetonates/Al(C$_2$H$_5$)$_3$	—	30/2	,,	196a
	Co 2-ethylhexanoate/n-BuLi	Cyclohexane	50/3	,, (30)	204

Note: References 303–344 are on pp. 185–186.

103

TABLE XV. HYDROCARBONS—INTERNAL ACYCLIC OLEFINS (Continued)

Substrate	Catalyst	Solvent	Temperature (°)/Pressure (atm)	Product (Yield %)	Refs.
	RhCl(PPh$_3$)$_3$	C$_6$H$_6$	22–25/1	—C$_3$H$_{7-n}$ (70)	39a
C$_7$ 2-Heptene	RhCl(PPh$_3$)$_3$	C$_6$H$_6$, C$_6$H$_6$/C$_2$H$_5$OH	Room temp/1	hexanes (30) Heptane, heptenes	41a
cis-2-Heptene	Rh, IrCl(PPh$_3$)$_3$	C$_6$H$_5$CH$_3$	25/1	Heptane (K)	26b, 56b, 306
trans-2-Heptene	Rh, Ir(NO)(PPh$_3$)$_3$	C$_6$H$_5$CH$_3$, CH$_2$Cl$_2$	25/1	" (K)	56a, 56b
3-Heptene	RuHCl(PPh$_3$)$_3$	C$_6$H$_6$	50/1	Heptenes	207a, b
	Cr, Co, Fe, Ni acetyl-acetonates/Al(C$_2$H$_5$)$_3$	—	30/2	Heptane	196a
	TiCl$_2$(C$_5$H$_5$)$_2$/Al(C$_2$H$_5$)$_3$				
	Rh, IrCl(CO)(PPh$_3$)$_3$	C$_6$H$_5$CH$_3$	30/2	", heptenes	197
trans-3-Heptene	RhX(CO)L$_2$, X = Cl, Br, I L = SCN, PPh$_3$, P(OPh)$_3$, P(C$_6$H$_{11}$)$_3$	C$_6$H$_5$CH$_3$	80/1	" (K)	77
		"	80/1	" (K)	80d, f
	Rh, IrH(CO)(PPh$_3$)$_3$	"	25/1	" (K)	56b, 83a
3-Ethyl-2-pentene	RhCl(PPh$_3$)$_3$	C$_6$H$_6$	Room temp/1	3-Ethylpentane (K)	26a

104

	Substrate	Catalyst	Solvent	Conditions	Products	Ref.
C$_8$	2-Octene	RhCl(PPh$_3$)$_3$	C$_6$H$_6$	Room temp/1	Octane	6b
		RuHCl(PPh$_3$)$_3$	"	50/1	Octenes	207a, b
		Mn$_2$(CO)$_{10}$	", dioxane, methyl-cyclohexane	160/20	Octane	265
	cis-2-Octene	RhCl(PPh$_3$)$_3$	C$_6$H$_6$	22/1	" (K)	31a
	trans-2-Octene	RhCl(PPh$_3$)$_3$	C$_6$H$_6$	22/1	" (K)	31a
	2,4,4-Trimethyl-2-pentene	RhCl(PPh$_3$)$_3$	C$_6$H$_6$, C$_2$H$_5$OH	22/1	2,2,4-Trimethyl-pentane	6b, 31a
	(internal olefin portion of diisobutylene)	Co$_2$(CO)$_8$	—	200/ 300(H$_2$ + CO)	C$_8$H$_{18}$ (63), C$_9$H$_{20}$O (25)	163c
C$_{10}$	cis-2-Decene	RhCl(PPh$_3$)$_3$	C$_6$H$_6$	Room temp/ 1(D$_2$)	Decane	32b
	trans-2-Decene	RhCl(PPh$_3$)$_3$	C$_6$H$_6$	Room temp/ 1(D$_2$)	"	32b
	cis-3-Decene	RhCl(PPh$_3$)$_3$	C$_6$H$_6$	Room temp/ 1(D$_2$)	"	32b
	cis-4-Decene	RhCl(PPh$_3$)$_3$	C$_6$H$_6$	Room temp/ 1(D$_2$)	"	32b
	cis-5-Decene	RhCl(PPh$_3$)$_3$	C$_6$H$_6$	Room temp/ 1(D$_2$)	"	32b
	trans-5-Decene	RhCl(Ph$_3$)$_3$	C$_6$H$_6$	Room temp/ 1(D$_2$)	"	32b
C$_{12}$	2-Dodecene	Ti(C$_5$H$_5$)$_2$(1-methylallyl)	Cyclohexane	Room temp/1	Dodecane	217b

Note: References 303–344 are on pp. 185–186.

TABLE XVI. HYDROCARBONS—CYCLOHEXENE TO CYCLOHEXANE

Catalyst	Solvent	Temperature (°)/Pressure (atm)	Yield (%)	Refs.
RhCl(PPh$_3$)$_3$	C$_6$H$_6$, C$_6$H$_6$/C$_2$H$_5$OH	22–60/0–2(T$_2$) (K)	—	6b, 8, 25, 26a, 28, 31a, 35, 36, 40a, b, 52a, 310
RhCl(PPh$_3$)$_3$/O$_2$, H$_2$O$_2$	C$_6$H$_6$	25/1	—	42
RhCl(PPh$_3$)$_3$, polymer supported	C$_6$H$_6$, C$_2$H$_5$OH, C$_6$H$_6$/C$_2$H$_5$OH	25/1	—	5a, d
RhCl (tertiary phosphine)$_3$	C$_6$H$_6$, C$_6$H$_6$/C$_2$H$_5$OH	25/1	—	35, 52a
Rh(NO)(PPh$_3$)$_3$	CH$_2$Cl$_2$	25/1,4(D$_2$)	—	14, 56a
Rh(O$_2$CCH$_3$)(PPh$_3$)$_3$	C$_6$H$_6$	25/1 (K)	—	19b
RhH(CO)(PPh$_3$)$_3$	C$_6$H$_6$	50/30 (K)	—	311
IrCl(CO)(PPh$_3$)$_2$	C$_6$H$_6$	50/1	—	69
[RhH$_2$(PPh$_3$)$_2$(solvent)$_2$]$^+$	THF	25/1	—	94
Rh$_2$, Ir$_2$HCl$_3$[C$_5$(CH$_3$)$_5$]$_2$	i-PrOH	24/1	—	237
RhCl$_2$(2-methylallyl)/phosphine, amine, sulfide	C$_2$H$_5$OH	—/1 (K)	—	241
RhCl$_3$(N-formylpiperidine)$_3$	DMF	—/—	—	240
Rh salts of organic acids	DMF	—/—	—	243
RhCl$_3$/CO [Rh$_6$(CO)$_{16}$?]	None	200/240 (H$_2$+CO)	—	318
[RhCl(pyridine)$_2$(DMF)(BH$_4$)]$^+$	DMF	Room temp/1	—	57, 60
RuCl$_3$/PPh$_3$	CH$_3$OH	—/—	—	312
RuHCl(PPh$_3$)$_3$	C$_6$H$_6$, C$_2$H$_5$OH	Room temp, 50/1	—	207a, b, c

Catalyst	i-PrOH and other alcohols	Solvent	Conditions	References
$H_2PtCl_6/SnCl_2$	—			111
$Co_2(CO)_8$	$(75)^a$	—	200/ 300 (H_2 + CO)	163c
$[Co(CO)_2(n\text{-}Bu_3P)]_3$	—	C_7H_{16}	66/15 (K)	180a, b
$CoH(CO)_2(n\text{-}Bu_3P)_2$	—	C_6H_{14}	130/30 (K)	179c
$CoH(CO)(PPh_3)_2$	—	C_6H_6	30–150/1–50	179a
$CoH_3(PPh_3)_3$	—	C_6H_6	Room temp, 40/ 0–2	179a, 262
$CoHN_2(PPh_3)_3$/Na naphthalide	—	C_6H_6/THF	20/1	263a
$Ti(C_5H_5)_2$(1-methylallyl)	—	C_6H_6; THF, dimethoxyethane, cyclohexane	Room temp/1	217b
$TiCl_2(C_5H_5)_2$/Na, RLi, RMgCl, AlR_3	—	THF	Room temp–30/ 1, 2	190, 194, 195, 196b
$TiCl_2(C_5H_5)_2$/BuLi; Na naphthalide; free and polymer supported	—	THF, C_6H_{14}	—/1	187a
Ziegler catalysts	—	$C_6H_5CH_3$, C_6H_5Cl, THF, C_7H_{16}, cyclohexane, decalin	0–60/1–3.5	182, 192, 193, 195, 196b, 198, 204, 206d, e, f, h
$NiCl_2/NaBH_4$	—	DMF	25/1	270
Metal stearates	—	C_2H_5OH	20–70/1–100	274, 277a, c, 313
$n\text{-}Bu_3B$, $i\text{-}Bu_3B$	—	None	220/67, 170	279, 281a

[a] The product is cyclohexylcarbinol.

Note: References 303–344 are on pp. 185–186.

TABLE XVII. HYDROCARBONS—CYCLIC OLEFINS OTHER THAN CYCLOHEXENE

Substrate	Catalyst	Solvent	Temperature (°)/ Pressure (atm)	Product (Yield %)	Refs.
C_5 Cyclopentene	RhCl(PPh$_3$)$_3$	C$_6$H$_6$	22/1	Cyclopentane (K)	26a, 31a
	RhCl[PPh(piperidyl)$_2$]$_3$	C$_6$H$_6$	20/1	"	11
	Rh(NO)(PPh$_3$)$_3$	CH$_2$Cl$_2$	25/1	"	56a
	Rh(O$_2$CCH$_3$)(PPh$_3$)$_3$	C$_6$H$_6$	25/1	" (K)	19b
	Rh$_2$, Ir$_2$HCl$_3$[C$_5$(CH$_3$)$_5$]$_2$	i-PrOH	24/1	"	237
	[RhCl(pyridine)$_2$(DMF)(BH$_4$)]$^+$	DMF	Room temp/1	"	57, 60
	RuCl$_3$/PPh$_3$	CH$_3$OH	—/—	"	312
	RuHCl(PPh$_3$)$_3$	C$_2$H$_5$OH	Room temp/1	" (K)	207c
	Pd$_2$(Ph$_2$PCH$_2$PPh$_2$)$_3$	C$_6$H$_5$CH$_3$	Room temp/6	"	258a
	Co, Ni, Fe, Ti salts/ LiAlH(OR)$_3$	THF/C$_7$H$_{16}$	20–40/1–15	"	206h
	Mn(II) stearate	Liquid paraffin	20–60/1	"	274, 314
C_6	RhCl(PPh$_3$)$_3$	C$_6$H$_6$	22–25/1	(93)	39a
C_7 Cycloheptene	RhCl(PPh$_3$)$_3$	C$_6$H$_6$, C$_6$H$_5$CH$_3$	22, 25/1	Cycloheptane	26, 31a
	Rh, IrCl(CO)(PPh$_3$)$_3$	C$_6$H$_5$CH$_3$	80/1	" (K)	77
	Rh, IrH(CO)(PPh$_3$)$_3$	"	25/1	" (K)	83a
	RhX(CO)L$_2$, X = Cl, Br, I, L = SCN, PPh$_3$, P(OPh)$_3$, P(C$_6$H$_{11}$)$_3$	"	80/1	" (K)	80f
	[RhCl(pyridine)$_2$(DMF)(BH$_4$)]$^+$	DMF	Room temp/1	" (K)	57, 60
	RuHCl(PPh$_3$)$_3$	C$_2$H$_5$OH	Room temp/1	" (K)	207c
	Co 2-ethylhexanoate/n-BuLi	Cyclohexane	50/3	"	204
1-Methylcyclohexene	RhCl(tertiary phosphine)$_3$	C$_6$H$_6$, various other solvents	25/1(D$_2$)	Methylcyclohexane	26a, 35, 36, 40a
	Rh(NO)(PPh$_3$)$_3$	CH$_2$Cl$_2$	25/1	"	56a

Na naphthalide

Substrate	Catalyst	Solvent	Temp/Press	Product	Refs
3-Methylcyclo-hexene	Co 2-ethylhexanoate/n-BuLi	Cyclohexane	50/3	'' (44)	204
	RhCl(PPh$_3$)$_3$	—	—/—	''	36
4-Methylcyclo-hexene	RhCl(PPh$_3$)$_3$	—	—/—(T$_2$)	''	24, 36
(Norbornene)	RhCl(PPh$_3$)$_3$	C$_6$H$_6$	25/1	Norbornane	35
	Rh(NO)(PPh$_3$)$_3$	CH$_2$Cl$_2$	25/1	''	56a
	RhX(CO)L$_2$, X = Cl, I, L = PPh$_3$, AsPh$_3$, P(OPh)$_3$, P(C$_6$H$_{11}$)$_3$	C$_6$H$_5$CH$_3$	70/1	''	80d
	[RhCl(pyridine)$_2$(DMF)(BH$_4$)]$^+$	DMF	Room temp/1	''	57, 60
	RuHCl(PPh$_3$)$_3$	C$_2$H$_5$OH	Room temp/1	''	207c
	Pd$_2$(Ph$_2$PCH$_2$PPh$_2$)$_3$	C$_6$H$_5$CH$_3$	Room temp/6	''	258a
C$_8$ Cyclooctene	RhCl(PPh$_3$)$_3$	C$_6$H$_6$	22/1	Cyclooctane (K)	31a
	RhCl(PPh$_3$)$_3$, polymer supported	C$_6$H$_6$, C$_6$H$_6$/C$_2$H$_5$OH	21, 25/1	''	5a, d
	Rh(NO)(PPh$_3$)$_3$	CH$_2$Cl$_2$	25/1	''	56a
	Rh(O$_2$CCH$_3$)(PPh$_3$)$_3$	C$_6$H$_6$	25/1	'' (K)	19b
	RhX(CO)L$_2$, X = Cl, I, L = PPh$_3$, AsPh$_3$, P(OPh)$_3$, P(C$_6$H$_{11}$)$_3$	C$_6$H$_5$CH$_3$	70, 80/1	''	80d
	[RhCl(pyridine)$_2$(DMF)(BH$_4$)]$^+$	DMF	Room temp/1	'' (K)	57, 60
	RuHCl(PPh$_3$)$_3$	C$_2$H$_5$OH	Room temp/1	'' (K)	207c
	RuCl$_3$/PPh$_3$	CH$_3$OH	—/—	'' (K)	312
	Co 2-ethylhexanoate/(alkyl), (aryl)Li	Cyclohexane	50/3	''	204
1,2-Dimethyl-cyclohexene	[Ir(cyclooctene)$_2$Cl]$_2$	DMA	Room temp/1	''	244
	TiCl$_2$(C$_5$H$_5$)$_2$/BuLi, Na naphthalide (free and polymer supported)	C$_6$H$_{14}$	—/1	1,2-Dimethyl-cyclohexane	187a

109

Note: References 303–344 are on pp. 185–186.

TABLE XVII. HYDROCARBONS—CYCLIC OLEFINS OTHER THAN CYCLOHEXENE (*Continued*)

Substrate	Catalyst	Solvent	Temperature (°)/ Pressure (atm)	Product (Yield %)	Refs.
1,4-Dimethyl-cyclohexene	RhCl(PPh$_3$)$_3$	C$_6$H$_6$, C$_6$H$_6$/C$_2$H$_5$OH	25/1(D$_2$)	1,4-Dimethyl-cyclohexane (*cis*, 50) (K)	35, 40a
2,3-Dimethyl cyclohexene	RhCl(PPh$_3$)$_3$	C$_6$H$_6$/C$_2$H$_5$OH	25/1(D$_2$)	1,2-Dimethyl-cyclohexane (*cis*, 50) (K)	35
2,4-Dimethyl-cyclohexene	RhCl(PPh$_3$)$_3$	C$_6$H$_6$/C$_2$H$_5$OH	25/1(D$_2$)	1,3-Dimethyl-cyclohexane (*cis*, 48) (K)	35
4,4-Dimethyl-cyclohexene	RhCl(PPh$_3$)$_3$	C$_6$H$_6$/C$_2$H$_5$OH	25/1	1,1-Dimethyl-cyclohexane	35
C$_9$ (Indene)	RhCl(PPh$_3$)$_3$	C$_6$H$_6$	22/1	Indane	31a
C$_{10}$	RhCl(PPh$_3$)$_3$	C$_6$H$_6$, C$_6$H$_6$/C$_2$H$_5$OH	25/1(D$_2$)		40a
C$_{12}$ Cyclododecene	RhCl(PPh$_3$)$_3$, polymer supported	C$_6$H$_6$	25/1	Cyclododecane (*trans*, 70)	5a
	Ti(C$_5$H$_5$)$_2$(1-methylallyl)	Cyclohexane	Room temp/1	"	217b

Note: References 303–344 are on pp. 185–186.

TABLE XVIII. HYDROCARBONS—ACYCLIC DI- AND HIGHER OLEFINS

Substrate	Catalyst	Solvent	Temperature (°)/Pressure (atm)	Product (Yield %)	Refs.
A. Nonconjugated					
C$_5$ 1,4-Pentadiene	RhCl(PPh$_3$)$_3$	C$_6$H$_6$	22/1	Pentane, pentenes	31a
C$_6$ Hexadiene	Pd$_2$(Ph$_2$PCH$_2$PPh$_2$)$_3$	C$_6$H$_5$CH$_3$	Room temp/7	Pentenes	258a
	Cr acetylacetonate/i-Bu$_3$Al	Decalin	20/—	—	193
1,4-Hexadiene	Cr (C$_6$H$_5$CO$_2$CH$_3$)(CO)$_3$	C$_5$H$_{12}$	175/30	2- and 3-Hexene	220
cis-1,4-Hexadiene	RuHCl(PPh$_3$)$_3$	C$_6$H$_6$	25/1	cis- and trans-2-Hexene (K)	207a
trans-1,4-Hexadiene	RuHCl(PPh$_3$)$_3$	C$_6$H$_6$	25/1	", "	207a
1,5-Hexadiene	RhCl(PPh$_3$)$_3$	C$_6$H$_6$	22/1	Hexane, hexenes	26a, 31a
	Rh(NO)(PPh$_3$)$_3$	CH$_2$Cl$_2$	25/1	", "	56a
	RhH(CO)(PPh$_3$)$_3$	C$_6$H$_6$	25/0.7	", " (K)	99
	RhH(DBP)$_4$	C$_6$H$_6$	20/0.1	—	100c
	PtCl$_2$(PPh$_3$)$_2$/SnCl$_2$	CH$_2$Cl$_2$	90–105/33, 39	Hexenes	105, 315
	Pd$_2$(Ph$_2$PCH$_2$PPh$_2$)$_3$	C$_6$H$_5$CH$_3$	Room temp/7	1-Hexene (mainly)	258a
C$_7$ 1,5-Heptadiene	PtCl$_2$(PPh$_3$)$_2$/SnCl$_2$	CH$_2$Cl$_2$	90–105/33, 39	Heptenes	105, 315
2-Methyl-1,5-hexadiene	RuHCl(PPh$_3$)$_3$	C$_6$H$_6$	25/1	2- and 5-Methyl-1-hexene	207a
3,3-Dimethyl-1,4-pentadiene	RuCl$_2$(CO)$_2$(PPh$_3$)$_2$/PPh$_3$	C$_6$H$_6$	140/10–15	3,3-Dimethyl-pentane, 3,3,-dimethyl-pentenes	215

Note: References 303–344 are on pp. 185–186.

111

TABLE XVIII. HYDROCARBONS—ACYCLIC DI- AND HIGHER OLEFINS (Continued)

Substrate	Catalyst	Solvent	Temperature (°)/Pressure (atm)	Product (Yield %)	Refs.
A, Nonconjugated (Continued)					
C$_8$ 1,7-Octadiene	RhCl(PPh$_3$)$_3$	C$_6$H$_6$	Room temp/1	Octane, octenes	26a
	PtCl$_2$(PPh$_3$)$_2$/SnCl$_2$	C$_6$H$_6$/CH$_3$OH, CH$_2$Cl$_2$	90/33	Octenes	105
C$_{10}$ 1,9-Decadiene	Pd$_2$(Ph$_2$PCH$_2$PPh$_2$)$_3$	C$_6$H$_5$CH$_3$	Room temp/7	Decenes	258a
1,4,9-Decatriene	PtCl$_2$(PPh$_3$)$_2$/SnCl$_2$	C$_6$H$_6$/CH$_3$OH, CH$_2$Cl$_2$	90/33–37	"	105
B. Conjugated (Including Allenes)					
"Diene hydrocarbons"	RuCl$_2$(PPh$_3$)$_3$, RuH(O$_2$CCF$_3$)(PPh$_3$)$_2$	—	—/—	"cis-β-Olefins"	209
C$_4$ 1,3-Butadiene	RhH(PPh$_3$)$_4$	Cyclohexane	50/15	1-Butene	100f
	[Rh(CO)$_2$(PPh$_3$)]$_2$.2 C$_6$H$_6$	Cyclohexane, dioxane	65/15	"	100f
	Pd$_2$(Ph$_2$PCH$_2$PPh$_2$)$_3$, PdCl$_2$(Ph$_2$PCH$_2$PPh$_2$), and similar complexes	C$_6$H$_5$CH$_3$, CH$_2$Cl$_2$, AcOH, diglyme, C$_6$H$_5$CN	Room temp/7	Butenes	258a
	[Co(CN)$_5$]$^{3-}$	H$_2$O, H$_2$O/alcohols, glycerol	20–30/1(D$_2$)	cis- and trans-2-Butene, 1-butene (K)	123–128, 130, 131
	[Co(CN)$_3$(amine)]$^-$	CH$_3$OH/H$_2$O	20/1	cis- and trans-2-Butene, 1-butene	144b

	Catalyst	Solvent	Conditions	Products	Ref.
	$[Ni_2(CN)_6]^{4-}$	H_2O	22, 25/1	cis- and trans-2-Butene, 1-butene (K)	266
	Rh carbonyls				
	$Co_2(CO)_8$	—	—/—	Butane	304
		Ether	145–175/ 330(H_2 + CO)	n-C_3H_7CHO, nonan-5-one	164
	$CoH(CO)_4$	C_5H_{12}, $C_{10}H_{22}$	Room temp/ (CO only)	Butenes	159
	$CoH(CO)_2(n$-$Bu_3P)_2$ $CoH(CO)(n$-$Bu_3P)_3$	C_7H_{16}	45, 115/30	Butane, butenes	179b
	$[Co(L)(PR_3)_2]^+$, L = bipyridine, phenanthroline	CH_3OH	—/—	1-Butene (mainly)	260
	Ziegler catalysts	C_6H_6	40–45/1	Butenes, butane	200
	$RhCl(PPh_3)_3$	C_6H_6	Room temp/1	— (slow reaction)	26a
	$RhH(PPh_3)_4$	Cyclohexane	46/15	1-Pentene	100f
	$[Rh(CO)_2(PPh_3)]_2 \cdot 2\,C_6H_6$	Dioxane	65/15	''	100f
	$RuHCl(PPh_3)_3$	C_6H_6	25/1	2-Pentene	207a, b
	$RuCl_2(CO)_2(PPh_3)_2/PPh_3$	C_6H_6	140/10–15	Pentenes	215
	$CoH(CO)_2(n$-$Bu_3P)_2$, $CoH(CO)(n$-$Bu_3P)_3$	C_7H_{16}	45, 115/30	Pentane, pentenes	179b
	$[Cr(C_5H_5)(CO)_3]_2$	C_6H_6	70/90	cis- and trans-2-Pentene	225
C_5 1,3-Pentadiene	$Ti(C_5H_5)_2$(1-methylallyl)	Cyclohexane	Room temp/1	Pentane (+ π-allyl complex)	217b
cis-1,3-Pentadiene	$RhCl(PPh_3)_3$	C_6H_6	22/1	2-Pentene	31a

Note: References 303–344 are on pp. 185–186.

TABLE XVIII. HYDROCARBONS—ACYCLIC DI- AND HIGHER OLEFINS (*Continued*)

Substrate	Catalyst	Solvent	Temperature (°)/ Pressure (atm)	Product (Yield %)	Refs.
		B, Conjugated (Including Allenes) (Continued)			
trans-1,3-Pentadiene	Pd₂(Ph₂PCH₂PPh₂)₃	C₆H₅CH₃	Room temp/7	Pentenes	258a
	Cr(CO)₃(CH₃CN)₃	None	40/1.3	*cis*-2-Pentene	229f
	RhCl(PPh₃)₃	C₆H₆	22/1	2-Pentene	31a
Isoprene (2-methyl-1,3-butadiene)	Pd₂(Ph₂PCH₂PPh₂)₃	C₆H₅CH₃	Room temp/7	Pentenes	258a
	Cr(CO)₆/*hv*	C₆H₆, isooctane	10/1	*cis*-2-Pentene	229a
	Cr(CO)₃(CH₃CN)₃	None	40/1.3	″	229f
	Rh(NO)(PPh₃)₃	CH₂Cl₂	25/1	2-Methylbutenes	56a
	RhH(PPh₃)₄	Cyclohexane	92/15	″	100f
	[Rh(CO)₂(PPh₃)]₂,2 C₆H₆	Cyclohexane	92/15	″	100f
	Pd₂(Ph₂PCH₂PPh₂)₃	C₆H₅CH₃	Room temp/7(D₂)	″	258
	[Co(CN)₅]³⁻	H₂O	Room temp, 30/1	″ (K)	123, 125, 132, 316
	[Co(CN)₃(amine)]⁻	CH₃OH/H₂O	20/1	2-Methylbutenes	144b
	Co₂(CO)₈	Ether	145–175/ 330 (H₂ + CO)	Hexanals	164
	Ziegler catalysts	C₆H₆, THF, THF/ C₇H₁₆	0–100/1	2-Methylbutane, 2-methyl-butenes	200, 206d,h
	Ti(C₅H₅)₂(1-methylallyl)	Cyclohexane	Room temp/1	2-Methylbutane (+ π-allyl complex)	217b

114

Substrate	Catalyst	Solvent	Conditions	Product	Ref.
	Cr(CO)$_6$/hv	C$_6$H$_6$, isooctane	10/1	2-Methyl-2-butene	229a
C$_6$ Hexadiene	Cr(CO)$_3$(CH$_3$CN)$_3$	None	40/1.3	''	229f
1,3-Hexadiene	Cr acetylacetonate/i-Bu$_3$Al	Decalin	20/—	—	193
2,4-Hexadiene	Cr(PhCO$_2$CH$_3$)(CO)$_3$	C$_5$H$_{12}$	160/30	cis-2-Hexene	220
	Rh(NO)(PPh$_3$)$_3$	CH$_2$Cl$_2$	25/1	Hexenes	56a
	Cr(PhCO$_2$CH$_3$)(CO)$_3$	C$_5$H$_{12}$	160/30	cis-3-Hexene	220
	Cr(CO)$_6$/hv	C$_5$H$_{12}$	Room temp/0.5(D$_2$)	''	229b
trans,trans-2,4-Hexadiene	Cr(norbornadiene)(CO)$_4$/hv	C$_5$H$_{12}$	Room temp/0.5	''	229b
	Cr(CO)$_6$/hv	C$_6$H$_6$, isooctane	10/1(D$_2$)	cis-3-Hexene	229a, b
cis,trans-2,4-Hexadiene	Cr(norbornadiene)(CO)$_4$/hv	C$_5$H$_{12}$/(CH$_3$)$_2$CO	Room temp/0.5	''	229e
	Cr(CO)$_3$(CH$_3$CN)$_3$	None	40/1.3	''	229f
	Cr(CO)$_6$/hv	C$_5$H$_{12}$	Room temp/0.5	cis-3-Hexene	229b
1,3,5-Hexatriene	Cr(PhCO$_2$CH$_3$)(CO)$_3$	C$_5$H$_{12}$	170/30	Mixture	219
2-Methyl-1,3-pentadiene	Cr(PhCO$_2$CH$_3$)(CO)$_3$	C$_5$H$_{12}$	160/30	2-Methyl-2-pentene	220
4-Methyl-1,3-pentadiene	Cr(PhCO$_2$CH$_3$)(CO)$_3$	C$_5$H$_{12}$	160/30	2-Methyl-2-pentene	220
	[Cr(C$_5$H$_5$)(CO)$_3$]$_2$	C$_6$H$_6$	70/90	2- and 4-Methyl-2-pentene	225
2,3-Dimethyl-1,3-butadiene	RhX(CO)L$_2$, X = Cl, I, L = PPh$_3$, AsPh$_3$, P(OPh)$_3$, P(C$_6$H$_{11}$)$_3$	C$_6$H$_5$CH$_3$	70/1	—	80d
	Cr(PhCO$_2$CH$_3$)(CO)$_3$	C$_5$H$_{12}$	160/30	2,3-Dimethyl-2-butene	220
	Cr(CO)$_6$/hv	Decalin	10/1	''	228, 229a
	Cr(CO)$_3$(CH$_3$CN)$_3$	None	40/1.3	''	229f

Note: References 303–344 are on pp. 185–186.

115

TABLE XVIII. HYDROCARBONS—ACYCLIC DI- AND HIGHER OLEFINS (*Continued*)

B. Conjugated (Including Allenes) (Continued)

Substrate	Catalyst	Solvent	Temperature (°)/ Pressure (atm)	Product (Yield %)	Refs.
	$Co_2(CO)_8$	Ether	145–175/ 330 (H_2 + CO)	2,3-Dimethyl-butane, heptanals	164
C₇ 1,3,5-Heptatriene	$Pd_3(Ph_2PCH_2PPh_2)_3$	$C_6H_5CH_3$	Room temp/7	Heptenes	258a
3-Ethyl-1,2-pentadiene	$RhCl(PPh_3)_3$	C_6H_6	60/1	3-Ethyl-2-pentene	31b
2,4-Dimethyl-2,3-pentadiene	$RhCl(PPh_3)_3$	C_6H_6	60/1	2,4-Dimethyl-2-pentene	31b
(tetramethyl-allene)	$Ti(C_5H_5)_2(1\text{-methylallyl})$	Cyclohexane	Room temp/1	''	217b
C₈ 1,3-Octadiene	$RhCl(PPh_3)_3$	C_6H_6	22/1	—	31a
1,3,6-Octatriene	$PtCl_2(PPh_3)_2/SnCl_2$	C_6H_6/CH_3OH	90/33	Octenes (91)	105
2,4,6-Octatriene	$PtCl_2(PPh_3)_2/SnCl_2$	C_6H_6/CH_3OH	90/33	Octenes (90)	105
2,5-Dimethyl-2,4-hexadiene	$Cr(PhCO_2CH_3)(CO)_3$	C_5H_{12}	175/30	2,5-Dimethyl-3-hexene	220
C₉ 1,2-Nonadiene	$RhCl(PPh_3)_3$	C_6H_6	60/1	*cis*-2-Nonene	31b
4,5-Nonadiene	$RhCl(PPh_3)_3$	C_6H_6	60/1	*cis*-4-Nonene	31b
C₁₀ 1-Phenyl-1,3-butadiene	$[Co(CN)_5]^{3-}$	H_2O, CH_3OH/ glycerol, H_2O/ $HOCH_2CH_2OH$	20/1(D_2)	1-Phenylbutenes	133
C₁₆ 1,4-diphenyl-1,3-butadiene	$Ti(CO)(C_5H_5)_2(PhC{=}CPh)$	C_7H_{16}	Room temp/1	1,4-Diphenyl-butane	217c

Note: References 303–344 are on pp. 185–186.

TABLE XIX. HYDROCARBONS—CYCLIC DI- AND HIGHER OLEFINS

Substrate	Catalyst	Solvent	Temperature (°)/Pressure (atm)	Product (Yield %)	Refs.
		A. Nonconjugated			
C_6 1,4-Cyclo-hexadiene	$IrCl(CO)(PPh_3)_2$	None, DMA	$83/2(D_2)$	Cyclohexene	79b
	$Pd_2(Ph_2PCH_2PPh_2)_3$	$C_6H_5CH_3$	Room temp/7	”	258a
	$Cr(PhCO_2CH_3)(CO)_3$	C_6H_{14}	$160/30(D_2)$	”	220, 221
	$MoH_2(C_5H_5)_2$	None	140–180/160	”	227
C_7 Norbornadiene	$Rh(NO)(PPh_3)_3$	CH_2Cl_2	25/1	Norbornene, norbornane	56a
	$Rh_2HCl_3[C_5(CH_3)_5]_2$	i-PrOH	24/100	—	237
	$[RhH_2(PPh_3)_2(solvent)_2]^+$	THF	25/1	—	94
	$RuHCl(PPh_3)_3$	C_2H_5OH	Room temp/1	— (K)	207c
	$RuCl_2(CO)_2(PPh_3)_2/PPh_3$	C_6H_6	140/10–15	Norbornene,	215
	$PtHCl(PPh_3)_2/SnCl_2$, $RuCl_3/PPh_3$	C_6H_6, CH_3OH	—/—	Norbornane	312
	$Pd_2(Ph_2PCH_2PPh_2)_3$ and similar complexes	$C_6H_5CH_3$	Room temp/7	Norbornene (mainly)	258a
	$Cr(norbornadiene)(CO)_4/h\nu$	C_5H_{12}	$23/0.5(D_2)$	Norbornene, nortricyclene	228, 229c

Note: References 303–344 are on pp. 185–186.

117

TABLE XIX. HYDROCARBONS—CYCLIC DI- AND HIGHER OLEFINS (*Continued*)

A. Nonconjugated (*Continued*)

Substrate	Catalyst	Solvent	Temperature (°)/Pressure (atm)	Product (Yield %)	Refs.
C$_8$ 1,4-Cyclooctadiene	MoH$_2$(C$_5$H$_5$)$_2$	None	140–180/160	Norbornene	227
1,5-Cyclo-octadiene	Ir$_2$H$_2$Cl$_2$(1,5-COD)(PPh$_3$)$_2$	C$_6$H$_6$	22, 68/1	Cyclooctene	100h
	RhCl(PPh$_3$)$_3$	C$_6$H$_6$	22, 25/1	—(slow reaction)	8, 26a, 31a
	Rh(O$_2$CCH$_3$)(PPh$_3$)$_3$	C$_6$H$_6$	25/1	—	19b
	Rh$_2$HCl$_3$[C$_5$(CH$_3$)$_5$]$_2$	i-PrOH	24/100	—	237
	[Rh, IrH$_2$(PPh$_3$)$_2$(solvent)$_2$]$^+$	THF, acetone	25/1	Cyclooctene, cyclooctadienes	94
	[Rh, Ir(1,5-COD)(CH$_3$CN)$_2$]$^+$	THF, acetone, AcOH	—/—	Cyclooctene	242
	[Ir(1,5-COD)$_2$]$^+$	Acetone	30/1	∴	97
	Ir$_2$H$_2$Cl$_2$(1,5-COD)(PPh$_3$)$_2$	C$_6$H$_6$	22, 68/1	∴	100h
	RuHCl(PPh$_3$)$_3$	C$_2$H$_5$OH	Room temp/1	—(K)	207c
	RuCl$_2$(CO)$_2$(PPh$_3$)$_2$/PPh$_3$	C$_6$H$_6$	140/10–15	Cyclooctene	215
	PtCl$_2$(PPh$_3$)$_2$/SnCl$_2$, other Pt, Pd complexes	CH$_2$Cl$_2$	90–105/37–47	∴	105, 315
	Co$_2$(CO)$_8$/n-Bu$_3$P	—	—/—	∴	177a
	Cr(PhCO$_2$CH$_3$)(CO)$_3$	C$_6$H$_{14}$	160, 170/30	1,3-Cyclo-octadiene, cyclooctene	220
	Ti(C$_5$H$_5$)$_2$(1-methylallyl)	Cyclohexane	Room temp/1	Cyclooctane	217b
	TiCl$_2$(C$_5$H$_5$)$_2$/BuLi, Na naphthalide; (polymer supported)	C$_6$H$_{14}$	—/1	—	187a

4-Vinylcyclo-hexene	RhCl(PPh$_3$)$_3$	C$_6$H$_6$	22/1	Ethylcyclo-hexane (K)	31a
	Rh(NO)(PPh$_3$)$_3$	CH$_2$Cl$_2$	25/1	;	56a
	RhH(CO)(PPh$_3$)$_3$	C$_6$H$_6$	25/0.7	4-Ethylcyclo-hexene (K)	99
	Rh$_2$HCl$_3$[C$_5$(CH$_3$)$_5$]$_2$	i-PrOH	24/100	—	237
	RuHCl(PPh$_3$)$_3$	C$_6$H$_6$	25/1	4-Ethylcyclo-hexene (K)	207a
	PtCl$_2$(PPh$_3$)$_2$/SnCl$_2$	C$_6$H$_6$/C$_2$H$_5$OH, CH$_2$Cl$_2$	90/33–37	4-Ethylcyclo-hexene (85) (K) "Monoene"	105
	Cr acetylacetonate/i-Bu$_3$Al	Decalin	20/—	4-Ethylcyclo-hexene, ethyl-cyclohexane	193
C$_9$	RhCl(PPh$_3$)$_3$	C$_6$H$_6$	Room temp/1		45
C$_{10}$	RhCl(PPh$_3$)$_3$	C$_6$H$_6$/C$_2$H$_5$OH	Room temp/1		38
	RhCl(PPh$_3$)$_3$	C$_6$H$_6$, C$_6$H$_6$/C$_2$H$_5$OH	Room temp/1		38, 45
(dicyclo-pentadiene)	RhCl$_3$, RuCl$_3$, PdCl$_2$, K$_2$PtCl$_4$	DMF	Room temp/1	—	119

Note: References 303–344 are on pp. 185–186.

119

TABLE XIX. HYDROCARBONS—CYCLIC DI- AND HIGHER OLEFINS (Continued)

Substrate	Catalyst	Solvent	Temperature (°)/Pressure (atm)	Product (Yield %)	Refs.
		A. Nonconjugated (Continued)			
C_{12} 1,5,9-Cyclo-dodecatriene	$Co_2(CO)_8/PPh_3$, n-Bu_3P	C_6H_6	160–180/30	cis- and trans-Cyclododecene	177a
	$[Co(CO)_3(PR_3)]_2$ R = alkyl, aryl	C_6H_6	110–155/25	Cyclododecene	177b, 178
	$RuCl_2(CO)_2(PPh_3)_2$/ Lewis base (and other Ru complexes)	C_6H_6, DMF, others	125–160/7–15	''	178, 215
trans,trans,trans-1,5,9-Cyclo-dodecatriene	$CoH_3(PPh_3)_3$	C_6H_6	80/50	Cyclododecane (16), cyclo-dodecenes (41), cyclododeca-dienes (27), cyclododeca-trienes (16)	179a
	$Ti(C_5H_5)_2$(1-methylallyl)	Cyclohexane	Room temp/1	π-Dienyl com-plex, no hydrogenation	217b
	$H_2PtCl_6/SnCl_2$	i-PrOH, butanone	25–50/1(D_2)	(cis and trans)	117
C_{14}	$RhCl(PPh_3)_3$	C_6H_6/C_2H_5OH	Room temp/1		45

120

B. Conjugated (Including Cyclic Allenes)

	Substrate	Catalyst	Solvent	Temp/pressure	Product	Ref.
C$_5$	Cyclopentadiene	[Co(CN)$_5$]$^{3-}$	H$_2$O	Room temp/1	Cyclopentene	125
		MoH$_2$(C$_5$H$_5$)$_2$	None	180/160	; ;	227
C$_6$	1,3-Cyclohexadiene	RhCl(PPh$_3$)$_3$	C$_6$H$_6$	Room temp/1	— (slow reaction)	26a
		RhX(CO)L$_2$, X = Cl, I, L = PPh$_3$, AsPh$_3$, P(OPh)$_3$, P(C$_6$H$_{11}$)$_3$	C$_6$H$_5$CH$_3$	70/1	—	80d
		IrCl(CO)(PPh$_3$)$_2$	None, DMA	83/2(D$_2$)	Cyclohexene	79b
		IrCl(CO)(PPh$_3$)$_2$/hv	C$_6$H$_5$CH$_3$	50/1	; ; (K)	79d
		Pd$_2$(Ph$_2$PCH$_2$PPh$_2$)$_3$	C$_6$H$_5$CH$_3$	Room temp/7	; ;	258a
		[Co(CN)$_5$]$^{3-}$	H$_2$O	Room temp/1	; ;	125, 145, 316
		Co(CN)$_2$(pyridine)$_2$	H$_2$O	Room temp/1	; (?)	145
		Cr(PhCO$_2$CH$_3$)(CO)$_3$	C$_6$H$_{14}$	160/30(D$_2$)	; ;	220, 221
		[Cr(C$_5$H$_5$)(CO)$_3$]$_2$	C$_6$H$_6$	70/90	; ;	225
		Cr(CO)$_6$/hv	Decalin	Room temp/1	; ;	228
		Cr(norbornadiene)(CO)$_4$/hv	—	Room temp/0.5	; ;	229b
		MoH$_2$(C$_5$H$_5$)$_2$	None	140–180/160	; ;	227
		MoH$_2$(C$_5$H$_5$)$_2$	None	140–180/160	; ;	227
C$_7$	1,3-Cycloheptadiene	PtCl$_2$(PPh$_3$)$_2$/SnCl$_2$	CH$_2$Cl$_2$	90/37	Cycloheptene	105
	1,3,5-Cycloheptatriene	Pd$_2$(Ph$_2$PCH$_2$PPh$_2$)$_3$	C$_6$H$_5$CH$_3$	Room temp/7	—	258a
		Cr(PhCO$_2$CH$_3$)(CO)$_3$	C$_6$H$_{14}$	160, 175/30(D$_2$)	1,3-Cycloheptadiene, cycloheptene	219

Note: References 303–344 are on pp. 185–186.

TABLE XIX. HYDROCARBONS—CYCLIC DI- AND HIGHER OLEFINS (Continued)

B. Conjugated (Including Cyclic Allenes) (Continued)

Substrate	Catalyst	Solvent	Temperature (°)/ Pressure (atm)	Product (Yield %)	Refs.
	$[Cr(C_5H_5)(CO)_3]_2$	C_6H_6	70/90	1,4-Cyclo-heptadiene, cycloheptene	225
	$MoH_2(C_5H_5)_2$	None	140, 180/160	1,3-Cyclohepta-diene, then cycloheptene	227
C_8 1,3-Cycloocta-diene	$RhCl(PPh_3)_3$	C_6H_6	22/1	—	31a
	$Rh(NO)(PPh_3)_3$	CH_2Cl_2	25/1	Cyclooctene, cyclooctane	56a
	$Rh(O_2CCH_3)(PPh_3)_3$	C_6H_6	25/1	—	19b
	$RhX(CO)L_2$, X = Cl, Br, I, L = SCN, PPh_3, AsPh_3, P(OPh)_3, P(C_6H_{11})_3	$C_6H_5CH_3$	70, 80/1	— (K)	80d, f
	Rh, IrH(CO)(PPh_3)_3	¨	25/1	— (K)	83a
	$[RhH_2(PPh_3)_2(solvent)_2]^+$	THF	25/1	—	94
	$Rh_2HCl_3[C_5(CH_3)_5]_2$	i-PrOH	24/100	Cyclooctane (?)	237
	$PtCl_2(AsPh_3)_2/SnCl_2$ and similar complexes	CH_2Cl_2	90/37	Cyclooctene	105
	$Co_2(CO)_8/n$-Bu_3P	—	—/—	¨	177a
	$Cr(PhCO_2CH_3)(CO)_3$	C_6H_{14}	160/30	¨	220

	Cyclohexane C_6H_{14}	Room temp/1		
1,3,5-Cyclo-octatriene	Ti(C_5H_5)(1-methylallyl) TiCl$_2$(C_5H_5)$_2$/BuLi, Na naphthalide; (polymer supported) MoH$_2$(C_5H_5)$_2$ Ziegler catalysts [Cr(C_5H_5)(CO)$_3$]$_2$	Cyclohexane C_6H_{14} None C_7H_{16} C_6H_6	Room temp/1 —/1 140–180/160 20/1 70/90	Cyclooctane — Cyclooctene Cyclooctane (?) 1,4- and 1,5-Cyclooctadiene	217b 187a 227 206h 225
Cyclooctatetraene	PtCl$_2$(PPh$_3$)$_2$/SnCl$_2$ [Cr(C_5H_5)(CO)$_3$]$_2$	CH$_2$Cl$_2$ C_6H_6	90/37 70/90	Cyclooctene ", 1,4- and 1,5-cyclooctadiene	105 225
	PtCl$_2$(PPh$_3$)$_2$/SnCl$_2$	CH$_2$Cl$_2$	90/37	"Monoene"	105
	[Cr(C_5H_5)(CO)$_3$]$_2$	C_6H_6	70/90		225
C$_9$ 1,2-Cyclonona-diene	RhCl(PPh$_3$)$_3$	C_6H_6	60/1	cis-Cyclononene	31b
1,2,6-Cyclo-nonatriene	RhCl(PPh$_3$)$_3$	C_6H_6	60/1	cis,cis-1,5-Cyclononadiene	31b
C$_{13}$ 1,2-Cyclotri-decadiene	RhCl(PPh$_3$)$_3$	C_6H_6	60/1	Cyclotridecene	31b

Note: References 303–344 are on pp. 185–186.

TABLE XX. SATURATED ALDEHYDES AND KETONES (INCLUDING AROMATICS)

	Substrate	Catalyst	Solvent	Temperature (°)/Pressure (atm)	Product (Yield %)	Refs.
			A. Aldehydes			
	RCH$_2$CHO (from cracked gasoline)	Co$_2$(CO)$_8$	Petroleum spirit	100–240/130–200 (H$_2$ + CO)	RCH$_2$CH$_2$OH	317
C$_3$	C$_2$H$_5$CHO	RhCl$_3$/CO [Rh$_6$(CO)$_{16}$?]	None	175/300 (H$_2$ + CO)	1-Propanol	318
		Co$_2$(CO)$_8$	C$_6$H$_5$CH$_3$	150/100–300 (H$_2$ + CO)	'' (K)	154
C$_4$	n-C$_3$H$_7$CHO	RhCl$_3$/CO [Rh$_6$(CO)$_{16}$?]	None	170–200/300 (H$_2$ + CO)	1-Butanol	318, 319
		IrH$_3$(PPh$_3$)$_3$	AcOH	50/1	''	93
		[IrH$_2$(PPh$_3$)$_2$(acetone)$_2$]$^+$	Dioxane	50/1	''	94
		Co$_2$(CO)$_8$	C$_6$H$_5$CH$_3$, C$_6$H$_{14}$, cyclohexane	100–240/130–250 (H$_2$ + CO)	''	169, 317, 320
		CoH(CO)$_2$(n-Bu$_3$P)$_2$, CoH(CO)(n-Bu$_3$P)$_3$	C$_7$H$_{16}$	60, 130/30	''	179b
	i-C$_3$H$_7$CHO	Co$_2$(CO)$_8$	C$_6$H$_5$CH$_3$	130–185/250 (H$_2$ + CO)	(CH$_3$)$_2$CHCH$_2$OH	320
		CoH(CO)$_2$(n-Bu$_3$P)$_2$, CoH(CO)(n-Bu$_3$P)$_3$	C$_7$H$_{16}$	60, 130/30	''	179b
C$_5$	n-C$_4$H$_9$CHO	CoH(CO)$_2$(n-Bu$_3$P)$_2$, CoH(CO)(n-Bu$_3$P)$_3$	C$_7$H$_{16}$	60, 130/30	n-C$_4$H$_9$CH$_2$OH	179b
	t-C$_4$H$_9$CHO	CoH(CO)$_2$(n-Bu$_3$P)$_2$, CoH(CO)(n-Bu$_3$P)$_3$	C$_7$H$_{16}$	60, 130/30	t-C$_4$H$_9$CH$_2$OH	179b

B. Ketones — Aldehydes (continued)

C	Substrate	Catalyst	Solvent	Conditions (°/atm)	Product	Refs.
	(2-thienyl)CHO	Co$_2$(CO)$_8$	C$_6$H$_{14}$	180/ 150 (H$_2$ + CO)	(2-thienyl)CH$_2$OH; (2-methylthiophene)	169a
C$_7$	n-C$_6$H$_{13}$CHO	RhCl(PPh$_3$)$_3$, [RhCl(SnCl$_3$)$_2$]$_2$	C$_6$H$_6$/C$_2$H$_5$OH	110/50	1-Heptanol	96
		Co$_2$(CO)$_8$	C$_6$H$_{14}$	180/ 200 (H$_2$ + CO)	`"`	169a
	(cyclohexyl)CHO	Co$_2$(CO)$_8$	C$_6$H$_{14}$	100–240/ 130–300 (H$_2$ + CO)	(cyclohexyl)CH$_2$OH	174b, 317
	Benzaldehyde	RhCl$_3$/CO [Rh$_6$(CO)$_{16}$?]	None	200/ 300 (H$_2$ + CO)	PhCH$_2$OH	318
		IrCl$_4$/HP(O)(OCH$_3$)$_2$	i-PrOH/H$_2$O	Reflux/(no H$_2$)	`"`	87a
		[Co(CN)$_5$]$^{3-}$	H$_2$O	Room temp/1	`"`	125
		FeCl$_3$/Grignard reagent	THF	—/—	`"`	206a
C$_8$	2-Ethylhexanal	Co$_2$(CO)$_8$	Di-n-hexyl ether	160/ 100 (H$_2$ + CO)	2-Ethylhexanol	155
	Butyraldehyde dimer	Co$_2$(CO)$_8$, Fe(CO)$_5$	Cyclohexane	150/ 200 (H$_2$ + CO)	C$_8$-alcohols	170
C$_9$	PhCH(CH$_3$)CHO	Co$_2$(CO)$_8$	C$_6$H$_{14}$	180/ 133 (H$_2$ + CO)	PhCH(CH$_3$)-CH$_2$OH	169a

B. Ketones

C	Substrate	Catalyst	Solvent	Conditions	Product	Refs.
C$_3$	Acetone	{RhH$_2$[PPh(CH$_3$)$_2$]$_2$(solvent)$_2$}$^+$	H$_2$O	25/1(D$_2$)	(CH$_3$)$_2$CHOH	95
		Co$_2$(CO)$_8$	C$_6$H$_{14}$	180/ 200 (H$_2$ + CO)	`"`	169a
C$_4$	CH$_3$COC$_2$H$_5$	Ziegler catalysts	THF	35/—	—	206f
		{RhH$_2$[PPh(CH$_3$)$_2$]$_2$(solvent)$_2$}$^+$	—	25/1	2-Butanol	95

Note: References 303–344 are on pp. 185–186.

125

TABLE XX. SATURATED ALDEHYDES AND KETONES (INCLUDING AROMATICS) (Continued)

Substrate	Catalyst	Solvent	Temperature (°)/Pressure (atm)	Product (Yield %)	Refs.
		B. Ketones (Continued)			
C$_4$ (Contd.)	{Rh(norbornadiene)-[P*Ph(CH$_3$)(CH$_2$Ph)]$_2$}$^+$	C$_6$H$_5$CH$_3$	Room temp/1	Optically active 2-butanol	294
Biacetyl	Co(dimethylglyoximato)$_2$	CH$_3$OH, C$_2$H$_5$OH	Room temp/1	CH$_3$CHOH-COCH$_3$ (acetoin)	231a
C$_5$ C$_2$H$_5$COC$_2$H$_5$	FeCl$_3$/LiAlH$_4$	THF	—/—	3-Pentanol	306c
C$_6$ Cyclohexanone	{RhH$_2$[PPh(CH$_3$)$_2$]$_2$(solvent)$_2$}$^+$	—	25/1	Cyclohexanol	95
	IrCl$_3$/R'R''SO	i-PrOH	90/(no H$_2$)	''	89a
	RuCl$_2$(PPh$_3$)$_3$	Various alcohols	84–140/1	''	212a
	Organic Ni, Fe, Co salts	—	—/—	'' (?)	321
2-Acetylthiophene	Co$_2$(CO)$_8$	None	180/— 250 (H$_2$ + CO)	2-Ethylthiophene	146
[furan–COR structure]	Co$_2$(CO)$_8$	—	130/—	[2-CH$_2$R furan structure]	166
C$_7$ 2-Methylcyclohexanone	IrCl$_4$/P(OCH$_3$)$_3$	i-PrOH/H$_2$O	Reflux/(no H$_2$)	[2-methylcyclohexanol structure, OH]	87a
4-Methylcyclohexanone	IrCl$_4$/HP(O)(OCH$_3$)$_2$	''	Reflux/(no H$_2$)	[4-methylcyclohexanol structure, OH]	87a
C$_8$ Acetophenone	{RhH$_2$[PPh(CH$_3$)$_2$]$_2$(solvent)$_2$}$^+$	—	25/1	1-Phenylethanol	95
	{Rh(norbornadiene)-[P*Ph(CH$_3$)(CH$_2$Ph)]$_2$}$^+$	C$_6$H$_5$CH$_3$	Room temp/1	Optically active 1-phenyl-ethanol	294

126

Substrate	Catalyst	Solvent	Conditions	Product	Refs.
	[Rh(norbornadiene)(diop)]+	CH₃OH, C₂H₅OH, i-PrOH	30/1	Optically active 1-phenyl-ethanol	290c
	RhCl₃(pyridine)₃/NaBH₄, Co₂(CO)₈	DMF, None	Room temp/1, 180/250 (H₂ + CO)	1-Phenylethanol, Ethylbenzene	62b, 146
2,2-Dimethyl-cyclohexanone	IrCl₄/H₃PO₃	i-PrOH/H₂O	Reflux/(no H₂)		87a
	{RhH₂[PPh(CH₃)₂]₂(solvent)₂}+	—	25/1	(?)	95
C₉ p-Methoxyaceto-phenone	Co₂(CO)₈	C₆H₆	180/250 (H₂ + CO)	1-Ethyl-4-methoxy-benzene	146
3,5,5-Trimethyl-cyclohexanone	H₂IrCl₆/P(OCH₃)₃, IrCl₄/HP(O)(OCH₃)₂	i-PrOH/H₂O	Reflux/(no H₂)		85, 87a, 88
C₁₀ PhCOCO₂C₂H₅	Co(dimethylglyoximato)₂	CH₃OH, C₂H₅OH	Room temp/1	PhCHOHCO₂C₂H₅	231a
3-t-Butylcyclo-hexanone	IrCl₄, H₂IrCl₆/P(OCH₃)₃	i-PrOH/H₂O	Reflux/(no H₂)		85, 87a, 88

Note: References 303–344 are on pp. 185–186.

127

TABLE XX. SATURATED ALDEHYDES AND KETONES (INCLUDING AROMATICS) (Continued)

Substrate	Catalyst	Solvent	Temperature (°)/Pressure (atm)	Product (Yield %)	Refs.
		B. Ketones (Continued)			
4-t-Butylcyclo-hexanone	{RhH$_2$[PPh(CH$_3$)$_2$]$_2$(solvent)$_2$}$^+$	—	25/1	[4-t-butylcyclohexanol, OH] (86)	95
	IrCl$_3$/R'R"SO	i-PrOH	90/(no H$_2$)	″	89a
	IrCl$_4$/H$_3$PO$_3$	i-PrOH/H$_2$O	Reflux/(no H$_2$)	[4-t-butylcyclohexanol, OH]	87a
	Ir(H)Cl[P[O(O)(OCH$_3$)$_2$]-[P(OH)(OCH$_3$)$_2$]$_3$	i-PrOH/H$_2$O	—/(no H$_2$)	″	87c
C$_{13}$ Benzophenone	RhCl$_3$(pyridine)$_3$/NaBH$_4$ Co$_2$(CO)$_8$	DMF C$_6$H$_6$	Room temp/1 180/ 250 (H$_2$ + CO)	Ph$_2$CHOH Ph$_2$CH$_2$	62b 146
[fluorenone structure]	Co$_2$(CO)$_8$	C$_6$H$_6$	180/ 250 (H$_2$ + CO)	[fluorene structure]	146
C$_{14}$ Benzoin	RhCl$_3$(pyridine)$_3$/NaBH$_4$	DMF	Room temp/1	PhCHOH·CHOHPh	62b
Benzil	[Co(CN)$_5$]$^{3-}$	H$_2$O	Room temp/1	Benzoin	316

128

	Substrate	Catalyst	Solvent	Conditions	Product	Refs.
C₁₇	[polycyclic ketone structure]	Co(dimethylglyoximato)₂ Co(dimethylglyoximato)₂/quinine	CH_3OH, C_2H_5OH C_6H_6/CH_3OH, THF	Room temp/1 Room temp/1	'' S(+)-Benzoin	231a 295
		$Co_2(CO)_8$	C_6H_6	180/ 250 (H_2 + CO)	[polycyclic structure]	146
C₁₉	2-Nonadecanone	$IrCl_4$/H_3PO_3	i-PrOH/H_2O	Reflux/(no H_2)	2-Nonadecanol	87a

C. Reductive Amination of Aldehydes and Ketones

	Substrate	Catalyst	Solvent	Conditions	Product	Refs.
C₃	CH_3COCO_2H	$[Co(CN)_5]^{3-}$	H_2O/NH_3	40–70/50	$CH_3CH(NH_2)$-CO_2H	142
		$[Co(NH_3)_5Cl]^{2+}$	H_2O	40/50	$CH_3CH(NH_2)$-CO_2H	142, 143
	$CH_3COCO_2C_2H_5$	$[Co(CN)_5]^{3-}$	H_2O/NH_3	40/50	$CH_3CH(NH_2)$-CO_2H	142, 143
C₄	$(CH_3)_2CHCHO$	$Co_2(CO)_8$	C_6H_6/piperidine	150–180/ 100–300 (H_2 + CO)	$(CH_3)_2CHCH_2$-NC_5H_{10}	162b
		$Rh_6(CO)_{16}$	C_6H_{14}/$PhNHCH_3$	110–160/ 100–300 (H_2 + CO)	$PhN(CH_3)$-$CH_2CH(CH_3)_2$	162b

Note: References 303–344 are on pp. 185–186.

129

TABLE XX. SATURATED ALDEHYDES AND KETONES (INCLUDING AROMATICS) (*Continued*)

C. Reductive Amination of Aldehydes and Ketones (Continued)

Substrate	Catalyst	Solvent	Temperature (°)/Pressure (atm)	Product (Yield %)	Refs.
C_5 $HO_2CCH_2CH_2$-$COCO_2H$	$[Co(CN)_5]^{3-}$	H_2O/NH_3	40/50	$HO_2CCH_2CH_2$-$CH(NH_2)CO_2H$	142, 143
Furfural	$Co_2(CO)_8/PBu_3$	C_2H_5OH/aniline	150–200/100–300	$C_4H_4OCH_2NHPh$	162b
C_6 Cyclohexanone	$Rh_6(CO)_{16}$	C_6H_{14}/i-$PrNH_2$	110–160/ 100–300 (H_2 + CO)	$C_6H_{11}NHPr$-i	162b
C_7 PhCHO	$Co_2(CO)_8/PBu_3$	C_2H_5OH/n-$BuNH_2$	150–200/100–300	$PhCH_2NHBu$-n	162b
	$Rh_6(CO)_{16}$	C_6H_{14}/piperidine	110–160/ 100–300 (H_2 + CO)	$PhCH_2NC_5H_{10}$	162b
2-Methylcyclo-hexanone	$[Co(CN)_5]^{3-}$	$H_2O/C_2H_5OH/NH_3$	70/50		144a
3-Methylcyclo-hexanone	$[Co(CN)_5]^{3-}$	$H_2O/C_2H_5OH/NH_3$	70/50		144a
4-Methylcyclo-hexanone	$[Co(CN)_5]^{3-}$	$H_2O/C_2H_5OH/NH_3$	70/50		144a

	Substrate	Catalyst	Solvent	Conditions	Product	Refs.
C$_8$	PhCOCH$_3$	Co$_2$(CO)$_8$/PBu$_3$	C$_2$H$_5$OH/NH$_3$	150–200/100–300	PhCH(CH$_3$)NH$_2$, [PhCH(CH$_3$)]$_2$-NH	162b
	n-BuCH-(C$_2$H$_5$)CHO	Rh$_6$(CO)$_{16}$	C$_6$H$_{14}$/i-PrNH$_2$	110–160/100–300 (H$_2$ + CO)	n-BuCH(C$_2$H$_5$)-CH$_2$NHPr-i	162b
C$_9$	PhCH$_2$COCO$_2$H	[Co(CN)$_5$]$^{3-}$	H$_2$O/NH$_3$	40–70/50	PhCH$_2$CH-(NH$_2$)CO$_2$H	142, 143
		[Co(CN)$_5$]$^{3-}$	H$_2$O/C$_2$H$_5$OH/RNH$_2$, R = H, CH$_3$, t-Bu, C$_6$H$_{11}$, Ph	70/50	PhCH$_2$CH-(NHR)CO$_2$H	144a
		[Co(NH$_3$)$_5$Cl]$^{2+}$	H$_2$O	40/50	PhCH$_2$CH(NH$_2$)-CO$_2$H	142, 143
	PhCH$_2$COCO$_2$-C$_2$H$_5$	[Co(CN)$_5$]$^{3-}$	H$_2$O/NH$_3$	40/50	PhCH$_2$CH(NH$_2$)-CO$_2$H	142
C$_{10}$	PhCH=CHCOCH$_3$	[Co(CN)$_5$]$^{3-}$	H$_2$O/C$_2$H$_5$OH/NH$_3$	70/50	PhCH$_2$CH$_2$CH-(NH$_2$)CH$_3$	144a
C$_{14}$	Benzil	[Co(CN)$_5$]$^{3-}$	H$_2$O/NH$_3$	–/50	Benzoin, PhCHOHCH-(NH$_2$)Ph	140
		[Co(NH$_3$)$_5$Cl]$^{2+}$	H$_2$O/C$_2$H$_5$OH	–/50	Benzoin, PhCHOHCH-(NH$_2$)Ph	140
C$_{15}$	PhCH=CHCOPh	[Co(CN)$_5$]$^{3-}$	H$_2$O/C$_2$H$_5$OH/NH$_3$	70/50	PhCH$_2$CH$_2$CH-(NH$_2$)Ph	144a

Note: References 303–344 are on pp. 185–186.

TABLE XXI. Unsaturated Aldehydes and Ketones

Substrate	Catalyst	Solvent	Temperature (°)/Pressure (atm)	Product	Refs.
A. α,β-Unsaturated Aldehydes					
C_3 Acrolein	$RhCl(PPh_3)_3$	C_6H_6	25/1	CH_3CH_2CHO	47
	$CO_2(CO)_8$	C_6H_6	120/ 300 (H_2 + CO)	"	162a
C_4 Crotonaldehyde	$RhCl(PPh_3)_3$	C_6H_6	25/1, 55	$n\text{-}C_3H_7CHO$	47
	$RhCl(PPh_3)_3$, polymer supported	—	25/35	"	5b
	$[Co(CN)_5]^{3-}$	H_2O	Room temp/1	"	125
	$CO_2(CO)_8$	C_6H_6, C_6H_{14}	120, 180/ 220, 300 (H_2 + CO)	"	162a, 169a
C_5 2-Methyl-2-butenal	$MoH_2(C_5H_5)_2$	None	140–150/160	"	227
$(CH_3)_2C{=}CHCHO$ (tiglic aldehyde)	$[Co(CN)_5]^{3-}$	H_2O	Room temp/1	2-Methylbutanal	125
	$[Co(CN)_5]^{3-}$	H_2O	Room temp/1	3-Methylbutanal	125, 316
C_6 2-Methyl-2-pentenal	$[Co(CN)_5]^{3-}$	H_2O	Room temp/1	2-Methyl-pentanal	125
trans-2-Methyl-2-pentenal	$RhCl(PPh_3)_3$	C_6H_6	25/1	"	47
	$RhCl(CO)(PPh_3)_2$	C_6H_6	80/80	2-Methyl-pentanol	47
B. Nonconjugated, Monounsaturated Ketones					
"Unsaturated	$[RhH_2(PPh_3)_2(solvent)_2]^+$	THF	25/1	Saturated	94

132

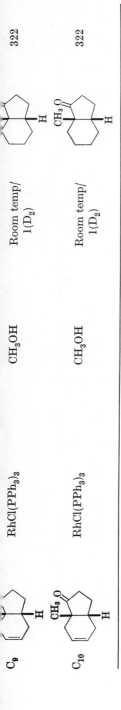

	Catalyst	Solvent	Temp/Pressure	Product	Ref.
C_9 [structure]	$RhCl(PPh_3)_3$	CH_3OH	Room temp/ 1(D_2)	[structure] H	322
C_{10} [structure, CH_3O]	$RhCl(PPh_3)_3$	CH_3OH	Room temp/ 1(D_2)	[structure] CH_3O, H	322

C. α,β-Unsaturated Ketones

	Catalyst	Solvent	Temp/Pressure	Product	Ref.
"Unsaturated ketones"	$[RhH_2(PPh_3)_2(solvent)_2]^+$	THF	25/1	Saturated ketones	94
C_4 $CH_3COCH=CH_2$	$Co_2(CO)_8$	C_6H_6	120/ 300 (H_2 + CO)	Butanone	162a
	$Rh_2HCl_3[C_5(CH_3)_5]_2$	i-PrOH	24/100	—	237
C_6 $(CH_3)_2C=CHCOCH_3$ (mesityl oxide)	$PtCl_2(PPh_3)_2/SnCl_2$	C_6H_6/CH_3OH	90/33	2-Methyl-4-pentanone	105
	$Co_2(CO)_8$	C_6H_6	120/ 300 (H_2 + CO)	''	162a
	$MoH_2(C_5H_5)_2$	None	140–150/160	''	227
Cyclohexen-2-one	$PtCl_2(PPh_3)_2/SnCl_2$	C_6H_6/CH_3OH	90/33	Cyclohexanone	105
C_7 [structure]	$RhCl[P*Ph(CH_3)Pr]_3$	C_6H_6/CH_3OH	60/27	3-Methylcyclo- hexanone (optically active)	287
C_8 Furalacetone	$Co_2(CO)_8$	—	<130/ −(H_2 + CO)	[structure] $CH_2CH_2COCH_3$	166

Note: References 303–344 are on pp. 185–186.

133

TABLE XXI. Unsaturated Aldehydes and Ketones (Continued)

Substrate	Catalyst	Solvent	Temperature (°)/Pressure (atm)	Product	Refs.
C. α,β–Unsaturated Ketones (Continued)					
(cyclohexadienone with CH_3, CCl_3)	$RhCl(PPh_3)_3$	C_6H_6	Room temp/1	(cyclohexenone with CH_3, CCl_3)	45
C_9 (cyclohexadienone with CH_3, $OCOCH_3$)	$RhCl(PPh_3)_3$	C_6H_6	Room temp/1	p-Cresol	45
C_{10} (methyl isopropenyl cyclohexenone)	$Co(dimethylglyoximato)_2$	CH_3OH	Room temp/1	(methyl isopropenyl cyclohexanone) *cis* 30%, *trans* 70%	231c
C_{13} $(CH_3)_3COCH{=}CHPh$	$H[IrCl_4(DMSO)_2]\cdot 2\ DMSO$	i-PrOH	Reflux/(no H_2)	$(CH_3)_3COCH_2CH_2Ph$	342, 92b
C_{15} $PhCOCH{=}CHPh$ $PhCOC(Ph){=}CH_2$	$H[IrCl_4(DMSO)_2]\cdot 2\ DMSO$ $Co(dimethylglyoximato)_2$	i-PrOH $CH_3OH,\ C_2H_5OH$	Reflux/(no H_2) Room temp/1(D_2)	$PhCOCH_2CH_2Ph$ $PhCOCH(Ph)CH_3$	342, 92b 231b
C_{17} $PhCO(CH{=}CH)_2Ph$	$H[IrCl_4(DMSO)_2]\cdot 2\ DMSO$	i-PrOH	Reflux/(no H_2)	$PhCO(CH_2)_4Ph$	342, 92b

134

D. Quinones

	(PhCH=CH)$_2$CO	H[IrCl$_4$(DMSO)$_2$]·2 DMSO	i-PrOH	Reflux/(no H$_2$)	(PhCH$_2$CH$_2$)$_2$CO	342, 92b
C$_6$	p-Benzoquinone	PdCl$_2$	DMF	Room temp/1	—	119
		[Co(CN)$_5$]$^{3-}$	H$_2$O	Room temp/1	Hydroquinone	125, 316
		(CuOAc)$_2$	Quinoline	Room temp/1	''	323
C$_8$		RhCl(PPh$_3$)$_3$	C$_6$H$_6$	Room temp/1		45, 324
C$_{10}$	1,2-Naphtho-quinone	PdCl$_2$	DMF	Room temp/1	—	119
	1,4-Naphtho-quinone	RhCl(PPh$_3$)$_3$	C$_6$H$_6$	Room temp/1		45, 324
		PdCl$_2$	DMF	Room temp/1	—	119
	 (Juglone)	RhCl(PPh$_3$)$_3$	C$_6$H$_6$	Room temp/1		324

Note: References 303–344 are on pp. 185–186.

135

TABLE XXII. Unsaturated Carboxylic Acids and Derivatives (Including Fatty Acids)

Substrate	Catalyst	Solvent	Temperature (°)/ Pressure (atm)	Product	Refs.
A. Mono-unsaturated, Nonconjugated					
C$_6$ Methyl 3-hexenoate	Fe(methyl sorbate)(CO)$_3$	C$_6$H$_6$	175/47	Methyl hexanoate	272
Methyl 4-hexenoate	Fe(methyl sorbate)(CO)$_3$	C$_6$H$_6$	175/47	"	272
C$_{18}$ Oleic acid	Cu, Cd soaps	None	220–380/240–400	Oleyl alcohol	275, 276
	Organic Ni, Fe, Co salts	—	—/—	"	321
Methyl oleate	RhCl(PPh$_3$)$_3$	C$_6$H$_6$	Room temp/ 1 (D$_2$)	Methyl stearate	28
	Fe(C$_{18}$-dienoate)(CO)$_3$	Ether	190/30	"	325
	NiCl$_2$/NaBH$_4$	DMF	25/1	—	270
	Cu, Cd soaps	None	315/200	Oleyl alcohol	276
Ricinoleic acid	Cu, Cd soaps	None	220–250/240–270	Ricinoleyl alcohol	275
B. α,β-Unsaturated					
"Olefinic substrates" (probably maleic and fumaric acids)	RuCl$_3$(HDMA), Ru$_2$Cl$_3$(HDMA) (HDMA = protonated DMA)	DMA	60/1	Saturated compounds	253a
C$_3$ "Unsaturated esters"	[RhH$_2$(PPh$_3$)$_2$(solvent)$_2$]$^+$	THF	25/1	Saturated esters	94
Acrylic acid	IrH$_3$(PPh$_3$)$_3$	AcOH	50/1	C$_2$H$_5$CO$_2$H	93
	[RuCl$_4$]$^{2-}$	—	—/—	" (?)	254a
	[Co(CN)$_5$]$^{3-}$	Water	> Room temp/1	"	125, 316
Methyl acrylate	IrH$_3$(PPh$_3$)$_3$	AcOH	50/1	C$_2$H$_5$CO$_2$CH$_3$	93
	Fe(CO)$_5$	C$_6$H$_6$	160/155	"	168

Substrate	Catalyst	Solvent	Conditions	Product	References
Ethyl acrylate	MoH₂(C₅H₅)₂	None	140–150/160	″	227
	Rh, IrCl(PPh₃)₃	C₆H₅CH₃	25/1	C₂H₅CO₂C₂H₅ (K)	26b, 56b, 306
	IrCl(PPh₃)₃/H₂O₂	C₆H₆, C₆H₅CH₃	50/1	″ (K)	76b
	IrCl(CO)(PPh₃)₂/hv	C₆H₅CH₃	50/1	″ (K)	79c
	Rh, Ir(NO)(PPh₃)₃	C₆H₅CH₃	25/1	″ (K)	56b
	Rh, IrCl(CO)(PPh₃)₃	″	80/1	″ (K)	77
	Rh, IrX(CO)L₂, X = Cl, Br, I, L = SCN, PPh₃, P(OPh)₃, P(C₆H₁₁)₃	″	80/1	″ (K)	80f, g, h, i
	IrCl(CO)(PPh₃)₂/H₂O₂	C₆H₅CH₃, C₆H₆	50/1	″ (K)	76b
	Rh, IrH(CO)(PPh₃)₃	C₆H₅CH₃	25/1	″ (K)	56b, 83a
	RhH(CO)(PPh₃)₃/hv	C₆H₅CH₃	25/1	″ (K)	86c
Acrylamide	RhCl(PPh₃)₃	C₆H₆	Room temp/1	C₂H₅CONH₂	26a
Methyl crotonate	RhCl(PPh₃)₃	—	—/— (T₂)	n-C₃H₇CO₂CH₃	24
	MoH₂(C₅H₅)₂	None	140–150/160	″	227
Ethyl crotonate	PdCl₂/metal ions	H₂O	30/1	n-C₃H₇CO₂C₂H₅	326
Methacrylic acid	[Co(CN)₅]³⁻	H₂O	Room temp/1	(CH₃)₂CHCO₂H	125, 316
Methyl methacrylate	RhCl(PPh₃)₃	C₆H₆	22/1	(CH₃)₂CHCO₂CH₃	31a
	[Co(CN)₅]³⁻	H₂O	Room temp/1	″	316
	Co(dimethylglyoximato)₂	CH₃OH, C₂H₅OH	Room temp/1	″	231a, b
Maleic acid	RhCl(PPh₃)₃	C₆H₆/C₂H₅OH	25/1 (D₂)	meso-1,2-Dideuterio-succinic acid	8
	[RhCl(cyclooctene)₂]₂	DMA	60/1	Succinic acid (K)	247
	RhCl₃	DMA	80/1	″ (K)	245, 252

C₄

Note: References 303–344 are on pp. 185–186.

Substrate	Catalyst	Solvent	Temperature (°)/ Pressure (atm)	Product	Refs.
		B. α,β-Unsaturated (Continued)			
C$_4$ Maleic acid (Contd.) (Contd.)	RhCl$_3$(SR$_2$)$_3$	DMA	50, 80/1(D$_2$)	Succinic acid (K)	23, 235, 246
	IrX(CO)(PPh$_3$)$_2$	DMA	80/1	,, (K)	69
	RuCl$_2$(AsPh$_3$)$_3$	C$_6$H$_6$	Room temp/1	,,	214a
	RuCl$_3$	DMA	80/1	,,	253b
	[RuCl$_4$]$^{2-}$, [RuCl$_4$(bi-pyridine)]$^{2-}$	H$_2$O/HCl	80/1	,,	254, 255
	[Co(CN)$_5$]$^{3-}$	H$_2$O. H$_2$O/C$_2$H$_5$OH	—/50	,,	140
Dimethyl maleate	Rh, IrCl(PPh$_3$)$_3$	C$_6$H$_5$CH$_3$	25/1	Dimethyl succinate (K)	26b, 56b, 306
	Rh, Ir(NO)(PPh$_3$)$_3$	,, , CH$_2$Cl$_2$	25/1(D$_2$)	,, (K)	56a, 56b
	Rh, IrCl(CO)(PPh$_3$)$_3$	,,	80/1(D$_2$, T$_2$)	,, (K)	75, 77
	Rh, IrH(CO)(PPh$_3$)$_3$	,,	25, 50, 60/1	,, (K)	56b, 83
	Rh, IrX(CO)L$_2$ X = Cl, Br, I, L = SCN, PPh$_3$, AsPh$_3$, P(OPh)$_3$, P(C$_6$H$_{11}$)$_3$	,, , DMF	80/1	,, (K)	70, 71, 80a, d, f, g, h, i
	CoD(CO)$_4$	..	26/(CO only)	Deuterated dimethyl succinate (K)	160
	Co(dimethylglyoximato)$_2$	CH$_3$OH	Room temp/1	Dimethyl succinate	231c

Diethyl maleate	RhCl(PPh₃)₃	C₆H₆	22/1	Diethyl succinate (K)	31a
	RhCl₃	DMA	80/1	"	245a
	Co(dimethylglyoximato)₂	CH₃OH, C₂H₅OH	Room temp/1	"	231a, b
Malonamic acid	RhCl₃	DMA	80/1	Succinamic acid	245a
Maleic anhydride	IrX(CO)L₂, X = Cl, Br, I; L = PPh₃, P(OPh)₃, P(C₆H₁₁)₃	C₆H₅CH₃	80/1	— (K)	80g, h, i
Fumaric acid	RhCl₃	DMA	80/1	—	245a
	RhCl(PPh₃)₃	C₆H₆/C₂H₅OH	25/1(D₂)	Succinic acid (K)	8, 51
	RhCl₃	DMA	80/1	"	245a
	IrX(CO)(PPh₃)₂	DMA	80/1	" (K)	69
	RuCl₃	DMA	80/1	" (K)	253b
	[RuCl₄]²⁻	H₂O/HCl	80/1(D₂)	" (K)	254
Dimethyl fumarate	RhCl(PPh₃)₃	C₆H₅CH₃	25/1(T₂)	Dimethyl succinate	24, 26b, 306
	Rh(NO)(PPh₃)₃	CH₂Cl₂	25/1	"	56a
	Rh, IrCl(CO)(PPh₃)₃	C₆H₅CH₃	80/1(D₂, T₂)	" (K)	75, 77
	RhX(CO)L₂, X = Cl, Br, I,; L = SCN, PPh₃, AsPh₃, P(OPh)₃, P(C₆H₁₁)₃	"	80/1	" (K)	80d, f
	IrCl(CO)(PPh₃)₂	DMF	80/1	" (K)	71
Diethyl fumarate	Co(dimethylglyoximato)₂	CH₃OH	Room temp/1	Dimethyl succinate	231c
	RhCl(PPh₃)₃	C₆H₆	22/1	Diethyl succinate (K)	31a
	Co(dimethylglyoximato)₂	CH₃OH, C₂H₅OH	Room temp/1	"	231a, b

Note: References 303–344 are on pp. 185–186.

139

TABLE XXII. Unsaturated Carboxylic Acids and Derivatives (Including Fatty Acids) (Continued)

Substrate	Catalyst	Solvent	Temperature (°)/ Pressure (atm)	Product	Refs.
		B. α,β-Unsaturated (Continued)			
C₅ CH₃CH=C(CH₃)CO₂H (Tiglic acid)	[Co(CN)₅]³⁻/1,2-propanediamine	H₂O	Room temp/1	CH₃CH₂CH(CH₃)CO₂H	293
CH₃O₂C—CO₂CH₃ / H—CH₃ (Dimethyl citraconate)	Co(dimethylglyoximato)₂	CH₃OH	Room temp/1	Dimethyl methyl-succinate	231c
CH₃O₂C—CH₃ / H—CO₂CH₃ (Dimethyl mesaconate)	Co(dimethylglyoximato)₂	CH₃OH	Room temp/1	Dimethyl methyl-succinate	231c
CH₂CO₂H / CH₂=CCO₂H (Itaconic acid)	RhCl[P*PhPr(CH₃)]₃	C₆H₆/CH₃OH, C₂H₅OH	25, 60/20–27	CH₂CO₂H | CH₃CHCO₂H * (optically active)	284a, 287
	RhH(CO)[PPh₂CH₂CH₂CH-(CH₃)C₂H₅]₃ *	C₆H₆/C₂H₅OH	Room temp/1	CH₂CO₂H | CH₃CHCO₂H * (optically active)	284b

	Substrate	Catalyst	Solvent	Temp/Pressure	Product	Ref.
					CH₃CHCO₂H	
		[Co(CN)₅]³⁻/1,2-propanediamine	H₂O	Room temp/1	;	293
	Dimethyl itaconate	Co(dimethylglyoximato)₂	CH₃OH, C₂H₅OH	Room temp/1	CH₂CO₂CH₃ / CH₃CHCO₂CH₃	231a
	NHCOCH₃ — CH₂=CCO₂H	RhCl(chiral phosphine)₂	CH₃OH	25/1	CH₃CHCO₂CH₃ NHCOCH₃	288
		RhCl(chiral diphosphine)	C₆H₆/C₂H₅OH	Room temp/1	(optically active) NHCOCH₃ — CH₃CHCO₂H *	290b, 284b
		Co(dimethylglyoximato)₂	CH₃OH, C₂H₅OH	Room temp/1	(optically active) NHCOCH₃ — CH₃CHCO₂H *	231b
C₆	Methyl 2-hexenoate	Fe(methyl sorbate)(CO)₃, Fe(dimethyl fumarate)(CO)₄	C₆H₆	175/47	CH₃CHCO₂H Methyl hexanoate	272
C₇	furan—CH=CHCO₂C₂H₅	Co₂(CO)₈	C₆H₆	120/ 300 (H₂ + CO)	furan—CH₂CH₂CO₂C₂H₅	162a
C₉	Cinnamic acid	RhCl(PPh₃)₃ RhCl₃(SR₂) Ir(1,5-COD)(tertiary phosphine)₂	C₆H₆/C₂H₅OH DMA CH₃OH	Room temp/1, 55, 80/1, 100/5	C₆H₅CH₂CH₂CO₂H ; (K) ;	51 235, 246 55

Note: References 303–344 are on pp. 185–186.

141

Substrate	Catalyst	Solvent	Temperature (°)/ Pressure (atm)	Product	Refs.
		B. α,β-Unsaturated (Continued)			
C$_9$ Cinnamic acid (Contd.)	[Co(CN)$_5$]$^{3-}$	H$_2$O	Room temp, 70/ 1, 50	C$_6$H$_5$CH$_2$CH$_2$CO$_2$H	125, 138, 142, 316, 327
Ethylcinnamate	Co$_2$(CO)$_8$	C$_6$H$_6$	120/300	C$_6$H$_5$CH$_2$CH$_2$CO$_2$C$_2$H$_5$	162a
Benzyl cinnamate	RhCl(PPh$_3$)$_3$	C$_6$H$_6$/C$_2$H$_5$OH	Room temp/1	C$_6$H$_5$CH$_2$CH$_2$CO$_2$CH$_2$-C$_6$H$_5$	51
Cinnamoyl chloride	RhCl(PPh$_3$)$_3$	C$_6$H$_6$	Room temp/1	C$_6$H$_5$CH$_2$CH$_2$COCl and other products	28
Ph $\mid$ CH$_2$=CCO$_2$H (Atropic acid)	RhCl(chiral phosphine)$_2$	C$_6$H$_6$/CH$_3$OH, C$_2$H$_5$OH	60/20–27	2-Phenylpropionic acid (optically active)	284a, 287, 289
	RhCl(chiral diphosphine)	C$_6$H$_6$/C$_2$H$_5$OH	Room temp/1	" (optically active)	290a
	[Co(CN)$_5$]$^{3-}$	H$_2$O	Room temp/1	2-Phenylpropionic acid (optically active)	125, 316
	[Co(CN)$_5$]$^{3-}$/optically active amine	H$_2$O	Room temp/1	" (optically active)	293
Methyl atropate	RhCl(chiral diphosphine), free and polymer supported	C$_6$H$_6$/C$_2$H$_5$OH	Room temp/1	Methyl 2-phenyl-propionate (optically active)	290a, 344
	Co(dimethylglyoximato)$_2$	CH$_3$OH, C$_2$H$_5$OH	Room temp/1	Methyl 2-phenyl-propionate	231b
Ethyl atropate	Co(dimethylglyoximato)$_2$	CH$_3$OH, C$_2$H$_5$OH	Room temp/1	Ethyl 2-phenyl-propionate	231a

Substrate	Catalyst	Solvent	Conditions	Product	Ref.
$\begin{array}{c}\text{COCH}_3\\ |\\ (\text{CH}_3)_3\text{CCH}=\text{CCO}_2\text{C}_2\text{H}_5\end{array}$ α-Methylcinnamic acid	Co(dimethylglyoximato)$_2$	CH$_3$OH, C$_2$H$_5$OH	Room temp/1	$\begin{array}{c}\text{COCH}_3\\ |\\ (\text{CH}_3)_3\text{CCH}_2\text{CHCO}_2\text{C}_2\text{H}_5\end{array}$ 2-Methyl-3-phenyl-propionic acid (optically active)	231b
C$_{10}$ α-Methylcinnamic acid	RhCl[P(neomenthyl)$_3$]$_3$	C$_6$H$_6$/C$_2$H$_5$OH	60/20	2-Methyl-3-phenyl-propionic acid (optically active)	289
	Ir(1,5-COD)(tertiary phosphine)$_2$	CH$_3$OH	100/5	2-Methyl-3-phenyl-propionic acid	55
Ethyl α-cyano-cinnamate	Co(dimethylglyoximato)$_2$	CH$_3$OH, C$_2$H$_5$OH	Room temp/1	Ethyl 2-cyano-3-phenylpropionate	231a, b
3-Phenyl-2-butenoic acid	RhCl[P(neomenthyl)$_3$]$_3$	C$_6$H$_6$/C$_2$H$_5$OH	60/20	3-Phenylbutyric acid (optically active)	289
Methyl 3-phenyl-2-butenoate (cis and trans)	[RhCl(pyridine)$_2$(DMF)(BH$_4$)]$^+$	DMF	Room temp/1(D$_2$)	Methyl 3-phenyl-butanoate	57
	RhCl$_3$(pyridine)$_3$/NaBH$_4$/optically active amide	None, diethylene glycol mono-ethyl ether	Room temp/1	″ (optically active)	58,59
$\begin{array}{c}\text{CH}_2\text{Ph}\\ |\\ \text{CH}_2=\text{CCO}_2\text{H}\end{array}$	RhCl[P*PhPr(CH$_3$)]$_3$	C$_6$H$_6$/CH$_3$OH	60/27	$\begin{array}{c}\text{CH}_2\text{Ph}\\ |\\ \text{CH}_3\text{CHCO}_2\text{H}\\ *\end{array}$	287
C$_{11}$ $\begin{array}{c}\text{NHCOCH}_2\text{Ph}\\ |\\ \text{CH}_2=\text{CCO}_2\text{H}\end{array}$	RhCl(chiral diphosphine)	C$_6$H$_6$/C$_2$H$_5$OH	Room temp/1	$\begin{array}{c}\text{NHCOCH}_2\text{Ph}\\ |\\ \text{CH}_3\text{CHCO}_2\text{H}\\ *\end{array}$ (optically active)	290a

Note: References 303–344 are on pp. 185–186.

143

TABLE XXII. UNSATURATED CARBOXYLIC ACIDS AND DERIVATIVES (INCLUDING FATTY ACIDS) (Continued)

Substrate	Catalyst	Solvent	Temperature (°)/ Pressure (atm)	Product	Refs.
B. α,β-Unsaturated (Continued)					
C₁₁ (Contd.)					
CH₂=CCO₂CH₃	Co(dimethylglyoximato)₂	CH₃OH, C₂H₅OH	Room temp/1	CH₃CHCO₂CH₃	231b
NHCOCH₂Ph				NHCOCH₂Ph	
	RhCl (diphosphine)	C₆H₆/C₂H₅OH	Room temp/1.1		52b
PhCH=CCO₂H	RhCl(chiral phosphine)₂	CH₃OH	25/1	PhCH₂CHCO₂H	288
NHCOCH₃				NHCOCH₃	
	RhCl(chiral diphosphine) [Rh(norbornadiene)- (diop)]⁺	C₆H₆/C₂H₅OH CH₃OH, C₂H₅OH, i-PrOH	Room temp/1 30/1	PhCH₂CHCO₂H * (optically active) " (optically active) NHCOCH₃ PhCH₂CHCO₂H * (optically active)	290a, b 290c
PhCH=CCO₂CH₃	RhCl(chiral diphosphine)	C₆H₆/C₂H₅OH	Room temp/1	PhCH₂CHCO₂CH₃ * (optically active)	290b
NHCOCH₃				NHCOCH₃	

Substrate	Catalyst	Solvent	Conditions	Product	Ref.
PhCH=CCONH₂ $\|$ NHCOCH₃	RhCl(chiral diphosphine)	C₆H₆/C₂H₅OH	Room temp/1	PhCH₂CHCONH₂ $\|$ NHCOCH₃ * (optically active)	290b
C₁₂ p-HOC₆H₄CH=CCO₂H $\|$ NHCOCH₃	RhCl(chiral diphosphine)	C₆H₆/C₂H₅OH	Room temp/1	p-HOC₆H₄CH₂CHCO₂H * $\|$ NHCOCH₃ (optically active)	290b
3,4-CH₂O₂C₆H₃CH=CCO₂H $\|$ NHCOCH₃	RhCl(chiral diphosphine)	C₆H₆/C₂H₅OH	Room temp/1	3,4-CH₂O₂C₆H₃CH₂CHCO₂H * $\|$ NHCOCH₃ (optically active)	290b
C₁₃ i-PrCH=CCO₂H $\|$ NHCOPh	RhCl(chiral diphosphine)	C₆H₆/C₂H₅OH	Room temp/1	i-C₃H₇CH₂CHCO₂H * $\|$ NHCOPh (optically active)	290b
C₁₄ CH₃O, AcO—C₆H₃—CH=CCO₂H $\|$ NHCOCH₃	RhCl(chiral phosphine)₂	CH₃OH	25/1	CH₃O, AcO—C₆H₃—CH₂—CHCO₂H * $\|$ NHCOCH₃ (optically active)	288

145

Note: References 303–344 are on pp. 185–186.

TABLE XXII. Unsaturated Carboxylic Acids and Derivatives (Including Fatty Acids) (Continued)

Substrate	Catalyst	Solvent	Temperature (°)/ Pressure (atm)	Product	Refs.
			B. α,β-Unsaturated (Continued)		

C_{16} NHCOPh
      |
PhCH=CCO₂H
	RhCl(chiral phosphine)₂	CH₃OH	25/1	NHCOPh
				   |
				PhCH₂CHCO₂H
				       *
					288

	RhCl(chiral diphosphine)	C₆H₆/C₂H₅OH	Room temp/1	" (optically active)	290b

p-HOC₆H₄CH=CCO₂H
         |
       NHCOPh
	RhCl(chiral diphosphine)	C₆H₆/C₂H₅OH	Room temp/1	p-HOC₆H₄CH₂CHCO₂H
				            *
				         |
				        NHCOPh
				(optically active)
					290b

(maleic anhydride with Ph, Ph substituents)	[Co(CN)₅]³⁻	H₂O, H₂O/ C₂H₅OH	—/50	Diphenylsuccinic acid	140

C_{17} CH₃O
            NHCOPh
             |
HO   CH=CCO₂H
	RhCl(chiral phosphine)₂	CH₃OH	25/1	CH₃O
				         NHCOPh
				         |
				HO CH₂-CHCO₂H
				            *
				(optically active)
					288

146

C. Polyunsaturated

Substrate	Catalyst	Solvent	Conditions (temp/pressure)	Products	Refs.
"Unsaturated esters"	$[RhH_2(PPh_3)_2(solvent)_2]^+$	THF	25/1	Saturated esters	94
"Unsaturated fatty acids and esters"	$Co_2(CO)_8$	—	120–190/25–30 (H_2 + CO)	Monounsaturated fatty acids	165a
	$Co_2(CO)_8/n\text{-}Bu_3P$	—	—/—	"	177a
	Cu, Cd soaps	None	—/—	Unsaturated fatty alcohols	328
C_6 $CH_3(CH{=}CH)_2CO_2H$ (Sorbic acid)	$[Co(CN)_5]^{3-}$	H_2O, CH_3OH	0–70/1, 50	2-, 3-, and 4-Hexenoic acids (ratios depend on conditions)	134–137, 144a, 145, 316
Methyl sorbate	$Co(CN)_2(pyridine)_2$	H_2O	Room temp/1	— (catalyst poisoning)	145
	Cr(methyl benzoate)(CO)$_3$; other Cr, Mo, W(arene)(CO)$_3$	Cyclohexane	120–200/15–47(D_2)	Methyl 3-hexenoate	218, 221, 222
	Co(dimethylglyoximato)$_2$	CH_3OH	Room temp/1	"	231c
	$MoH_2(C_5H_5)_2$	None	140–150/160	Mixture	227
	$NiCl_2/NaBH_4$	DMF	25/1(D_2)	Methyl 2-hexenoate	270
	Fe(diene)(CO)$_3$; Fe(monoene)(CO)$_4$	C_6H_6	165/10–33	Methyl 2-, 3-, and 4-hexenoates, methyl hexanoate (K)	272
CH_3O_2C‑‑‑CO_2CH_3 (Dimethyl muconate)	$MoH_2(C_5H_5)_2$	None	140–150/160	$CH_3O_2CCH_2CH_2$‑‑‑CO_2CH_3	227

Note: References 303–344 are on pp. 185–186.

147

C. Polyunsaturated (Continued)

Substrate	Catalyst	Solvent	Temperature (°)/ Pressure (atm)	Product	Refs.
C$_8$	H$_2$PtCl$_6$/SnCl$_2$	i-PrOH	25/1		111
C$_{11}$ Ph(CH=CH)$_2$CO$_2$H	[Co(CN)$_5$]$^{3-}$	H$_2$O	Room temp/1	—	145
C$_{18}$ Methyl 9,11-octa-decadienoate	Cr(C$_6$H$_6$)(CO)$_3$	Cyclohexane	160/30(D$_2$)	Methyl octadecenoates	221
Methyl cis-9, trans-11-octadecadienoate	Cr(arene)(CO)$_3$	"	125–175/30(D$_2$)	Methyl cis-octa-decenoates (mainly)	220, 221
Methyl trans-9, trans-11-octa-decadienoate	Cr(arene)(CO)$_3$	"	125–175/30(D$_2$)	"	220, 221
Methyl 10,12-octa-decadienoate	Cr(C$_6$H$_6$)(CO)$_3$	"	160/30(D$_2$)	Methyl octadecenoates	221
Methyl linoleate (9,12-octa-decadienoate)	RhCl(PPh$_3$)$_3$	C$_6$H$_6$	Room temp/ 1(D$_2$)	Methyl stearate	28
	RhCl$_3$(pyridine)$_3$/NaBH$_4$	DMF	Room temp/1	trans Monoene esters	269
	H$_2$PtCl$_6$, PtCl$_2$(PPh$_3$)$_2$/ SnCl$_2$, and similar	C$_6$H$_6$/CH$_3$OH	90/39	"	103, 104, 107, 116, 329

Substrate	Catalyst	Solvent	Temp/Pressure	Products	References
	Co$_2$(CO)$_8$	Cyclohexane	75/200 (H$_2$ + CO)	Monoene esters	165b
	Ziegler catalysts	C$_6$H$_{14}$	150/150	Mixed saturated, mono-, and di-ene esters	199
	NiX$_2$(PPh$_3$)$_2$	C$_6$H$_6$, C$_6$H$_5$CH$_3$, THF	90, 140/39	trans Monoene esters	116, 268
	Ni(acetylacetonate)$_3$	CH$_3$OH	100–180/67	Mono- and di-ene esters	205
	NiCl$_2$/NaBH$_4$	DMF	25/1, 20	cis Monoene esters	269, 270
	Cr(arene)(CO)$_3$	Cyclohexane	165, 175/30(D$_2$)	''	220, 221
	Fe(CO)$_5$,Fe(diene)(CO)$_3$	''	150–180/27	Monoene esters	273b
Alkali-conjugated linoleate	Cr(arene)(CO)$_3$	Cyclohexane	125–175/30	cis Monoenes	220
Dehydrated methyl ricinolate	Cr(arene)(CO)$_3$	Cyclohexane	150/47	Monoene and non-conjugated diene esters	218a
Methyl α- and β-eleostearates (9,11,13-octa-decatrienoates)	Cr(arene)(CO)$_3$	Cyclohexane	120–175/30, 47	Monoene and non-conjugated diene esters	218a, 223
Linolenic acid (9,12,15-octa-decatrienoic acid)	Cu, Cd soaps	None	260–380/100–300	Linolenyl alcohol	276
Methyl linolenate	H$_2$PtCl$_6$/SnCl$_2$ and similar	C$_6$H$_6$/CH$_3$OH	30–100/1–33	Mixed monoene and diene esters	118
	Co$_2$(CO)$_8$	Cyclohexane	75/200 (H$_2$ + CO)	trans Mono- and di-ene esters	165b

Note: References 303–344 are on pp. 185–186.

149

TABLE XXII. UNSATURATED CARBOXYLIC ACIDS AND DERIVATIVES (INCLUDING FATTY ACIDS) (Continued)

Substrate	Catalyst	Solvent	Temperature (°)/Pressure (atm)	Product	Refs.
			C. Polyunsaturated (Continued)		
C_{18} (Contd.)					
Methyl linolenate (Contd.)	Ziegler catalysts	C_6H_{14}	150/150	Mono- and di-ene esters	199
	Ni(acetylacetonate)$_3$	CH_3OH	100–180/67	,,	205
	Cr($C_6H_5CO_2CH_3$)(CO)$_3$	Cyclohexane	165, 175/30	Monoene and nonconjugated diene esters	223
	Fe(CO)$_5$	None	180/27	Mono-, di-, and tri-ene esters and Fe(CO)$_3$ complexes	273c
Soybean oil	NiCl$_2$/NaBH$_4$	DMF	25/40	cis Monoene esters	270
	Fe(CO)$_5$	None	180/3–24	trans Monoene and conjugated diene esters	273a
Soybean oil methyl esters	H$_2$PtCl$_6$, PtCl$_2$(PPh$_3$)$_2$/ SnCl$_2$ and similar	C_6H_6/CH_3OH	30–105/27–70	trans Monoene esters	103, 107, 116, 315, 329
	Co$_2$(CO)$_8$	Cyclohexane	75/ 200 (H_2 + CO)	Monoene esters	165b

	Catalyst	Solvent	Conditions	Products	Ref.
	Ziegler catalysts	C_6H_{14}	150/150	Mono- and di-ene esters	199
	Metal acetylacetonates	CH_3OH	100–180/67	Mono- and di-ene esters	205
	$Fe(CO)_5$	None, cyclohexane	180/8–27	*trans* Monoene and conjugated diene esters	273a
	$Cr(arene)(CO)_3$	Cyclohexane	175/30	*cis* Monoene and nonconjugated diene esters	222
Coconut oil	Cu, Cd soaps	None	250/100–300	Fatty alcohols	276
Cottonseed oil methyl esters	$Fe(diene)(CO)_3$	—	180–200/1–25	Monoene esters and $Fe(CO)_3$ complexes	271a
Linseed oil	Cu, Cd soaps	None	250/100–300	Fatty alcohols	276
	$Cr(C_6H_5CO_2CH_3)(CO)_3$	None	175/15	*cis* Monoene and nonconjugated diene esters	223
Linseed oil methyl esters	Metal acetylacetonates	CH_3OH	100–180/67	Mono- and di-ene esters	205
Olive oil	Cu, Cd soaps	None	250/100–300	Fatty alcohols	276
Sperm oil	Cu, Cd soaps	None	250/100–300	Fatty alcohols	276
Tung oil	$Cr(arene)(CO)_3$	C_6H_{14}, cyclohexane	115–170/15, 47	*cis* Monoene and nonconjugated diene esters	218a, 223

Note: References 303–344 are on pp. 185–186.

TABLE XXIII. STYRENES

Substrate	Catalyst	Solvent	Temperature (°)/Pressure (atm)	Product (Yield %)	Refs.
C$_8$ Styrene	RhCl(PPh$_3$)$_3$	C$_6$H$_6$, CHCl$_3$, CH$_2$Cl$_2$	22–60/1(D$_2$)	C$_6$H$_5$C$_2$H$_5$ (K)	11, 12 26, 31a, 37, 56b, 306
	RhCl(tertiary phosphine)$_3$	None, C$_6$H$_6$	40, 60/1	" (K)	11, 12
	RhCl(diphosphine)	C$_6$H$_6$	Room temp/1.1	" (K)	52a
	IrCl(PPh$_3$)$_3$	C$_6$H$_5$CH$_3$	25/1	" (K)	56b, 306
	IrCl(PPh$_3$)$_3$/H$_2$O$_2$	C$_6$H$_6$, C$_6$H$_5$CH$_3$	50/1(D$_2$)	" (K)	76b
	Rh, Ir(NO)(PPh$_3$)$_3$	C$_6$H$_5$CH$_3$	25/1	" (K)	56b
	Rh, IrCl(CO)(PPh$_3$)$_3$	"	80/1	" (K)	77
	Rh, IrH(CO)(PPh$_3$)$_3$	"	25/1	" (K)	56b, 83a, 99
	RhH(DBP)$_4$	C$_6$H$_6$	20/0.1	" (K)	100c
	RhX(CO)L$_2$, X = Cl, Br, I, L = SCN, PPh$_3$, AsPh$_3$, P(OPh)$_3$, P(C$_6$H$_{11}$)$_3$	C$_6$H$_5$CH$_3$	80/1	" (K)	80d, f, g
	IrCl(CO)(PPh$_3$)$_2$/H$_2$O$_2$	C$_6$H$_6$, C$_6$H$_5$CH$_3$	50/1(D$_2$)	" (K)	76b
	RuH$_2$(PPh$_3$)$_4$	C$_6$H$_6$	—/—	"	251
	RuH(NO)(tertiary phosphine)$_3$	"	20/1	"	213
	PtCl$_2$(SnCl$_3$)$_2$	—	—/—	" (slow)	305

	Catalyst	Solvent	Temp/pressure	Product	Refs.
[Co(CN)₅]³⁻		H_2O	0–50/1	··	125, 139, 316
	$CoHN_2(PPh_3)_3$/ Na naphthalide	C_6H_6/THF	20/1	—	263a
	Ziegler catalysts	$C_6H_5CH_3$, heptane, THF, dioxane, various ethers	0–60/1	$C_6H_5C_2H_5$	198, 204, 206a, c–h, 330
	$Ti(C_5H_5)_2(CO)_2$	C_6H_6	50–65/50	··	188
	$Ti(CO)(C_5H_5)_2(PhC{\equiv}CPh)$	C_7H_{16}	Room temp/1	··	217c
	$Ti(C_5H_5)_2(\text{1-methylallyl})$	Cyclohexane	Room temp/1	··	217b
	$TiCl_2(C_5H_5)_2$/Na	THF	—/—	··	190
	$TiCl_2(C_5H_5)_2$/BuLi, Na naphthalide; (polymer supported)	C_6H_{14}	—/1	··, polystyrene	187a
ω-Bromostyrene	Rh, $IrCl(PPh_3)_3$	$C_6H_5CH_3$	25/1	— (K)	26b, 56b, 306
	Rh, $Ir(NO)(PPh_3)_3$	··	25/1	— (K)	56b
	Rh, $IrCl(CO)(PPh_3)_3$	··	80/1	— (K)	77
	Rh, $IrH(CO)(PPh_3)_3$	··	25/1	— (K)	56b, 83a
	$RhX(CO)L_2$, X = Cl, Br, I, L = SCN, PPh_3, $P(OPh)_3$, $P(C_6H_{11})_3$	··	80/1	— (K)	80f
ω-Nitrostyrene	$RhCl(PPh_3)_3$	C_6H_6	Room temp/1	$C_6H_5CH_2CH_2NO_2$	28
p-Fluorostyrene	$RhCl(PPh_3)_3$	C_6H_6	Room temp/1	$p\text{-}FC_6H_4C_2H_5$ (K)	26a
Pentafluorostyrene	$RhCl(PPh_3)_3$	—	—/—	$C_6F_5C_2H_5$	37

Note: References 303–344 are on pp. 185–186.

TABLE XXIII. Styrenes (Continued)

Substrate	Catalyst	Solvent	Temperature (°)/ Pressure (atm)	Product (Yield %)	Refs.
C$_9$ Indene	RhCl(PPh$_3$)$_3$	C$_6$H$_6$	22/1	Indane	31a
	Ziegler catalysts	THF	0/1	—	206d, e
α-Methylstyrene	[Co(CN)$_5$]$^{3-}$	H$_2$O	Room temp/1	C$_6$H$_5$CH(CH$_3$)$_2$	125, 316
	Co$_2$(CO)$_8$	C$_6$H$_5$CH$_3$, CH$_3$OH, pentane, acetone, ether	130/ 140 (H$_2$ + CO)	''; C$_{10}$-aldehyde	3a, 331
	RhCl(chiral diphosphine), polymer supported	C$_6$H$_6$	Room temp/1	C$_6$H$_5$CH(CH$_3$)$_2$	344
	CoH(CO)$_4$	—	15/(CO only)	''; C$_{10}$-aldehyde	3a
	FeCl$_3$/LiAlH$_4$	THF	0/1	—	206d
α-Methoxystyrene	RhCl[P*PhPr(CH$_3$)]$_3$	C$_6$H$_6$	Room temp/1	1-Methoxy-1-phenylethane (optically active)	286
ω-Cyanostyrene	[Co(CN)$_5$]$^{3-}$	H$_2$O	70, 90/1	C$_6$H$_5$CH$_2$CH$_2$CO$_2$H, C$_6$H$_5$CH$_2$CH$_2$CH$_2$CONH$_2$	141
PhCH=C(NO$_2$)CH$_3$	RhCl$_3$(pyridine)$_3$/NaBH$_4$	DMF	Room temp/1	PhCH$_2$CH(NO$_2$)CH$_3$	62b
p-CH$_3$OC$_6$H$_4$CH=CH$_2$	RhCl(PPh$_3$)$_3$	C$_6$H$_6$	Room temp/1	p-CH$_3$OC$_6$H$_4$C$_2$H$_5$ (K)	26a, 37
p-CH$_3$OC$_6$H$_4$CH= CHNO$_2$	RhCl(PPh$_3$)$_3$	C$_6$H$_6$	Room temp/1	p-CH$_3$OC$_6$H$_4$CH$_2$- CH$_2$NO$_2$	45
C$_{10}$ α-Ethylstyrene	RhCl[P*PhPr(CH$_3$)]$_3$	C$_6$H$_6$	Room temp/1	2-Phenylbutane (optically active)	286
	RhCl[P(neomenthyl)$_3$]$_3$	C$_6$H$_6$/C$_2$H$_5$OH	60/20	''	289
	RhCl(chiral diphosphine), polymer supported	C$_6$H$_6$	Room temp/1	''	344
α-Acetoxystyrene	RhCl[P*PhPr(CH$_3$)]$_3$	C$_6$H$_6$/CH$_3$OH	60/27	1-Acetoxy-1-phenyl-ethane (optically active)	287

154

Substrate	Catalyst	Solvent	Conditions	Product	Refs.
CH₃OOC₆H₄OH—CHCH₃ (fragment)					
3,4-CH₂O₂C₆H₃CH=CHCH₃	FeCl₃/LiAlH₄	THF	0/1	—	206d
3,4-(CH₃O)₂C₆H₃CH=CHNO₂	RhCl₃(pyridine)₃/NaBH₄	DMF	Room temp/1	CH₃O, CH₃O—C₆H₃—CH₂CH₂NO₂	62b
C₁₂ (cyclohexene-OH-Ph structure)	RhCl(PPh₃)₃	C₆H₆, C₆H₆/C₂H₅OH	Room temp/1	(cyclohexanol-Ph structure) OH, Ph	332
(cyclohexene-OH, Ph substituted structure)	RhCl(PPh₃)₃	C₆H₆, C₆H₆/C₂H₅OH	Room temp/1	(cyclohexanol-Ph structure) OH, Ph	332
C₁₄ trans-Stilbene	RhCl(PPh₃)₃	C₆H₆	25/1(D₂)	PhCH₂CH₂Ph (K)	8
	Ti(OPr-i)₄/i-Bu₃Al	C₆H₅CH₃, C₇H₁₆	—/3.5	"	182
	Ti(CO)(C₅H₅)₂(PhC≡CPh)	C₇H₁₆	Room temp/1	"	217c
	CoH(CO)₄	C₅H₁₂	25/(CO only)	1,1-Diphenylethane	161
(C₆H₅)₂C=CH₂					
C₁₅ (phenanthrene CH₂OH, CH₃O structure)	Rh(substrate)(PPh₃)₃	C₆H₆	50/7	(octahydrophenanthrene CH₂OH, CH₃O, H structure)	40c
C₁₆ PhCH=C(N=CPh—O—C=O structure)	Co(dimethylglyoximato)₂	CH₃OH, C₂H₅OH	Room temp/1	PhCH₂CH(NHCOPh)CO₂CH₃	231a

Note: References 303–344 are on pp. 185–186.

TABLE XXIV. SUBSTITUTED OLEFINS OTHER THAN STYRENES

Substrate	Catalyst	Solvent	Temperature (°)/Pressure (atm)	Product	Refs.
		A. Monoolefins			
C₃ Acrylonitrile	$RhCl(PPh_3)_3$	C_6H_6	22/1	C_2H_5CN	6b, 31a
	$RuCl_2(acrylonitrile)_4$	C_2H_5OH, acetone	110, 150/10–40	"; acrylonitrile dimers	256, 257
	$Fe(CO)_5$, $Co_2(CO)_8$/base, borohydride	C_6H_6	110/100	"	3h
	$FeCl_3$, $CoCl_2/LiAlH_4$	THF	0/1	—	206d
Allyl alcohol	$RhCl(PPh_3)_3$	C_6H_6	22/1	$n\text{-}C_3H_7OH$ (?)	26a, 31a
	$RhCl(PPh_3)_3$, polymer supported	C_6H_6/C_2H_5OH	25/1	"	5d
	$Rh(NO)(PPh_3)_3$	CH_2Cl_2	25/1	"	56a
	$RhH(CO)(PPh_3)_3$	C_6H_6	25/0.8	" (?)	99
	$RhH(DBP)_4$	C_6H_6	20/0.1	" (?)	100c
C₄ Vinyl acetate	$RhCl(PPh_3)_3$	C_6H_6	22/1	$CH_3CO_2C_2H_5$	31a
	$RhCl(PPh_3)_3$, polymer supported	—	25/35	"	5b
	$TiCl_2(C_5H_5)_2/BuLi$, Na naphthalide; (polymer supported)	C_6H_{14}	—/1	" (?)	187a
$CH_2\text{=}CHOC_2H_5$	$FeCl_3$, $CoCl_2/LiAlH_4$	THF	0/1	" (?)	206d
	$RhCl(PPh_3)_3$	C_6H_6	22/1	$(C_2H_5)_2O$ (?)	31a
	$RhCl(PPh_3)_3$, polymer supported	—	110/40	" (?)	5b

156

Substrate	Catalyst	Solvent	Temp/Pressure	Product	Ref.
Allyl cyanide	FeCl$_3$, CoCl$_2$/LiAlH$_4$	THF	0/1	.. (?)	206d
	RhCl(PPh$_3$)$_3$	C$_6$H$_6$	22/1	n-C$_3$H$_7$CN (?)	26a, 31a
	RhH(CO)(PPh$_3$)$_3$	C$_6$H$_6$	25/0.8	.. (?)	99
	RhH(DBP)$_4$	C$_6$H$_6$	20/0.1	.. (?)	100c
2-Butene-1,4-diol	RuCl$_2$(PPh$_3$)$_3$	C$_6$H$_6$/C$_2$H$_5$OH	20/—	Butane-1,4-diol	210
α-Methylacrylonitrile	Co(dimethylglyoximato)$_2$	CH$_3$OH, C$_2$H$_5$OH	Room temp/1	i-C$_3$H$_7$CN	231b
C$_5$ Allyl acetate	RhCl(PPh$_3$)$_3$	C$_6$H$_6$	22/1	Propyl acetate (?)	31a
1-Penten-4-ol	RhCl(PPh$_3$)$_3$, polymer supported	C$_6$H$_6$/C$_2$H$_5$OH	25/1	1-Pentanol	5d
C$_6$ 2,3-Dihydropyran	RhCl(PPh$_3$)$_3$	C$_6$H$_6$	Room temp/1	Tetrahydropyran	28
CH$_2$=CHOC$_4$H$_9$-n	RhX(CO)L$_2$, X = Cl, I, L = PPh$_3$, AsPh$_3$, P(OPh)$_3$, P(C$_6$H$_{11}$)$_3$	C$_6$H$_5$CH$_3$	80/1	C$_2$H$_5$OC$_4$H$_9$-n (?)	80d
CH$_2$=CHOC$_4$H$_9$-i	RhCl(PPh$_3$)$_3$	C$_6$H$_6$	22/1	C$_2$H$_5$OC$_4$H$_9$-i (?)	31a
C$_7$ (OH, 2-methylcyclohexene)	RhCl(PPh$_3$)$_3$	C$_6$H$_6$, C$_6$H$_6$/C$_2$H$_5$OH	Room temp/1	o-Cresol	332
C$_7$ (OH, methylcyclohexene)	RhCl(PPh$_3$)$_3$	C$_6$H$_6$, C$_6$H$_6$/C$_2$H$_5$OH	Room temp/1	m-Cresol	332
C$_8$ 2-Butene-1,4-diol diacetate	RuCl$_2$(PPh$_3$)$_3$	C$_6$H$_6$/C$_2$H$_5$OH	20/—	Butane-1,4-diol diacetate	210

Note: References 303–344 are on pp. 185–186.

157

TABLE XXIV. SUBSTITUTED OLEFINS OTHER THAN STYRENES (Continued)

A. Monoolefins (Continued)

Substrate	Catalyst	Solvent	Temperature (°)/Pressure (atm)	Product	Refs.
C₉ Allyl phenyl sulfide	RhCl(PPh₃)₃	C₆H₆	Room temp/1	n-PrSC₆H₅	308
Cinnamyl alcohol	[Co(CN)₅]³⁻	H₂O	Room temp/1	C₆H₅(CH₂)₃OH	125
[structure: cyclohexene with OH, gem-dimethyl and methyl]	RhCl(PPh₃)₃	C₆H₆, C₆H₆/C₂H₅OH	Room temp/1	[structure: cyclohexanol with dimethyl, OH]	332
C₁₂ [structure: cyclohexene with OH and Ph]	RhCl(PPh₃)₃	C₆H₆, C₆H₆/C₂H₅OH	Room temp/1	[structure: cyclohexanol with OH and Ph]	332
[structure: cyclohexene with OH and Ph]	RhCl(PPh₃)₃	C₆H₆, C₆H₆/C₂H₅OH	Room temp/1	[structure: cyclohexanol with OH and Ph]	332

B. Di-, Tri-, and Higher Olefins

Substrate	Catalyst	Solvent	Temperature (°)/Pressure (atm)	Product	Refs.
C₇ [structure: methoxy-cyclohexadiene, OCH₃]	RhCl(PPh₃)₃	C₆H₆	Room temp/1	[structure: methoxy-cyclohexene, OCH₃]	28, 45

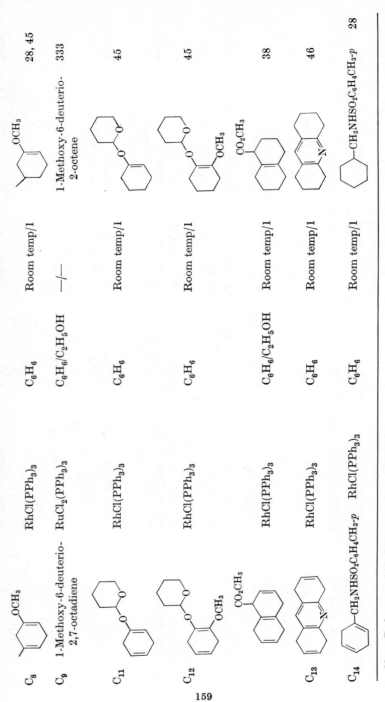

						Refs.
C$_8$	(OCH$_3$, methylcyclohexadiene)	RhCl(PPh$_3$)$_3$	C$_6$H$_6$	Room temp/1	(OCH$_3$, methylcyclohexene)	28, 45
C$_9$	1-Methoxy-6-deuterio-2,7-octadiene	RuCl$_2$(PPh$_3$)$_3$	C$_6$H$_6$/C$_2$H$_5$OH	—/—	1-Methoxy-6-deuterio-2-octene	333
C$_{11}$	(tetrahydropyranyloxy-phenyl)	RhCl(PPh$_3$)$_3$	C$_6$H$_6$	Room temp/1	(tetrahydropyranyloxy-cyclohexene)	45
C$_{12}$	(tetrahydropyranyloxy, OCH$_3$)	RhCl(PPh$_3$)$_3$	C$_6$H$_6$	Room temp/1	(tetrahydropyranyloxy, OCH$_3$)	45
C$_{13}$	(CO$_2$CH$_3$ naphthalene)	RhCl(PPh$_3$)$_3$	C$_6$H$_6$/C$_2$H$_5$OH	Room temp/1	(CO$_2$CH$_3$)	38
C$_{13}$	(acridine)	RhCl(PPh$_3$)$_3$	C$_6$H$_6$	Room temp/1	(acridine)	46
C$_{14}$	—CH$_2$NHSO$_2$C$_6$H$_4$CH$_3$-p	RhCl(PPh$_3$)$_3$	C$_6$H$_6$	Room temp/1	—CH$_2$NHSO$_2$C$_6$H$_4$CH$_3$-p	28

Note: References 303–344 are on pp. 185–186.

159

TABLE XXV. ACETYLENES

Substrate		Catalyst	Solvent	Temperature (°)/Pressure (atm)	Product (Yield %)	Refs.
	"Acetylenes"	RhHCl₂(n-Pr₂-t-BuP)₂	i-PrOH/i-PrONa	20/1	Alkanes	238b
	"Terminal and disubstituted acetylenes"	CoCl[P(OC₂H₅)₃]₃&₄	C₂H₅OH	>75/"High"	Olefins, alkanes	230a
	"Internal acetylenes"	[Rh(diene)Lₓ]⁺	—	—/—	cis Alkenes	97
	"Disubstituted acetylenes"	RuCl₂(PPh₃)₃	—	20/—	cis Alkenes	210
C₂	Acetylene	Rh, IrX(CO)(PPh₃)₂,₃	C₆H₆, C₆H₅CH₃ CH₃OH	40–60/1	C₂H₄, C₂H₆	68b, 82
		H₂PtCl₆/SnCl₂	C₆H₅CH₃	20/1(D₂)	" , ",	101, 114
		Fe(acetylacetonate)₃/ Al(C₂H₅)₃	—	—/—	Ethylene, polymer	189
C₃	Propyne	Pd₂(Ph₂PCH₂PPh₂)₃	C₆H₅CH₃	Room temp/6	C₃H₆, C₃H₈	258a
C₄	2-Butyne	Pd₂(Ph₂PCH₂PPh₂)₃	C₆H₅CH₃	Room temp/6	cis-2-Butene	258a
	2-Butyne-1,4-diol	RuCl₂(PPh₃)₃	C₆H₆/C₂H₅OH	20/—	cis-2-Butene-1,4-diol, butane-1,4-diol	210
	Acetylenedicarboxylic acid	[Co(CN)₅]³⁻, [Co(NH₃)₅Cl]²⁺/CN⁻	H₂O, H₂O/ C₂H₅OH	20, 70/50	Fumaric, succinic acids	140
C₅	1-Pentyne	CoH(CO)₂(n-Bu₃P)₂, CoH(CO)(n-Bu₃P)₃	C₇H₁₆	40–60/30	1-Pentene	179b
	2-Pentyne	Ti(C₅H₅)₂(CO)₂	C₆H₆, C₇H₁₆	50–60/50	"	188
		CoH(CO)₂(n-Bu₃P)₂, CoH(CO)(n-Bu₃P)₃	C₁₃H₂₈	40/30	cis-2-Pentene	179b
	2-Pentyne-1,4-diol	RhCl(PPh₃)₃	C₆H₆/C₂H₅OH	20/1	Pentane-1,4-diol (?)	6a
	3-Methyl-1-butyn-3-ol	RhCl(PPh₃)₃	C₆H₆/C₂H₅OH	20/1	3-Methylbutan-3-ol (?)	6a
		RhH(PF₃)(PPh₃)₃	C₆H₆	25/1	" (?)	100b

C6	Rh, IrCl(PPh₃)₃	C₆H₆, C₆H₆/various co-solvents, C₆H₅CH₃	20–25/1(D₂)	C₆H₁₄ (K)	
1-Hexyne	Rh, IrCl(PPh₃)₃	C₆H₆, C₆H₆/various co-solvents, C₆H₅CH₃	20–25/1(D₂)	C₆H₁₄ (K)	6a, 8, 26b, 31a, 56b, 306
	Rh, Ir(NO)(PPh₃)₃	C₆H₅CH₃, CH₂Cl₂	25/1	" (K)	56a, 56b
	Rh, IrCl(CO)(PPh₃)₃	C₆H₅CH₃	70/1	" (K)	77
	Rh, IrH(CO)(PPh₃)₃	"	25/1	" (K)	56b, 83a
	RhX(CO)L₂, X = Cl, Br, I, L = SCN, PPh₃, P(OPh)₃, P(C₆H₁₁)₃	"	70/1	" (K)	80f
	[RhH₂(PPh₃)₂(solvent)₂]⁺	THF	25/1	—	94
	Ti(C₅H₅)₂(CO)₂	C₆H₆	50–60/50	1-Hexene (90)	188
	Ti(C₅H₅)₂Cl₂/BuLi, Na naphthalide; (polymer supported)	C₆H₁₄	—/1	—	187a
2-Hexyne	RhCl(PPh₃)₃	C₆H₆	25/1	cis-2-Hexene, C₆H₁₄	8
	Rh(NO)(PPh₃)₃	CH₂Cl₂	25/1	cis-2-Hexene	56a
	[RhH₂(PPh₃)₂(solvent)₂]⁺	THF	25/1	—	94
3-Hexyne	RhCl(PPh₃)₃	C₆H₆, C₆H₆/various co-solvents	22/1	—	31a
	Rh(NO)(PPh₃)₃	CH₂Cl₂	25/1	—	56a
	Ti(C₅H₅)₂Cl₂/BuLi, Na naphthalide; (polymer supported)	C₆H₁₄	—/1	—	187a

Note: References 303–344 are on pp. 185–186.

161

TABLE XXV. ACETYLENES (Continued)

Substrate	Catalyst	Solvent	Temperature (°)/Pressure (atm)	Product (Yield %)	Refs.
3-Methyl-1-pentyn-3-ol	$RhCl(PPh_3)_3$	C_6H_6/C_2H_5OH	20/1	3-Methylpentan-3-ol (?)	6a
3,3-Dimethyl-1-butyne	$Ti(C_5H_5)_2(CO)_2$	C_6H_6	50–60/50	3,3-Dimethyl-1-butene	188
C₇ 1-Heptyne	$RhCl(PPh_3)_3$	C_6H_6	22/1	—	31a
	$RuCl_2(PPh_3)_3$	C_6H_6/C_2H_5OH	25/1(D_2)	C_7H_{16}	307
	$Fe(acetylacetonate)_3/$ $Al(C_2H_5)_3$	—	—/—	"	189
C₈ 1-Octyne	$RhCl(PPh_3)_3$	C_6H_6	22/1	—	31a
Phenylacetylene	Rh, $IrCl(PPh_3)_3$	C_6H_6, $C_6H_5CH_3$, C_6H_6/C_2H_5OH	20–25/1	PhC_2H_5 (K)	6a, b, 26b, 31a, 56b, 306
	Rh, $Ir(NO)(PPh_3)_3$	$C_6H_5CH_3$	25/1	" (K)	56b
	Rh, $IrCl(CO)(PPh_3)_3$	"	80/1	" (K)	77
	Rh, $IrH(CO)(PPh_3)_3$	"	25/1	" (K)	56b, 83a
	Rh, $IrX(CO)L_2$, X = Cl, Br, I, L = SCN, PPh_3, $P(OPh)_3$, $P(C_6H_{11})_3$	"	80/1	" (K)	80f, i
	$Ti(C_5H_5)_2(CO)_2$	C_6H_6, C_7H_{16}	50–60/50	"	188
	Metal acetylacetonates/ AlR_3	$C_6H_3CH_3,C_7H_{16}$ C_6H_5Cl	0–60/—	", trimeric products	189, 198

162

The substrate heading shows a cyclohexane ring bearing C≡CH and OH (1-ethynylcyclohexanol); the product heading shows a cyclohexane ring bearing C₂H₅ and OH.

Substrate	Catalyst	Solvent	Conditions	Product	Ref.
(cyclohexane ring) C≡CH, OH	RhCl(PPh$_3$)$_3$	C$_6$H$_6$/C$_2$H$_5$OH	20/1	(cyclohexane ring) C$_2$H$_5$, OH (?)	6a
3,5-Dimethyl-1-hexyn-3-ol	RuCl$_3$/PPh$_3$	CH$_3$OH	—/—	3,5-Dimethylhexan-3-ol	312
2,5-Dimethyl-3-hexyne-2,5-diol	PtHCl(PPh$_3$)$_2$/SnCl$_2$	CH$_3$OH	—/—	2,5-Dimethylhexane-2,5-diol	312
2-Butyne-1,4-diol diacetate	RuCl$_2$(PPh$_3$)$_3$	C$_6$H$_6$/C$_2$H$_5$OH	20/1	cis-2-Butene-1,4-diol diacetate	210
C$_9$ 1-Phenylpropyne	IrCl(PPh$_3$)$_3$	C$_6$H$_5$CH$_3$	50/1	cis- and trans-1-Phenylpropene	76b
C$_{10}$ 3,7-Decadiyne	Ti(C$_5$H$_5$)$_2$(1-methylallyl)	Cyclohexane	Room temp/1	C$_{10}$H$_{22}$	217b
	RhCl(PPh$_3$)$_3$	C$_6$H$_6$/C$_2$H$_5$OH	20/1	PhCH$_2$CH$_2$Ph	6a
C$_{14}$ Diphenylacetylene (tolan)	Rh$_2$HCl$_3$[C$_5$(CH$_3$)$_5$]$_2$	i-PrOH	24/100	—	237
	RuCl$_3$/PPh$_3$	CH$_3$OH	—/—	cis-Stilbene	312
	IrHCl$_2$(DMSO)$_3$	i-PrOH/HCl	73/(no H$_2$)	''	343
	Co$_2$(CO)$_8$	Cyclohexane	180/ 300 (H$_2$ + CO)	PhCH$_2$CH$_2$Ph	167
	Ti(C$_5$H$_5$)$_2$(CO)$_2$	C$_6$H$_6$	50–60/50	''	188
	Ti(CO)(C$_5$H$_5$)$_2$(PhC≡CPh)	C$_7$H$_{14}$	Room temp/1	''	217c
	Ti(C$_5$H$_5$)$_2$Cl$_2$/BuLi, Na naphthalide; (polymer supported)	C$_6$H$_{14}$	—/1	—	187a
C$_{18}$ Stearolic acid (9-octadecynoic acid)	RuCl$_3$/PPh$_3$	C$_6$H$_6$, CH$_3$OH	—/—	Oleic acid	312

Note: References 303–344 are on pp. 185–186.

TABLE XXVI. Aromatic and Heteroaromatic Compounds

Substrate	Catalyst	Solvent	Temperature (°)/ Pressure (atm)	Product (Yield %)	Refs.
C$_4$ "Alkyl benzenes"	Co(C$_3$H$_5$)[P(OCH$_3$)$_3$]$_3$	—	25/1	Alkylcyclohexanes	206i
Furan	Co$_2$(CO)$_8$	C$_6$H$_{14}$	180/ 200 (H$_2$ + CO)	Tetrahydrofurfuryl alcohol	169a
	FeCl$_3$, CoCl$_2$/LiAlH$_4$	THF	0/1	—	206d
Thiophene	Co$_2$(CO)$_8$	C$_6$H$_6$, C$_6$H$_{14}$	180/210, 250 (H$_2$ + CO)	Thiolane (66)	169a, 334
C$_5$ Pyridine	[RhCl(pyridine)$_2$(DMF)-(BH$_4$)]$^+$	DMF	Room temp/1	Piperidine	57, 62a
	Ni 2-ethylhexanoate/ Al(C$_2$H$_5$)$_3$	C$_6$H$_{14}$	150–174/66	''	184
2-Methylthiophene	Co$_2$(CO)$_8$	C$_6$H$_6$	180/ 250 (H$_2$ + CO)	[structure: tetrahydrothiophene-CH$_3$] (51)	334
2-Thienylmethanol	Co$_2$(CO)$_8$	C$_6$H$_6$	180/ 250 (H$_2$ + CO)	2-Methylthiophene (24), [structure: tetrahydrothiophene-CH$_3$] (57)	334

164

C_6	Benzene	$Co(C_3H_5)[P(OCH_3)_3]_3$	—	25/1	Cyclohexane	206i, j
		Ni 2-ethylhexanoate/ $Al(C_2H_5)_3$	None	150–190/66	,,	184
		Various Ziegler catalysts	*n*-Octane	80–150/—	,,	185
		Organic Ni, Fe, Co salts	—	—/—	,, (?)	321
	Phenol	Ni 2-ethylhexanoate/ $Al(C_2H_5)_3$	C_7H_{16}	150–160/66	Cyclohexanol (92), cyclohexanone (5)	184
	2,5-Dimethylfuran	$Co_2(CO)_8$	C_6H_{14}	180/ 200 (H_2 + CO)	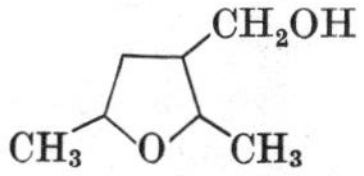	169a
	2-Ethylthiophene	$Co_2(CO)_8$	C_6H_6	180/250 (H_2 + CO)	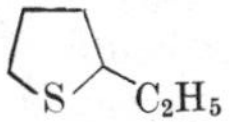 (82)	334
	2,5-Dimethyl-thiophene	$Co_2(CO)_8$	C_6H_6	180/250 (H_2 + CO)	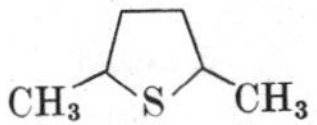 (22)	334
	2-Acetylthiophene	$Co_2(CO)_8$	C_6H_6	180/ 250 (H_2 + CO)	2-Ethylthiophene (52), 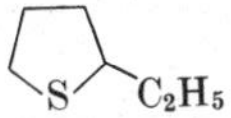 (26)	334

Note: References 303–344 are on pp. 185–186.

TABLE XXVI. AROMATIC AND HETEROAROMATIC COMPOUNDS (*Continued*)

Substrate	Catalyst	Solvent	Temperature (°)/ Pressure (atm)	Product (Yield %)	Refs.
C₇ Anisole	Co(C₃H₅)[P(OCH₃)₃]₃	—	25/1	Methoxycyclohexane	206i
Ethyl benzoate	Co(C₃H₅)[P(OCH₃)₃]₃	—	25/1	1-Carboethoxycyclo-hexene	206i
C₈ o-Xylene	Ni 2-ethylhexanoate/ Al(C₂H₅)₃	C₇H₁₆	150/66	1,2-Dimethylcyclo-hexane (*cis*, 66; *trans*, 34)	184
C₉ Quinoline	[RhCl(pyridine)₂(DMF)-(BH₄)]⁺	DMF	Room temp/1	1,2,3,4-Tetrahydro-quinoline	57, 62a
C₁₀ Naphthalene	Ni 2-ethylhexanoate/ Al(C₂H₅)₃	C₇H₁₆	210/66	Tetralin (84), decalin (13)	184
	Co₂(CO)₈	None	200/ 200 (H₂ + CO)	Tetralin (10)	173
Dimethyl phthalate	Ni 2-ethylhexanoate/ Al(C₂H₅)₃	C₇H₁₆	150/66		184
Dimethyl terephthalate	Ni 2-ethylhexanoate/ Al(C₂H₅)₃	C₇H₁₆	200/66		184

	Substrate	Catalyst	Conditions	Solvent	Products	References
C$_{11}$	2-Methylnaphthalene	Co$_2$(CO)$_8$	200/230 (H$_2$ + CO)	None	Methyltetralins (43)	173
C$_{12}$	Acenaphthene	Co$_2$(CO)$_8$	200/230 (H$_2$ + CO)	C$_6$H$_6$	2a,3,4,5-Tetrahydro-acenaphthene (45)	173
C$_{14}$	Anthracene	Co$_2$(CO)$_8$	135, 150/200 (H$_2$ + CO)	C$_6$H$_6$	9,10-Dihydroanthracene (99)	161, 173
	Phenanthrene	CoH(CO)$_4$	25/(CO only)	C$_6$H$_5$CH$_3$	''	161
		Co$_2$(CO)$_8$	180, 200/180, 230 (H$_2$ + CO)	C$_6$H$_6$, C$_6$H$_{14}$	Dihydro- and tetra-hydrophenanthrene	169a, 173
C$_{16}$	Pyrene	Co$_2$(CO)$_8$	150–200/200–240 (H$_2$ + CO)	C$_6$H$_6$	4,5-Dihydropyrene (69)	173
	(9,10-dimethylanthracene structure)	Co$_2$(CO)$_8$	150/200 (H$_2$ + CO)	C$_6$H$_6$	(9,10-dimethyl-9,10-dihydroanthracene structure, H CH$_3$)	161
C$_{18}$	Naphthacene	CoH(CO)$_4$	25/(CO only)	C$_6$H$_5$CH$_3$	''	161
		Co$_2$(CO)$_8$	140/200 (H$_2$ + CO)	C$_6$H$_6$	5,12-Dihydro-naphthacene (70)	173
	Chrysene	Co$_2$(CO)$_8$	150/230 (H$_2$ + CO)	C$_6$H$_6$	5,6-Dihydrochrysene (24)	173
C$_{20}$	Perylene	Co$_2$(CO)$_8$	150/200 (H$_2$ + CO)	C$_6$H$_6$	1,2,3,10,11,12-Hexa-hydroperylene (72)	173

Note: References 303–344 are on pp. 185–186.

TABLE XXVII. COMPOUNDS CONTAINING C:N OR N:N BONDS

Substrate		Catalyst	Solvent	Temperature (°)/ Pressure (atm)	Product (Yield %)	Refs.
C$_3$	Acetoxime	[Co(CN)$_5$]$^{3-}$	H$_2$O	70/50	CH$_3$CH(NH$_2$)CH$_3$	142
	CH$_3$C(=NOH)CO$_2$H	[Co(CN)$_5$]$^{3-}$	H$_2$O	40–70/50	CH$_3$CH(NH$_2$)CO$_2$H	142
C$_5$	CH$_2$CH$_2$CO$_2$H | C(=NOH)CO$_2$H	[Co(CN)$_5$]$^{3-}$	H$_2$O	70/50	CH$_2$CH$_2$CO$_2$H — CH(NH$_2$)CO$_2$H	142
C$_8$	PhCCH$_3$ || NOH	[Co(CN)$_5$]$^{3-}$	H$_2$O	70/50	PhCH(NH$_2$)CH$_3$	142
C$_9$	PhCH$_2$C(=NOH)CO$_2$H	[Co(CN)$_5$]$^{3-}$	H$_2$O	40–100/50–100	PhCH$_2$CH(NH$_2$)CO$_2$H	142
C$_{12}$	(cyclohexanone N-cyclohexyl imine)	Co$_2$(CO)$_8$	C$_6$H$_6$	135/ 240 (H$_2$ + CO)	(C$_6$H$_{11}$)$_2$NH	176b
	(cyclohexanone N-C$_6$H$_5$ imine)	Co$_2$(CO)$_8$	C$_6$H$_6$	135/ 240 (H$_2$ + CO)	C$_6$H$_{11}$NHPh	176b
	PhN=NPh	[RhCl(pyridine)$_2$(DMF)(BH$_4$)]$^+$	DMF	Room temp/1	PhNHNHPh	57, 62a
	PhN=NPh	Co$_2$(CO)$_8$	C$_6$H$_6$, C$_6$H$_6$/C$_2$H$_5$OH	130/ 200 (H$_2$ + CO)	PhNHCONHPh (15–20)	176, 335
	PhN=NPh	Co(dimethylglyoximato)$_2$	CH$_3$OH, C$_2$H$_5$OH	Room temp/1	PhNHNHPh	231a

	O^- / $\vert +$ / Ph—N=NPh	Co(dimethylglyoximato)$_2$	CH_3OH, C_2H_5OH	Room temp/1	PhN=NPh	231a
C_{13}	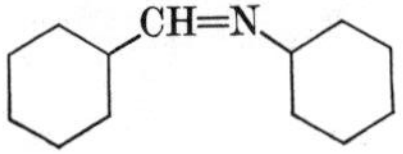	$Co_2(CO)_8$	C_6H_6	135/ 240 (H_2 + CO)	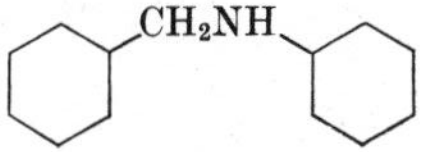	176b
	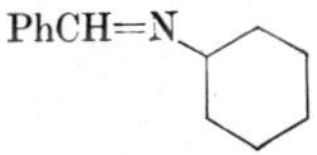	$Co_2(CO)_8$	C_6H_6	135/ 240 (H_2 + CO)	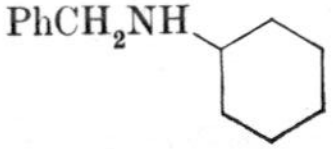	176b
	PhCH=NPh	$Co_2(CO)_8$	C_6H_6, C_6H_6/C_2H_5OH	120, 135/ 200 (H_2 + CO)	$PhCH_2NHPh$ (80)	176, 335
		[RhCl(pyridine)$_2$(DMF)-(BH_4)]$^+$	DMF	Room temp/1	"	57, 62a
	PhCH=NC_6H_4Cl-*p*	$Co_2(CO)_8$	C_6H_6, C_6H_6/C_2H_5OH	130, 150/ 200 (H_2 + CO)	$PhCH_2NHC_6H_4Cl$-*p* (79)	176, 335
	PhCH= $NC_6H_4NO_2$-*p*	$Co_2(CO)_8$	C_6H_6, C_6H_6/C_2H_5OH	120, 130/ 200 (H_2 + CO)	$PhCHNHC_6H_4NO_2$ (80)	176, 335
C_{14}	PhCH= $NC_6H_4CH_3$-*p*	$Co_2(CO)_8$	C_6H_6, C_6H_6/C_2H_5OH	120, 140/ 200 (H_2 + CO)	$PhCH_2NHC_6H_4$-CH_3-*p* (82)	176, 335
	PhCH=NC_6H_4-OCH_3-*p*	$Co_2(CO)_8$	C_6H_6, C_6H_6/C_2H_5OH	120, 130/ 200 (H_2 + CO)	$PhCH_2NHC_6H_4$-OCH_3-*p* (85)	176, 335
C_{15}	$PhCH_2N$=C(CH_3)Ph	[Rh(norbornadiene)-(diop)]$^+$	CH_3OH, C_2H_5OH, *i*-PrOH	30/1	$PhCH_2NHCH(CH_3)Ph$ * (optically active)	290c

Note: References 303–344 are on pp. 185–186.

TABLE XXVIII. Nitro Compounds

Substrate		Catalyst	Solvent	Temperature (°)/Pressure (atm)	Product (Yield %)	Refs.
C_2	Nitroethane	$[Co(CN)_5]^{3-}$	H_2O/C_2H_5OH	70/50	$C_2H_5NH_2$, C_2H_5OH, CH_3CHO	141
C_3	1-Nitropropane	CuCl	Ethylene-diamine	80–95/35	Propanal oxime	281c
	2-Nitropropane	CuCl	Ethylene-diamine	80–95/35	Acetone oxime	281c
C_4	$(CH_3)_3CNO_2$	$[Co(CN)_5]^{3-}$	H_2O/C_2H_5OH	70/50	$(CH_3)_3CNH_2$ (13)	141
C_5	Methyl 4-nitro-pentanoate	CuCl	Ethylene-diamine	80–95/35	Methyl 4-(hydroxy-imino)pentanoate	281c
C_6	Nitrocyclohexane	$RhCl_3(pyridine)_3/NaBH_4$	DMF	Room temp/1	Cyclohexylamine	62b
		CuCl	Ethylene-diamine	80–95/35	Cyclohexanone oxime	281c
	Nitrobenzene	$RhCl_3(pyridine)_3/NaBH_4$	DMF	Room temp/1	Aniline	57, 61, 62b
		$RhCl_3(N$-formyl-piperidine$)_3$	DMF	—/—	"	240
		$[Co(CN)_5]^{3-}$	H_2O, H_2O/C_2H_5OH	Room temp, 70j 1, 50	", PhN=NPh, PhNHNHPh	125, 141, 316

	Catalyst	Solvent	Conditions	Product	Ref.
	$Co_2(CO)_8$	C_6H_6, C_6H_6/C_2H_5OH	130/200 $(H_2 + CO)$	PhNHCONHPh (6)	176, 335
	Co(dimethylglyoximato)$_2$	CH_3OH, C_2H_5OH	Room temp/1	Aniline, PhNHNHPh	231a
o-Nitrophenol	$[Co(CN)_5]^{3-}$	H_2O/C_2H_5OH	70/50	o-Aminophenol (32)	141
p-Nitrophenol	$[Co(CN)_5]^{3-}$	H_2O/C_2H_5OH	70/50	p-Aminophenol (33)	141
C$_7$ o-Nitrotoluene	$[Co(CN)_5]^{3-}$	H_2O	Room temp/1	"Azoxy and azo compounds"	125
p-Nitrotoluene	RhCl$_3$(pyridine)$_3$/NaBH$_4$	DMF	Room temp/1	p-Toluidine	62b
α-Nitrotoluene	CuCl	Ethylenediamine	80–95/35	Benzaldoxime	281c
o-Nitroanisole	$[Co(CN)_5]^{3-}$	H_2O	Room temp/1	"Hydrazo compound"	125
C$_8$ p-Nitrobenzoic acid	RhCl$_3$(pyridine)$_3$/NaBH$_4$	DMF	Room temp/1	p-Aminobenzoic acid	62b
C$_{10}$ p-$(CH_3)_2NC_6H_4NO_2$	RhCl$_3$(pyridine)$_3$/NaBH$_4$	DMF	Room temp/1	p-$(CH_3)_2NC_6H_4NH_2$	62b
Methyl p-nitrocinnamate	Co(dimethylglyoximato)$_2$	CH_3OH, C_2H_5OH	Room temp/1	Methyl p-aminocinnamate	231a
C$_{12}$ Nitrododecanes	CuCl	Ethylenediamine	80–95/35	Dodecanone oximes	281c
C$_{15}$ α-(2-Cyanoethyl)-nitrododecanes	CuCl	Ethylenediamine	80–95/35	α-(3-Aminopropyl)-nitrosododecanes	281c

Note: References 303–344 are on pp. 185–186.

171

TABLE XXIX. STEROIDS

Substrate	Catalyst	Solvent	Temperature (°)/Pressure (atm)	Product (Yield %)	Refs.
"Steroid-4-en-3-ones"	[RhCl(pyridine)$_2$(DMF)-(BH$_4$)]$^+$	DMF	Room temp/1	5α, 5β-Dihydro derivatives	57
C$_{19}$ 5α-Androstan-3-one	H$_2$IrCl$_6$/P(OCH$_3$)$_3$	i-PrOH/H$_2$O	Reflux/(no H$_2$)	3α-Hydroxy-5α-androstane	90
17β-Hydroxy-5α-androstan-2-one	Na$_2$IrCl$_6$/P(OCH$_3$)$_3$; RhCl(PPh$_3$)$_3$/P(OCH$_3$)$_3$	i-PrOH/H$_2$O	Reflux/(no H$_2$)	2β,17β-Dihydroxy-5α-androstane	91
17β-Hydroxy-5α-androstan-3-one	H$_2$IrCl$_6$/P(OCH$_3$)$_3$	i-PrOH/H$_2$O	Reflux/(no H$_2$)	3β,17β-Dihydroxy-5α-androstane	90
17β-Hydroxy-5β-androstan-3-one	H$_2$IrCl$_6$/P(OCH$_3$)$_3$	i-PrOH/H$_2$O	Reflux/(no H$_2$)	3β,17β-Dihydroxy-5β-androstane	90
5α-Androstane-3,17-dione	Na$_2$IrCl$_6$/P(OCH$_3$)$_3$	i-PrOH/H$_2$O	Reflux/(no H$_2$)	3α-Hydroxy-5α-androstan-17-one	90, 91
5β-Androstane-3,17-dione	Na$_2$IrCl$_6$/P(OCH$_3$)$_3$; RhCl(PPh$_3$)$_3$/P(OCH$_3$)$_3$	i-PrOH/H$_2$O	Reflux/(no H$_2$)	3β-Hydroxy-5β-androstan-17-one	90, 91
5α-Androstane-3,11,17-trione	H$_2$IrCl$_6$/P(OCH$_3$)$_3$	i-PrOH/H$_2$O	Reflux/(no H$_2$)	3α-Hydroxy-5α-androstane-11,17-dione	90
Androst-4-ene-3,17-dione	RhCl(PPh$_3$)$_3$	Acetone	Room temp/1(D$_2$)	5α-Androstane-3,17-dione	336
Androsta-1,4-diene-3,17-dione	RhCl(PPh$_3$)$_3$	C$_6$H$_6$/CH$_3$OH, C$_2$H$_5$OH	Room temp/1	Androst-4-ene-3,17-dione	27, 336
	RuCl$_2$(tertiary phosphine)$_3$	C$_6$H$_6$	40/130	"	211
Androsta-4,6-diene-3,17-dione	RhCl(PPh$_3$)$_3$	C$_6$H$_6$/CH$_3$OH, C$_2$H$_5$OH	Room temp/1	Androst-4-ene-3,17-dione	27, 336

Substrate	Catalyst	Solvent	Conditions	Product(s)	Refs.
Testosterone	$RhCl(PPh_3)_3$	C_6H_6	Room temp/1(D_2)	5α-Dihydrotestosterone	28, 45
	$[RhCl(pyridine)_2(DMF)_2(BH_4)]^+$	DMF	Room temp/1	5α, 5β-Dihydro-testosterone	61
	$RhCl_3(N$-formyl-piperidine)$_3$	DMF	—/—	—	240
C_{20} 17α-Methyltesto-sterone	$[RhCl(pyridine)_2(DMF)_2(BH_4)]^+$	DMF	Room temp/1	5α-, 5β-17-Methyl-dihydrotestosterone	61
C_{21} 20α-Hydroxy-5α-pregnan-3-one	$H_2IrCl_6/P(OCH_3)_3$	i-PrOH/H_2O	Reflux/(no H_2)	3α,20α-Dihydroxy-5-pregnane	90
20β-Hydroxy-5α-pregnan-3-one	$H_2IrCl_6/P(OCH_3)_3$	i-PrOH/H_2O	Reflux/(no H_2)	3α,20β-Dihydroxy-5α-pregnane	90
20α-Hydroxy-5β-pregnan-3-one	$H_2IrCl_6/P(OCH_3)_3$	i-PrOH/H_2O	Reflux/(no H_2)	3β,20α-Dihydroxy-5β-pregnane	90
5α,17β-Pregnane-3,20-dione	$Na_2IrCl_6/P(OCH_3)_3$	i-PrOH/H_2O	Reflux/(no H_2)	3α-Hydroxy-5α,17α- and 17β-pregnan-20-one	90, 91
5β-Pregnane-3,20-dione	$H_2IrCl_6/P(OCH_3)_3$	i-PrOH/H_2O	Reflux/(no H_2)	3β-Hydroxy-5β-pregnan-20-one	90
5α-Pregnane-3,11-20-trione	$H_2IrCl_6/P(OCH_3)_3$	i-PrOH/H_2O	Reflux/(no H_2)	3α-Hydroxy-5α-pregnane-11,20-dione	90
5β-Pregnane-3,11,20-trione	$H_2IrCl_6/P(OCH_3)_3$	i-PrOH/H_2O	Reflux/(no H_2)	3β-Hydroxy-5β-pregnane-11,20-dione	90
5α-Pregnane-3,12,20-trione	$H_2IrCl_6/P(OCH_3)_3$	i-PrOH/H_2O	Reflux/(no H_2)	3α-Hydroxy-5α-pregnane-12,20-dione	90
5β-Pregnane-3,12,20-trione	$H_2IrCl_6/P(OCH_3)_3$	i-PrOH/H_2O	Reflux/(no H_2)	3β-Hydroxy-5β-pregnane-12,20-dione	90
5β-Pregn-1-ene-3,20-dione	$RhCl(PPh_3)_3$	C_6H_6/CH_3OH, C_2H_5OH	Room temp/1	"Saturated diketone" (13)	27

Note: References 303–344 are on pp. 185–186.

TABLE XXIX. Steroids (Continued)

Substrate	Catalyst	Solvent	Temperature (°)/ Pressure (atm)	Product (Yield %)	Refs.
C_{21} 9,11-Secopregna-1,4-*contd.* diene-3,20-dione	RhCl(PPh$_3$)$_3$	C_6H_6	Room temp/1	9,11-Secopregn-4-ene-3,20-dione	337
Progesterone	[RhCl(pyridine)$_2$(DMF)-(BH$_4$)]$^+$	DMF	Room temp/1	5α-, 5β-Dihydropro-gesterone	61
C_{23} 3β-Acetoxypregna-5,16-dien-20-one	RhCl(PPh$_3$)$_3$	C_6H_6	Room temp/1(D$_2$)	3β-Acetoxypregn-5-en-20-one	28, 45
21-Acetoxy-17α-hydroxypregna-1,4-diene-3,20-dione	RhCl(PPh$_3$)$_3$	C_6H_6	Room temp/1(D$_2$)	21-Acetoxy-17α-hydroxypregn-4-ene-3,20-dione	45
C_{25} 11α-Acetoxy-20-ethylenedioxypregna-1,4-dien-3-one	RhCl(PPh$_3$)$_3$	C_6H_6	Room temp/1	1,2-Dihydro-, 1,2,3,4-tetrahydro-derivatives	338
C_{27} 5α-Cholestan-3-one	H$_2$IrCl$_6$/P(OCH$_3$)$_3$	i-PrOH/H$_2$O	Reflux/(no H$_2$)	3α-Hydroxy-5α-cholestane	87a, 88
5α-Cholestane-3,6-dione	H$_2$IrCl$_6$/P(OCH$_3$)$_3$	i-PrOH/H$_2$O	Reflux/(no H$_2$)	3α-Hydroxy-5α-cholestan-6-one	90
Cholest-1-ene	RhCl(PPh$_3$)$_3$, RhI(PPh$_3$)$_3$	C_6H_6/CH$_3$OH, C_2H_5OH	Room temp/1(D$_2$)	5α-Cholestane	27
Cholest-2-ene	RhCl(PPh$_3$)$_3$, RhI(PPh$_3$)$_3$	C_6H_6, THF, acetone, C_6H_6/CH$_3$OH, C_2H_5OH	Room temp/1(D$_2$)	Cholestane	27, 29, 51, 336

174

Substrate	Catalyst	Solvent	Conditions	Product	Ref.
Cholest-3-ene	RhCl(PPh$_3$)$_3$, polymer supported	C$_6$H$_6$	25/1	"	5a
	RhCl(PPh$_3$)$_3$, RhI(PPh$_3$)$_3$	C$_6$H$_6$/CH$_3$OH, C$_2$H$_5$OH	Room temp/1	Cholestane	27
Cholest-4-ene	RhCl(PPh$_3$)$_3$	Acetone	Room temp/125	Cholestane (6)	336
Cholestenone	RhCl$_3$(N-formyl-piperidine)$_3$	DMF	—/—	—	240
	TiCl$_2$(C$_5$H$_5$/2)$_2$/BuLi, Na naphthalide (polymer supported)	C$_6$H$_{14}$	—/1	—	187a
Cholest-1-en-3-one	RhCl(PPh$_3$)$_3$	C$_6$H$_6$/CH$_3$OH, C$_2$H$_5$OH	Room temp/1	Cholestan-3-one	27
Cholest-4-en-3-one	RhCl(PPh$_3$)$_3$	Acetone	Room temp/ 1(D$_2$)	Cholestan-3-one	336
	RhCl(PPh$_3$)$_3$	CH$_3$OH	Room temp/1	Cholestan-3-one and its dimethyl acetal	50
	[RhCl(pyridine)$_2$(DMF)-(BH$_4$)]$^+$	DMF	Room temp/1	5β-Cholestan-3-one	61
Cholesta-1,4-dien-3-one	RhCl(PPh$_3$)$_3$	C$_6$H$_6$	Room temp/ 1(T$_2$)	Cholest-4-en-3-one	298
C$_{28}$ Ergosterol	RhCl(PPh$_3$)$_3$	C$_6$H$_6$	Room temp/1	5α,6-Dihydroergosterol	28, 308
Ergosta-1,4,22-trien-3-one	RhCl(PPh$_3$)$_3$	C$_6$H$_6$	Room temp/ 1(T$_2$)	Ergosta-4,22-dien-3-one	299
C$_{30}$ 22,23-Dihydro-ergosteryl acetate	RhCl(PPh$_3$)$_3$	C$_6$H$_6$	Room temp/ 1(D$_2$)	5α,6,22,23-Tetra-hydroergosteryl acetate	32a

Note: References 303–344 are on pp. 185–186.

175

TABLE XXX. Natural Products

Substrate		Catalyst	Solvent
C_{10}	Geraniol	$RhCl(PPh_3)_3$	C_6H_6/C_2H_5OH
	Linalool	$RhCl(PPh_3)_3$	C_6H_6
	Nerol	$RhCl(PPh_3)_3$	C_6H_6/C_2H_5OH
	(p-Menthene)	$RhCl(PPh_3)_3$	C_6H_6, $C_6H_6/$ C_2H_5OH
	(d-Limonene)	$Rh(NO)(PPh_3)_3$	CH_2Cl_2
		Ziegler catalysts	THF, decalin
	[(+)-Carvone]	$RhCl(PPh_3)_3$	C_6H_6
	(Juglone)	$RhCl(PPh_3)_3$	C_6H_6
	$3,4\text{-}CH_2O_2C_6H_3CH_2CH{=}CH_2$ (Safrole)	$FeCl_3/LiAlH_4$	THF
	$3,4\text{-}CH_2O_2C_6H_3CH{=}CHCH_3$ (Isosafrole)	$FeCl_3/LiAlH_4$	THF
	$4\text{-}CH_3OC_6H_4CH{=}CHCH_3$ (Anethole)	$FeCl_3/LiAlH_4$	THF
C_{12}	Neryl acetate	$RhCl(PPh_3)_3$	C_6H_6/C_2H_5OH
C_{14}		$RhCl(PPh_3)_3$	C_6H_6

Note: References 303–344 are on pp. 185–186.

Temperature (°)/ Pressure (atm)	Product (Yield %)	Refs.
Room temp/1	Dihydro-, tetrahydrogeraniol (also CO abstraction by catalyst)	28
Room temp/1	Dihydrolinalool	28
Room temp/1	Dihydro-, tetrahydronerol (also CO abstraction by catalyst)	28
25/1(D_2)	p-Menthane (K) (*cis*, 30, *trans*, 70)	35
25/1		56a
0, 20/1		193, 206d
Room temp/1		28
Room temp/1		324.
0/1	$3,4\text{-}CH_2O_2C_6H_3C_3H_7$	206d
0/1	$3,4\text{-}CH_2O_2C_6H_3C_3H_7$	206d
0/1	$4\text{-}CH_3OC_6H_4C_3H_7$	206d
Room temp/1	3,7-Dimethyloctyl acetate	28
Room temp/1		340

TABLE XXX. Natural Products

Substrate	Catalyst	Solvent

C_{14}
contd.

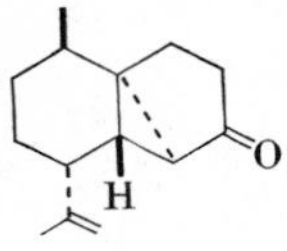

RhCl(PPh$_3$)$_3$ C$_6$H$_6$

C_{15}

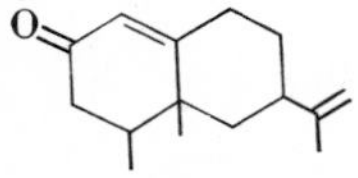

[(+)-Nootkatone]
RhCl(PPh$_3$)$_3$ C$_6$H$_6$

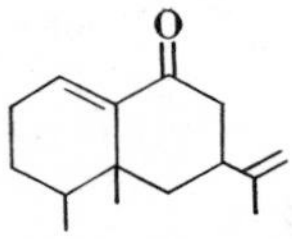

(Eremophilone)
RhCl(PPh$_3$)$_3$ C$_6$H$_6$

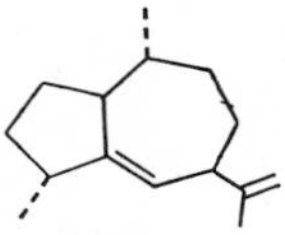

RhCl(PPh$_3$)$_3$ C$_6$H$_6$/C$_2$H$_5$OH

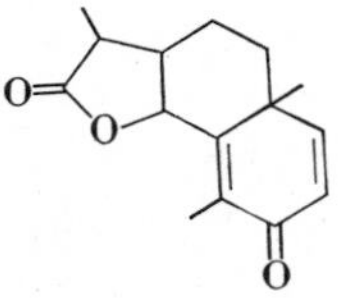

(Santonin)   RhCl(PPh$_3$)$_3$ C$_6$H$_6$/C$_2$H$_5$OH

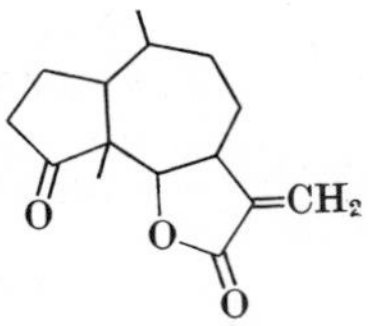

(Damsin)   RhCl(PPh$_3$)$_3$ —

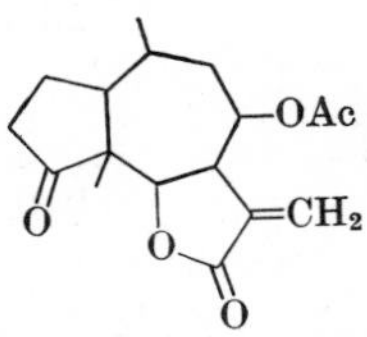

(Confertiflorin)   RhCl(PPh$_3$)$_3$ —

Note: References 303–344 are on pp. 185–186.

Temperature (°)/ Pressure (atm)	Product (Yield %)	Refs.
Room temp/1		340
Room temp/1		339
Room temp/1		30
Room temp/1(D₂)		29
Room temp/1		38
—/—		39b
—/—		39b

179

TABLE XXX. NATURAL PRODUCTS

Substrate	Catalyst	Solvent

C_{15}
contd.

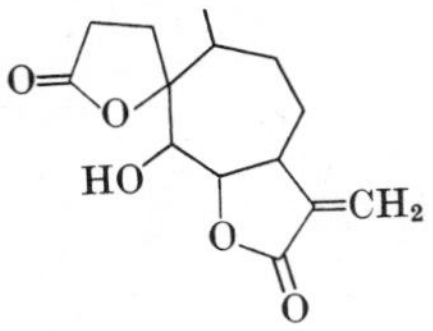

(Psilostachyine)

RhCl(PPh$_3$)$_3$ —

C_{17}

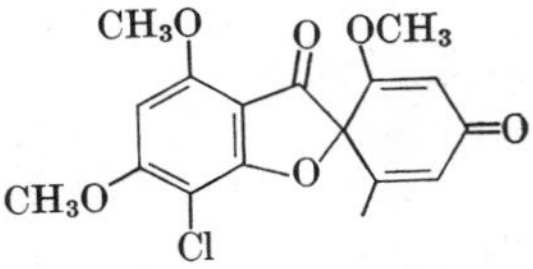

[(−)-Dehydrogriseofulvin)]

RhCl(PPh$_3$)$_3$ C$_6$H$_6$

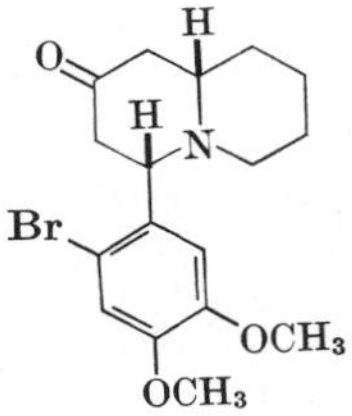

H$_2$IrCl$_6$/   *i*-PrOH/H$_2$O
P(OCH$_3$)$_3$

C_{19}

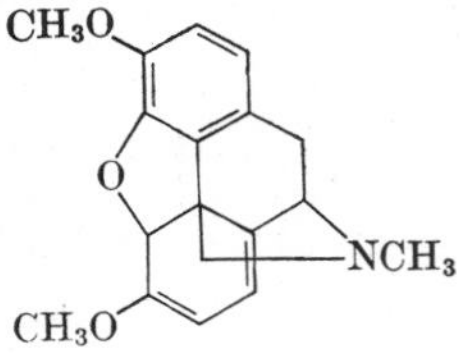

(Thebaine)

RhCl(PPh$_3$)$_3$ C$_6$H$_6$

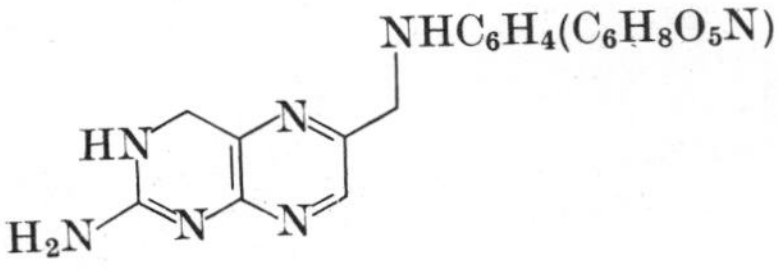

(Folic acid)

RhCl$_3$(pyri-   (+)-, (−)-N-
dine)$_3$/NaBH$_4$ (1-Phenyl-
ethyl)for-
mamide

Note: References 303–344 are on pp. 185–186.

Temperature (°)/ Pressure (atm)	Product (Yield %)	Refs.
—/—		39b
Room temp/1		45
Reflux/(no H$_2$)		92a
Room temp/1		28
—/—	Optically active tetrahydrofolic acid	291b

TABLE XXX. NATURAL PRODUCTS

Substrate	Catalyst	Solvent
C_{22} (Diospyrin)	RhCl(PPh$_3$)$_3$	C$_6$H$_6$
C_{32} AcO (Lupeol acetate)	RhCl(PPh$_3$)$_3$	C$_6$H$_6$/C$_2$H$_5$OH

Note: References 303–344 are on pp. 185–186.

Temperature (°)/ Pressure (atm)	Product (Yield %)	Refs.
Room temp/1		341
Room temp/1	Dihydrolupeol acetate	51

TABLE XXXI. POLYMERS

Substrate	Catalyst	Solvent	Temperature (°)/Pressure (atm)	Product	Refs.
1,2-Polybutadiene	i-Bu$_3$B	C$_6$H$_6$/cyclohexane	225/130	Liquid product	281a
1,4-Polybutadiene	Co 2-ethylhexanoate/BuLi	Cyclohexane	9–50/0.3–3	Saturated product	202a
	i-Bu$_3$B	C$_6$H$_6$	225/130	Degraded liquid product	281a
cis-1,4-Polybutadiene	Ziegler catalysts	Decalin	15–80/1	— (K)	192, 202b
	n-Bu$_3$B	"	180–220/67–100	Crystalline polyethylene	279, 280
	i-Bu$_3$B	C$_6$H$_6$	225/130	"	281a
	Triethylborazole	Decalin	180–220/100	"	280
Emulsion polybutadiene	n-Bu$_3$B	—	220/67	Saturated product	279
Butadiene rubbers	RhCl(PPh$_3$)$_3$	C$_6$H$_6$	50/1	Partially saturated product	45
Butadiene-styrene copolymer	Cr(acetylacetonate)$_3$/i-Bu$_3$Al	Decalin	20/1	Sequential saturation of 1,2 then 1,4 units	193
1,4-Polyisoprene	Co 2-ethylhexanoate/BuLi	Cyclohexane	50/3	Saturated product	202a
cis-1,4-Polyisoprene	i-Bu$_3$B	Cyclohexane	235/130	Semisolid product	281a
Polystyrene	Co 2-ethylhexanoate/BuLi	Cyclohexane	100–300/33–300	Saturation of aromatic rings	202a
Polypiperylene	i-Bu$_3$B	Cyclohexane/diglyme	240/130	Liquid product	281a
Neoprene-834	i-Bu$_3$B	C$_6$H$_6$	225/130	No reduction	281a
SBR polymer	n-Bu$_3$B	—	220/67	Saturated product	279
	i-Bu$_3$B	C$_6$H$_6$	225/130	Degraded semisolid product	281a

REFERENCES TO TABLES 14–31

[303] V. A. Avilov, O. N. Eremenko, and M. L. Khidekel, *Izv. Akad. Nauk. SSSR, Ser. Khim.*, **1967**, 2781 [*C.A.*, **68**, 117464z (1968)].

[304] N. S. Imyanitov and D. M. Rudkovskii, *Neftekhim.*, **3**, 198 (1963) [*C.A.*, **59**, 7396c (1963)].

[305] G. C. Bond and M. Hellier, *J. Catal.*, **7**, 217 (1967).

[306] W. Strohmeier and R. Endres, *Z. Naturforsch.*, *B*, **26**, 730 (1971).

[307] W. M. Moreau and K. Weiss, *Nature*, **208**, 1203 (1965).

[308] A. J. Birch and K. A. M. Walker, *Tetrahedron Lett.*, **1967**, 1935.

[309] S. Siegel and D. W. Ohrt, *Tetrahedron Lett.*, **1972**, 5155.

[310] S. Siegel and D. W. Ohrt, *Inorg. Nucl. Chem. Lett.*, **8**, 15 (1972).

[311] J. Hjortkjer and Z. Kulicki, *J. Catal.*, **27**, 452 (1972).

[312] I. Jardine and F. J. McQuillin, *Tetrahedron Lett.*, **1966**, 4871.

[313] (a) V. A. Tulupov, *Russ. J. Phys. Chem.*, **36**, 873 (1962); (b) V. A. Tulupov, *ibid.*, **37**, 365 (1963); (c) V. A. Tulupov and M. I. Gagarina, *ibid.*, **38**, 926 (1964); (d) V. A. Tulupov and T. I. Evlasheva, *ibid.*, **39**, 41 (1965).

[314] V. A. Tulupov, *Zh. Fiz. Khim.*, **32**, 727 (1958) [*C.A.*, **52**, 14302c (1958)].

[315] J. C. Bailar, Jr., H. Itatani, and H. Jayim, *Kagaku No Ryoiki*, **22**, 337 (1968) [*C.A.*, **69**, 44717t (1968)].

[316] J. Kwiatek, I. L. Mador, and J. K. Seyler, *J. Amer. Chem. Soc.*, **84**, 304 (1962).

[317] J. Berty and L. Marko, *Acta Chim. Acad. Sci. Hung.*, **3**, 177 (1953) [*C.A.*, **48**, 11294 (1954)].

[318] B. Heil and L. Marko, *Chem. Ber.*, **99**, 1086 (1966).

[319] B. Heil and L. Marko, *Acta Chim. Acad. Sci. Hung.*, **55**, 107 (1968) [*C.A.*, **68**, 77430b (1968)].

[320] K. A. Alekseeva, D. L. Libina, D. M. Rudkovskii, and A. G. Trifel, *Neftekhim.*, **6**, 458 (1966) [*C.A.*, **65**, 10451c (1966)].

[321] V. A. Tulupov, *Zhur. Fiz. Khim.*, **31**, 519 (1957) [*C.A.*, **51**, 17776i (1957)].

[322] B. Zeeh, G. Jones, and C. Djerassi, *Chem. Ber.*, **100**, 3204 (1967).

[323] (a) M. Calvin, *Trans. Faraday Soc.*, **34**, 1181 (1938); (b) S. Weller and G. A. Mills, *J. Amer. Chem. Soc.*, **75**, 769 (1953); (c) M. Calvin, *ibid.*, **61**, 2230(1939); (d) L. W. Wright and S. Weller, *ibid.*, **76**, 3345 (1954).

[324] A. J. Birch and K. A. M. Walker, *Tetrahedron Lett.*, **1967**, 3457.

[325] V. I. Ogata and A. Misono, *Bull. Chem. Soc. Jap.*, **37**, 900 (1964).

[326] E. B. Maxted and S. M. Ismail, *J. Chem. Soc.*, **1964**, 1750.

[327] (a) L. Simandi and F. Nagy, *Proc. Symp. Coord. Chem.*, Tihany, Hungary, **1964**, 83; (b) L. Simandi and F. Nagy, *Magy. Kem. Folyoirat*, **71**, 6 (1965) [*C.A.*, **62**, 13006b (1965)].

[328] H. W. van der Linden, B. Stouthamer, and I. C. Vlugter, *Chem. Weekblad*, **60**, 254 (1964) [*C.A.*, **61**, 9691f (1964)].

[329] J. C. Bailar and H. Itatani, *Proc. Symp. Coord. Chem.*, Tihany, Hungary, **1964**, 67.

[330] Y. Takegami, T. Ueno, T. Fujii, and T. Sakata, *Shokubai*, **8**, 54 (1966) [*C.A.*, **68**, 104625c (1968)].

[331] D. M. Rudkovskii and N. S. Imyanitov, *Zh. Prikl. Khim.*, **35**, 2719 (1962)[*C.A.*, **59**, 2689f (1963)].

[332] Y. Senda, T. Iwasaki, and S. Mitsui, *Tetrahedron*, **28**, 4059 (1972).

[333] S. Takahashi, H. Yamazaki, and N. Hagihara, *Bull. Chem. Soc. Jap.*, **41**, 254 (1968).

[334] H. Greenfield, S. Metlin, M. Orchin, and I. Wender, *J. Org. Chem.*, **23**, 1054 (1958).

[335] S. Murahashi and S. Horiie, *Ann. Rep. Sci. Works*, Osaka University, **7**, 89 (1959).

[336] W. Voelter and C. Djerassi, *Chem. Ber.*, **101**, 58 (1968).

[337] N. S. Crossley and R. Dowell, *J. Chem. Soc.*, *C*, **1971**, 2496.

[338] P. Wieland and G. Anner, *Helv. Chim. Acta*, **51**, 1698 (1968).

[339] (a) A. R. Pinder, *Tetrahedron Lett.*, **1970**, 413; (b) H. C. Odam and A. R. Pinder, *J. Chem. Soc. (Perkin I)*, **1972**, 2193.

[340] A. Tanaka, R. Tanaka, H. Uda, and A. Yoshikoshi, *J. Chem. Soc. (Perkin I)*, **1972**, 1721.

[341] M. Pardhasaradhi and G. S. Sidhu, *Tetrahedron Lett.*, **1972**, 4201.

[342] J. Trocha-Grimshaw and H. B. Henbest, *Chem. Commun.*, **1967**, 544.

[343] J. Trocha-Grimshaw and H. B. Henbest, *Chem. Commun.*, **1968**, 757.

[344] W. Dumont, J.-C. Poulin, T.-P. Dang, and H. B. Kagan, *J. Amer. Chem. Soc.*, **95**, 8295 (1973).

CHAPTER 2

ESTER CLEAVAGES VIA S$_N$2-TYPE DEALKYLATION

JOHN MC MURRY

University of California
Santa Cruz, California

CONTENTS

INTRODUCTION

The cleavage of esters to furnish carboxylic acids is a common organic transformation that is usually carried out in a routine manner by acidic or basic hydrolysis. It often happens however, particularly in the synthesis of natural products, that the substrate ester is sensitive to hydrolytic conditions. For such sensitive materials a number of mild, neutral methods of ester cleavage have been devised. Those methods that occur with displacement of carboxylate by S_N2 dealkylation are the subject of this review.

$$RCO_2R' + Nu^- \rightarrow RCO_2^- + R'Nu$$

As is expected for an S_N2 reaction, the ester cleavage works best when R' is unhindered (R' = methyl, ethyl) and when a powerful nucleophile such as iodide, cyanide, or mercaptide ion is used in a dipolar aprotic solvent.

Although there are scattered references to such ester cleavages in the older literature,[1,2] it was not until 1956 that Taschner and Liberek established the general synthetic value of the reaction.[3] Their results, however, were published in journals inaccessible to most chemists,[3,4] and it was not until 1960 that the method gained wide popularity through the work of Eschenmoser and his colleagues.[5,6]

[1] L. P. Hammett and H. L. Pfluger, J. Amer. Chem. Soc., 55, 4079 (1933).
[2] R. Willstätter and W. Kahn, Chem. Ber., 35, 2757 (1902).
[3] E. Taschner and B. Liberek, Rocz. Chem., 30, 323 (1956) [C.A., 51, 1039d (1957)].
[4] E. Taschner and B. Liberek, Bull. Acad. Pol. Sci., Ser. Sci., Chim., Geol. Geog., 7, 877 (1959) [C.A., 55, 16465e (1961)].
[5] F. Elsinger, J. Schreiber, and A. Eschenmoser, Helv. Chim. Acta, 43, 113 (1960).
[6] J. Schreiber, W. Leimgruber, M. Pesaro, P. Schudel, T. Threlfall, and A. Eschenmoser, Helv. Chim. Acta, 44, 540 (1961).

SCOPE AND LIMITATIONS

The Ester

One of the great values of ester cleavage by an S_N2 dealkylation is that the reaction is highly selective. Bimolecular nucleophilic substitution reactions are well known to be quite sterically sensitive, and thus only the esters of unhindered alcohols undergo cleavage. This fact was recognized immediately by Taschner and Liberek in their initial publication when they showed, for example, that treatment of methyl phenylacetate with lithium iodide for 15 hours in refluxing pyridine gave phenylacetic acid in 93 % yield, whereas the corresponding reaction with the ethyl ester gave phenylacetic acid in only 42 % yield after 27 hours.[4] This reactivity difference has occasionally been taken advantage of in complex syntheses. Thus the selective demethylation of the methyl ethyl diester 1 has been reported to take place in high yield.[7]

Similarly a selective cleavage of the methyl ester in the diester 2 has been noted,[8] and diester 3 has been demethylated in high yield.[9]

[7] R. F. Borch, C. V. Grudzinskas, D. A. Peterson, and L. D. Weber, J. Org. Chem., 37, 141 (1972).

[8] P. D. G. Dean, T. G. Halsall, and M. W. Whitehouse, J. Pharm. Pharmacol., 19, 682 1967).

[9] J. Meinwald and D. E. Putzig, J. Org. Chem., 35, 1891 (1970).

Although methyl esters are clearly the most reactive and by far the most common substrates, other esters also undergo cleavage. The studies of Taschner and Liberek demonstrated that ethyl esters undergo slow S_N2 cleavage and, more recently, cleavage of the ethyl ester 4 has been reported to take place in 70% yield.[10]

$$C_6H_5CH_2C(CH_3)[CON(CH_3)_2]CO_2C_2H_5 \xrightarrow[\text{Lutidine}]{\text{LiI}} C_6H_5CH_2CH(CH_3)CON(CH_3)_2$$

4

Ethyl esters are also cleaved during the concomitant dealkylationdecarboxylation procedure that has been used on substituted malonic esters.[11] Normally the decarbalkoxylation of a malonic ester involves a three-step sequence: saponification, thermal decarboxylation, and reesterification. On treatment of a diethyl malonate with sodium cyanide in dimethyl sulfoxide at 160°, however, decarbalkoxylation occurs in one step. The method has gained considerable popularity in recent years (Table V), but its mechanism is unclear and may not involve simple S_N2 attack by cyanide. Dimethyl malonates work equally well,[12] but the more readily available diethyl esters are normally used.

(75%)

A variation has recently been published whereby sodium chloride in hot aqueous dimethyl sulfoxide is used to effect decarbalkoxylation,[13] but it is not yet clear whether this reaction proceeds by S_N2 attack or by an entirely different mechanism.[14,15]

There are few reports of esters other than methyl and ethyl being successfully cleaved. It has recently been shown that isopropyl esters can be slowly cleaved by the use of sodium cyanide in hexamethylphosphoramide, but the result is probably due largely to acyl cleavage rather than S_N2 alkyl cleavage.[16] By contrast, success has been reported in effecting an apparent S_N2 cleavage of phenacyl esters in the penicillin series by sodium thiophenoxide in dimethylformamide.[17,18] Interestingly the methyl ester

[10] W. Sucrow, *Chem. Ber.*, **101**, 4230 (1968).

[11] A. P. Krapcho, G. A. Glynn, and B. J. Grenon, *Tetrahedron Lett.*, **1967**, 215.

[12] L. J. Dolby and H. Biere, *J. Org. Chem.*, **35**, 3843 (1970).

[13] A. P. Krapcho and A. J. Lovey, *Tetrahedron Lett.*, **1973**, 957.

[14] A. P. Krapcho, E. G. E. Jahngen, A. J. Lovey, and F. W. Short, *Tetrahedron Lett.*, **1974**, 1091.

[15] C. L. Liotta and F. L. Cook, *Tetrahedron Lett.*, **1974**, 1095.

[16] P. Müller and B. Siegfried, *Tetrahedron Lett.*, **1973**, 3565.

[17] J. C. Sheehan and G. D. Daves, *J. Org. Chem.*, **29**, 2006 (1964).

[18] P. Bamberg, B. Ekstrom, and B. Sjoberg, *Acta Chem. Scand.*, **21**, 2210 (1967).

corresponding to 5 is reported to give only a low yield of acid product upon cleavage with sodium thiophenoxide.[17]

5

The Nucleophile

A wide variety of nucleophiles have been used in the reaction, including halides, thiolates, t-butoxide, cyanide, and amines, but no careful comparison of their relative effectiveness has been made. Such a tabulation is certainly needed and would serve to remove some of the present confusion about the most suitable reagent for a given substrate.

Halides

Undoubtedly the single most commonly chosen nucleophile is iodide ion, usually introduced as the lithium salt (Tables II and VI). Taschner and Liberek reported brief trials with different lithium halides and found lithium iodide to be superior.[4] Their results are given in Table I. Similarly lithium iodide was better than other lithium halides when dimethylformamide was used as solvent.[19]

These results run counter to the nucleophilicity order usually found in dipolar aprotic solvents, and the situation seems to be still somewhat unsettled.[20] A reactivity order of $Cl^- > Br^- > I^-$ has been found for the

TABLE I. REACTION OF LITHIUM HALIDES WITH ESTERS

$$C_6H_5CH_2CO_2R \xrightarrow[\text{Pyridine}]{\text{LiX}} C_6H_5CH_2CO_2Li + RX$$

R	X	Time of Reflux (hr)	Yield (%)
CH_3	Cl	15	23
CH_3	Br	,,	65
CH_3	I	,,	93
C_2H_5	Cl	27	8
C_2H_5	Br	,,	19
$C_2^1H_5$	I	,,	42

[19] P. D. G. Dean, *J. Chem. Soc.*, **1965**, 6655.

[20] J. March, *Advanced Organic Chemistry: Reactions, Mechanisms, and Structure*, McGraw-Hill, New York, 1968, pp. 287–290.

demethylation of methyl tosylate in a pyridine-dimethylformamide solvent mixture.[21] However the order is concentration dependent, and changes to $Br^- > I^- > Cl^-$ at 0.35 M due to ion pairing. In hexamethylphosphoramide, where ion pairing is less, a reactivity order of $Cl^- > Br^-$ is found.[16] Clearly the exact nucleophilicity of an ion is both concentration and solvent dependent. Iodide ion seems to be the halide of choice for S_N2 ester dealkylations, judging from the extensive success it has enjoyed.

In addition to its use for cleavage of isolated ester functions (Table II), lithium iodide has also been much used to effect concomitant ester cleavage-decarboxylation of β-keto esters (Table V).[22] This method is clearly much milder than the strongly acidic conditions normally employed to effect decarbalkoxylation. This fact has been used to advantage in a recent synthesis of α-methylenebutyrolactones.[23] Simple heating of the Mannich salt 6 in dimethylformamide results in both decarbomethoxylation and elimination of trimethylamine to generate an α-methylenebutyrolactone.

$$CH_3O_2C \quad CH_2\overset{+}{N}(CH_3)_3 \; I^- \qquad \xrightarrow[\text{Dimethylformamide}]{\text{Heat}} \qquad$$

6

Thiolates

Thiolates have been used much less extensively than halides, although they seem to work quite well under mild conditions. Sodium thiophenoxide in dimethylformamide, for example, has been used to cleave phenacyl esters of penicillanic acid.[17] The reaction proceeds rapidly at room temperature. A brief study of this cleavage indicated that phenacyl esters are cleaved more rapidly than benzyl esters or p-bromophenacyl esters, but no extensive study has been done.

Undoubtedly the mildest method yet devised for effecting S_N2 ester dealkylation involves the use of lithium thiopropoxide in hexamethylphosphoramide.[24] A variety of highly hindered methyl esters have been cleaved in excellent yield by this reagent under very mild conditions. Methyl triisopropylacetate, for example, is converted into the corresponding acid in 99% yield after 1 hour at room temperature.

$$(i\text{-}C_3H_7)_3CCO_2CH_3 \xrightarrow[\substack{\text{Hexamethyl-}\\\text{phosphoramide}}]{\text{LiSC}_3H_7\text{-}n} (i\text{-}C_3H_7)_3CCO_2H$$

[21] P. Müller and B. Siegfried, *Helv. Chem. Acta,* **54,** 2675 (1971).
[22] F. Elsinger, *Org. Syntheses,* **45,** 7 (1965).
[23] E. S. Behare and R. B. Miller, *Chem. Commun.,* **1970,** 402.
[24] P. A. Bartlett and W. S. Johnson, *Tetrahedron Lett.,* **1970,** 4459.

Although the reaction is somewhat inconvenient to carry out, in that a standard solution of the thiolate must be carefully prepared and protected from oxygen, the method appears to be an excellent one and has found use in synthesis.[25,26]

Potassium t-Butoxide

Another method for effecting ester cleavage is through the use of potassium t-butoxide in dimethyl sulfoxide. This reaction was first discovered when an attempt was made to effect an elimination reaction of methyl desoxycholate dimesylate (7).[27] Rather than the expected dienoic ester, however, the corresponding acid was isolated in quantitative yield after 4 hours at 100°. The reaction has received considerable use with di- and tri-terpenes (Table IV).

The mechanism of the cleavage is of interest since one might question the likelihood of the bulky t-butoxide anion acting as an attacking nucleophile in an S_N2 cleavage step. A further cause for worry is the report that methyl 3α-acetoxy-12α-mesyloxycholanate (8) undergoes cleavage and elimination to yield the *hydroxy* acid 9.[27] Clearly the acetate in 8 cannot be cleaved by an S_N2 dealkylation.

[25] E. J. Corey, T. M. Brennan, and R. L. Carney, *J. Amer. Chem. Soc.*, **93**, 7316 (1971).
[26] G. Schneider, *Tetrahedron Lett.*, **1972**, 4053.
[27] F. C. Chang and G. F. Wood, *Steroids*, **4**, 55 (1964).

Nevertheless t-butoxy methyl ether was isolated in 50% yield from the cleavage of methyl triisopropylacetate.[28]

Thus it seems that the potassium t-butoxide-dimethyl sulfoxide reagent can cleave esters by two pathways. Although esters of hindered acids cleave by S_N2 alkyl attack, esters of unhindered acids cleave by another mechanism, presumably acyl attack by the dimethyl sulfoxide anion.

Cyanide

Largely through the work of Krapcho, cyanide ion in dimethyl sulfoxide has been shown to be effective in promoting ester cleavage.[11] The reagent has been used almost exclusively to effect decarbalkoxylation of substituted malonic esters. Its mechanism is unclear. One well-documented possibility is simple S_N2 attack by cyanide on the ethyl group. Several recent results, however, cast doubt on this explanation.[14,15] It has been shown, for example, that cyanide may not even be necessary. Simply heating diethyl phenylmalonate in pure, dry dimethyl sulfoxide at 178° causes decarbethoxylation.[15] One well-documented report claims the specific decarbethoxylation of a diethyl malonate in the presence of an isolated methyl ester group, but no details have been given.[29]

$$(C_2H_5O)_2CHCH_2C(CO_2C_2H_5)_2CH_2CH(C_2H_5)CO_2CH_3 \xrightarrow[\text{Dimethyl sulfoxide}]{\text{NaCN}}$$
$$(C_2H_5O)_2CHCH_2CH(CO_2C_2H_5)CH_2CH(C_2H_5)CO_2CH_3$$

One way to rationalize this result is to assume that sodium cyanide in dimethyl sulfoxide undergoes rapid and reversible acyl attack on an ester. If the equilibrium heavily favors ester, isolated ester groups should not be noticeably affected. In malonic esters, however, the equilibrium can be shifted by an irreversible loss of ethyl cyanoformate, thus accounting for the observed specificity.

$$RCO_2C_2H_5 \underset{\text{CN}^-}{\rightleftharpoons} R-\overset{O^-}{\underset{CN}{\overset{|}{\underset{|}{C}}}}-OC_2H_5 \overset{-OC_2H_5^-}{\rightleftharpoons} RCOCN$$

$$R_2C(CO_2C_2H_5)_2 \underset{\text{CN}^-}{\rightleftharpoons} R_2C(CO_2C_2H_5)-\overset{O^-}{\underset{CN}{\overset{|}{\underset{|}{C}}}}-OC_2H_5 \longrightarrow R_2\bar{C}CO_2C_2H_5 + NCCO_2C$$

$$\downarrow \text{Dimethyl sulfoxide}$$

$$R_2CHCO_2C_2H_5$$

[28] F. C. Chang and G. F. Wood, *Tetrahedron Lett.*, **1964**, 2969.
[29] J. Harley-Mason and A. Rahman, *Chem. Ind.* (London), **1968**, 1845.

More work needs to be done on this reaction to distinguish between these possibilities. Regardless of the mechanism of this malonate decarbalkoxylation, there are several clear-cut instances reported where cyanide does act as a nucleophile in promoting S_N2 ester cleavages.[16,30,31] There is, in fact, some evidence for believing that cyanide ion is a considerably more reactive nucleophile than iodide, and that it therefore deserves wider usage in synthesis. A reactivity order $CN^- \gg Cl^- > Br^-$ in hexamethylphosphoramide has been found,[16] and dealkylation of methyl benzoate in dimethylformamide has been shown to occur more readily with cyanide than with iodide.[32]

Amines

One of the earliest reported S_N2 ester dealkylation is that by Willstätter in 1902 using trimethylamine as a nucleophile.[2] In general, however, amines have not been much used until two recent reports appeared on the use of diazabicyclononene and diazabicycloundecene in refluxing xylene for cleaving hindered methyl esters.[33,34a] Methyl mesitoate, for example, is cleaved in 94 % yield by diazabicyclononene after 6 hours of refluxing in xylene. It is not clear whether these reagents offer any advantages over more commonly used ones.

Thiocyanate

Quite recently, potassium thiocyanate in refluxing dimethylformamide has been shown to cleave methyl and benzyl esters in moderate yield.[34b] Once again, however, there seems to be no particular advantage to this method since neither conditions nor yields make it preferable to the use of other reagent systems.

Side Reactions

Few side reactions are encountered in these dealkylations because of the relatively mild conditions employed. The side reactions that do occur are usually thermal rearrangements caused by the high temperature used, and could perhaps be avoided by choice of a more reactive nucleophile.

A retro-Diels-Alder isomerization 10 → 11 has been reported during

[30] T. Kappe, M. A. A. Chirazi, H. P. Stelzel, and E. Zielgler, *Monatsh. Chem.*, **103**, 586 (1972).
[31] P. Müller and B. Siegfried, *Helv. Chim. Acta*, **57**, 987 (1974).
[32] J. E. Mc Murry and G. B. Wong, *Syn. Commun.*, **2**, 389 (1972).
[33] D. H. Miles and E. J. Parish, *Tetrahedron Lett.*, **1972**, 3987.
[34a] E. J. Parish and D. H. Miles, *J. Org. Chem.*, **38**, 1223 (1973).
[34b] T.-L. Ho and C. M. Wong, *Syn. Commun.*, **5**, 305 (1975).

ester cleavage by lithium iodide in collidine.[35] Similarly, the diester **12** undergoes ester cleavage followed by thermal decarboxylation to give **13**.[36]

The epoxide **14** has been reported to undergo a lithium iodide-catalyzed ring opening followed by dehydration to give the diene acid **15** in 60 % yield.[37] The diene **16** undergoes double-bond isomerization to a mixture of dienes when subjected to demethylation by t-butoxide in dimethyl sulfoxide.[38]

[35] W. Herz, R. C. Blackstone, and M. G. Nair, *J. Org. Chem.* **32**, 2992 (1967).
[36] R. A. Eade, J. Ellis, and J. J. H. Simes, *Aust. J. Chem.*, **20**, 2737 (1967).
[37] W. Herz and H. J. Wahlborg, *J. Org. Chem.*, **30**, 1881 (1965).
[38] R. M. Magid, C. R. Grayson, and D. R. Cowsar, *Tetrahedron Lett.*, **1968**, 4877.

Other than these few reported cases, however, ester dealkylations by S$_N$2 reactions are unusually clean processes.

EXPERIMENTAL CONDITIONS

There is not yet enough information available to allow a knowledgeable selection of optimum reaction conditions for a given S$_N$2 ester dealkylation. The reason is simply that quantitative comparisons of the different reagents have not been carried out. In the absence of such studies, chemists faced with choosing a reagent system have usually selected one that has worked well in a similar, previously reported case. Thus lithium iodide in dimethylformamide is much used for cleavage of simple methyl esters and for decarbomethoxylation of β-keto esters (Tables II and VI), whereas sodium cyanide is dimethyl sulfoxide is generally favored for decarbalkoxylation of malonic esters (Table V). No logical reason appears to exist for these choices, however.

Solvent

A dipolar aprotic solvent appears preferable. In the only published comparison of solvents, which was carried out for dealkylations by lithium iodide, dimethylformamide was found to be most effective in promoting cleavage.[19] Hexamethylphosphoramide has not yet been used with lithium iodide, but is highly effective in promoting room-temperature ester dealkylations by lithium n-propyl mercaptide, and would probably work well with other nucleophiles also.[24] Dimethyl sulfoxide is generally used as solvent for the sodium cyanide-promoted cleavage of malonic esters,[11] but no study of other solvents has been reported.

Nucleophile

From what information is available, it seems that thiolate and cyanide nucleophiles are probably most reactive, iodide somewhat less reactive, and t-butoxide least reactive. There is work to show that cyanide ion effects ester cleavage more efficiently and at lower reaction temperatures than does lithium iodide.[16,32] Similarly the thiolate nucleophiles effect cleavage rapidly at low reaction temperatures. Lithium iodide, by contrast, requires temperatures of 140° in dimethylformamide to cleave methyl benzoate,[32] while potassium t-butoxide in dimethyl sulfoxide requires temperatures of 100° to 160° and is strongly basic. In view of the known ability of potassium t-butoxide in dimethyl sulfoxide to cause olefin isomerizations,[38] this reagent appears limited to use with stable, base-insensitive esters.

Other Considerations

The actual cleavage step is a straightforward S_N2 reaction. It has not previously been pointed out, however, that the cleavage may be either reversible or irreversible depending on the nucleophile used. For example, the reaction of an ester with either a thiolate or with cyanide is irreversible, leading to alkyl sulfide and alkyl cyanide, respectively. Reaction with iodide, however, is readily reversible and, in order to obtain an efficient cleavage, alkyl iodide must be removed to drive the equilibrium in the desired direction.

$$RCO_2CH_3 \begin{cases} \xrightarrow{\ CN^- \ } RCO_2^- + CH_3CN \\ \xrightarrow{\ R'S^- \ } RCO_2^- + CH_3SR' \\ \xrightarrow{\ I^- \ } RCO_2^- + CH_3I \end{cases}$$

In practice this removal of alkyl iodide undoubtedly occurs by evaporation from the hot reaction. It has been reasoned that if this removal could be made more efficient, reaction might occur at a lower temperature.[32] To accomplish removal various nucleophiles were added to the reaction to scavenge methyl iodide by competing with product carboxylate. Introduction of an equivalent amount of sodium acetate, for example, led to a 20° lowering (from 150° to 130°) of the temperature at which complete reaction occurred between lithium iodide and methyl benzoate in dimethylformamide. Addition of cyanide led to further drop in the reaction temperature, and addition of both acetate and cyanide reduced the reaction temperature still further. The mixture of lithium iodide plus sodium cyanide in dimethylformamide seems to be a potent reagent for S_N2 ester dealkylations.[32,39]

EXPERIMENTAL PROCEDURES

The procedures given below have been chosen to illustrate the use of different nucleophiles in carrying out S_N2 ester dealkylations.

3β-Acetoxy-Δ⁵-etiocholenic Acid. (Cleavage by Lithium Iodide in 2,6-Lutidine).[5] Methyl 3β-acetoxy-Δ⁵-etiocholenate (800 mg, 2.14 mmol) and 1.8 g (13.4 mmol) of lithium iodide were dissolved in 35 ml of 2,6-lutidine under a nitrogen atmosphere, and the solution was refluxed for 8 hours. The cooled mixture was then acidified with 2N hydrochloric acid and extracted with ether-methylene chloride (2:1). The extract was washed with 2N hydrochloric acid to remove lutidine, then dried and concentrated. Chromatography on silica gel gave 202 mg (25%) of

[39] B. M. Trost and T. J. Dietsche, J. Amer. Chem. Soc., 95, 8200 (1973).

starting material, followed by 380 mg (50%) of 3β-acetoxy-Δ^5-etiocholenic acid, mp 244–246°, identified by mixture melting point with an authentic sample.

Glycyrrhetic Acid. (Cleavage by Lithium Iodide in Dimethylformamide).[19] Methyl glycyrrhetate (100 mg) and lithium iodide (500 mg) were dissolved in 15 ml of dimethylformamide under a nitrogen atmosphere, and the solution was refluxed. After 2 hours the reaction mixture was cooled, poured into water, acidified with dilute hydrochloric acid, and extracted with ether. The ether extract was washed with water, dried, and concentrated to give the crude solid product. Crystallization from acetic acid gave a quantitative yield of glycyrrhetic acid, mp 303–305°, whose purity was established by thin-layer chromatography.

1-Benzyl-3-carboxy-4-ethoxycarbonyl-2(1H)-pyridone (Selective Cleavage of a Methyl Ester by Lithium Iodide in Pyridine).[7] To a refluxing solution of 6.15 g (46 mmol) of anhydrous lithium iodide in 50 ml of dry pyridine under nitrogen was added a solution of 3.67 g (11.65 mmol) of 1-benzyl-3-methoxycarbonyl-4-ethoxycarbonyl-2(1H)-pyridone. Refluxing was continued for 1 hour. The solution was cooled and the pyridine was removed under reduced pressure (bath temperature 40°). The residue was dissolved in 50 ml of water, acidified with 6N hydrochloric acid, and extracted with chloroform. The extracts were dried and evaporated. Crystallization of the residue from absolute ethanol gave 3.02 (86%) of 1-benzyl-3-carboxy-4-ethoxycarbonyl-2(1H)-pyridone; mp 96–100°; infrared (Nujol) cm^{-1}: 1740, 1630, 1450; mass spectrum m/e (relative intensity): 301 (P$^+$, 4).

Sodium Benzylpenicillin (Cleavage of a Phenacyl Ester by Sodium Thiophenoxide in Dimethylformamide).[17] A solution of 0.029 g of sodium thiophenoxide and 0.050 g of phenacyl benzylpenicillinate in 0.2 ml of dimethylformamide was allowed to stand at room temperature for 15 minutes. To this solution was added 20 ml of acetone. This solution was stirred for 10 minutes, during which time crystallization occurred. Filtration gave 0.033 g (84%) of sodium benzylpenicillin, mp 226–227°. The identity of the product was confirmed by mixture melting point with an authentic sample.

3β-Acetoxy-Δ^5-etiocholenic Acid (Cleavage by Lithium n-Propyl Mercaptide in Hexamethylphosphoramide).[24] The mercaptide reagent was prepared by adding freshly distilled n-propyl mercaptan (1.0 ml) to a suspension of finely ground lithium hydride (0.3 g) in 10 ml of dry, oxygen-free hexamethylphosphoramide under an argon atmosphere. After stirring at 25° for 1 hour, the mixture was filtered under argon and stored at 0°. The reagent was approximately 0.5 M.

To a solution of 173 mg of methyl 3β-acetoxy-Δ^5-etiocholenate in 0.9 ml of dry hexamethylphosphoramide under a nitrogen atmosphere was added 0.89 ml of 0.58 M mercaptide reagent. After 24 hours at 25° the reaction mixture was transferred into 100 ml of cold 0.1 N hydrochloric acid. Extraction with ether gave 166 mg of crude product that was purified by chromatography on 5 g of silica gel. Elution with 5% ether-benzene gave 152.9 mg (92%) of 3β-acetoxy-Δ^5-etiocholenic acid, mp 242–246°. One recrystallization from methanol gave material having mp 244–245°, which was identical with an authentic sample by mixture melting point and by infrared comparison.

Dehydroabietic Acid (Cleavage by Potassium t-Butoxide in Dimethyl Sulfoxide).[28] Methyl dehydroabietate (100 mg, 0.32 mmol) was added to 5 ml of a 1 N potassium t-butoxide solution in dimethyl sulfoxide at 25°. After 1 hour, no starting material could be detected by thin layer chromatography, and the reaction was poured into cold dilute hydrochloric acid. Filtration gave 90 mg (90%) of crude acid. Recrystallization from aqueous ethanol gave pure dehydroabietic acid, mp 168–170.6°, identified by infrared comparison and mixture melting point with an authentic sample.

N-Methyl-3-ethyl-4-methoxycarbonyl-2-piperidone (Cleavage and Decarboxylation of a Dimethyl Malonate by Sodium Cyanide in Dimethylformamide).[12] To a solution of 6.5 g of sodium cyanide (0.13 mol) in 100 ml of dimethylformamide was added 23 g (0.09 mol) of N-methyl-3-ethyl-4,4-dimethoxycarbonyl-2-piperidone. The solution was heated to reflux, and gas evolution was monitored with a gas burette. After 2 hours the solution was cooled, filtered, and concentrated under reduced pressure. The residue was taken up in chloroform and washed with water. The crude product was distilled under reduced pressure to yield 12.4 g (70%) of N-methyl-3-ethyl-4-methoxycarbonyl-2-piperidone, bp 110–120° (0.4 mm), infrared (CHCl$_3$) cm^{-1}; 1730, 1630; a suitable combustion analysis was obtained also.

2-Benzylcyclopentanone (Cleavage and Decarboxylation of a β-Keto Ester by Sodium Cyanide in Hexamethylphosphoramide).[16] To a solution of 0.57 g (11.6 mmol) of sodium cyanide in 100 ml of hexamethylphosphoramide under a nitrogen atmosphere was added 1.5 g (6.4 mmol) of 2-benzyl-2-methoxycarbonylcyclopentanone. The reaction was heated heated to 75° and stirred for 1 hour. After cooling, the solution was poured into 500 ml of 2 N hydrochloric acid (*caution!*) and extracted with carbon tetrachloride. Distillation of the extract gave 0.89 g (80%) of 2-benzyl-cyclopentanone, identified by spectral comparison with an authentic sample. The semicarbazone derivative melted at 198°.

2-Benzylcyclopentanone (Cleavage and Decarboxylation of a β-Keto Ester by Lithium Iodide in 2,4,6-Collidine).[22] A mixture of 30 g (0.177 mol) of lithium iodide dihydrate and 140 ml of 2,4,6-collidine was refluxed under a nitrogen atmosphere. When solution was complete, 30 g (0.129 mol) of 2-benzyl-2-methoxycarbonylcyclopentanone in 30 ml of 2,4,6-collidine was added. After 19 hours at reflux, the mixture was cooled, poured into 400 ml of cold 3 N hydrochloric acid, and extracted with ether. The extract was washed with 6 N hydrochloric acid, then dried, concentrated, and distilled to yield 17 g (76%) of 2-benzylcyclopentanone, bp 83–85° (0.3 mm); semicarbazone, mp 204–205°.

Triisopropylacetic Acid (Cleavage by Diazabicycloundecene in Xylene).[34] Methyl triisopropylacetate (156 mg, 0.83 mmol) and 1.202 g (8.3 mmol) of diazabicycloundecene were dissolved in 1 ml of *o*-xylene and heated to 165° for 48 hours. After it was cooled and acidified, the solution was extracted with ether to yield 130 mg (91%) of crude triisopropyl-acetic acid. Recrystallization from methanol-water gave the pure product, mp 136–137°, identified by mixture melting point with an authentic sample.

TABULAR SURVEY

Table I appears in the text. Tables II–VII, which follow, provide all examples known through December 1974.

Within each table, compounds are given in the order of increasing carbon numbers based on the parent acid. Compounds with the same numbers of carbon atoms are arranged by increasing complexity.

Abbreviations for solvents are as follows: DMF, N,N-dimethylform-amide; DMSO, dimethyl sulfoxide; and HMPA, hexamethylphosphor-triamide.

Other symbols used in formulas follow common conventions: Ac for COCH$_3$ (acetyl), Ms for OSO$_2$CH$_3$ (mesyl), and Ts for OSO$_2$C$_6$H$_4$CH$_3$-p (tosyl). Under reagents, DBN stands for diazabicyclononene and DBU represents diazabicycloundecene.

TABLE II. Ester Cleavage by Alkali Halides

	Reactant	Halide	Solvent	Temperature	Time (hr)	Product and Yield (%)	Refs.
C$_6$	Br CH$_2$Br, H, CO$_2$CH$_3$, CH$_3$O$_2$C, H	LiCl	DMF	—	—	Br, H, CH$_3$O$_2$C, CH$_2$O$_2$C, H (—)	40
		LiBr	"	—	—	H CH$_2$OH, CO$_2$H, HO$_2$C, H (—)	40
C$_8$	C$_6$H$_5$CH$_2$CO$_2$CH$_3$	LiCl	Pyridine	Reflux	15	C$_6$H$_5$CH$_2$CO$_2$H (23)	4
		LiBr	"	"	:	: (65)	4
		LiI	"	"	:	: (93)	4
	C$_6$H$_5$CH$_2$CO$_2$C$_2$H$_5$	LiCl	Pyridine	Reflux	27	: (8)	4
		LiBr	"	"	:	: (19)	4
		LiI	"	"	:	: (42)	4
	CO$_2$CH$_3$ OCOC$_6$H$_5$	LiI	Pyridine	Reflux	—	CO$_2$H OCOC$_6$H$_5$ (92)	29
	C$_6$H$_5$CH$_2$CONH	LiI	Pyridine	Reflux	2	C$_6$H$_5$CH$_2$CONH (—)	41

	Reactant	Reagent	Solvent	Conditions	Time	Product	Yield	Refs.
C_{10}			Collidine					42
C_{11}	CH_3O_2C ... H, CH_3, CH_2OH, CH_3O_2C	LiI	2,4,6-Collidine	Reflux	4	HO_2C ... H, CH_3, CH_2OH, HO_2C	(—)	43
C_{14}	NH_2, CO_2CH_3, N, O	LiI	Pyridine	Reflux	—	NH_2, CO_2H, N, O	(68)	44
	$NCCH(C_6H_5)C(C_2H_5)_2CO_2CH_3$	LiI	2,4,6-Collidine	Reflux	12	$NCCH(C_6H_5)C(C_2H_5)_2CO_2H$	(82)	45
	$CO_2C_2H_5$, CO_2CH_3, N=O, $C_6H_5CH_2$	LiI	Pyridine	Reflux	1	$CO_2C_2H_5$, CO_2H, N=O, $C_6H_5CH_2$	(86)	7
C_{17}	OR, CH_3O_2C, R = H	LiI	2,4,6-Collidine	Reflux	10	HO_2C	(25)	46
	R = CH_3	LiI	"	"	—		(92)	47

Note: References 40–88 are on pp. 223–224.

TABLE II. ESTER CLEAVAGE BY ALKALI HALIDES (*Continued*)

Reactant	Halide	Solvent	Temperature	Time (hr)	Product and Yield (%)	Refs.
C$_{17}$(*Contd.*)	LiI	Pyridine	Reflux	2	(85)	6
C$_{19}$	LiI	2,4,6-Collidine	Reflux	18	(—)	48
C$_{20}$	LiI	2,4,6-Collidine	Reflux	8	(50)	49
	LiI	2,4,6-Collidine	Reflux	—	(60)	37

204

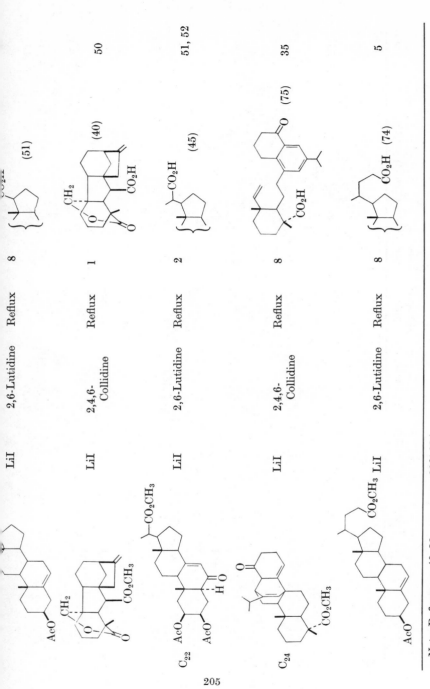

205

Note: References 40–88 are on pp. 223–224.

TABLE II. ESTER CLEAVAGE BY ALKALI HALIDES (Continued)

Reactant	Halide	Solvent	Temperature	Time (hr)	Product and Yield (%)	Refs.
C_{30}	LiI	2,4,6-Collidine	Reflux	8	(90)	53
	"	2,6-Lutidine	Reflux	8	(56)	5
		2,4,6-Collidine	"	"	(90)	5
	LiI	Pyridine	Reflux	8	(30)	19
	"	2,6-Lutidine	"	"	(100)	19
	"	2,4,6-Collidine	"	"	(100)	19

206

Substrate	Reagent	Solvent	Conditions		Product		Refs.
(steroid/triterpene with CO$_2$CH$_3$, HO)	LiF NaI	" "	" "	24	(product with CO$_2$H)	(0) (40)	19 19
	LiI	Pyridine	—	—	CO$_2$H (—)		54
(steroid with CO$_2$CH$_3$, RO, O)	LiI	DMF	Reflux	—	CO$_2$H (—)		55
R = Methyltriacetylglucosidur- onate							
(steroid CONHR, AcO, O)	LiI "	DMF "	Reflux "	7 "	R' = CONHCH$_2$CO$_2$H R' = CO$_2$H	(75) (—)	56 56
R = CH$_2$CO$_2$CH$_3$ R = CH(CO$_2$CH$_3$)— CH$_2$CH$_2$CO$_2$CH$_3$							

Note: References 40–88 are on pp. 223–224.

TABLE II. ESTER CLEAVAGE BY ALKALI HALIDES (*Continued*)

Reactant	Halide	Solvent	Temperature	Time (hr)	Product and Yield (%)	Refs.
C$_{35}$	LiI	DMF	Reflux	12	(—)	81
R = CH$_3$CH(CO$_2$CH$_3$)CH$_2$CO— R = CH$_3$CH(CO$_2$CH$_3$)CH$_2$CO—	LiI	2,4,6-Collidine	Reflux	8	R = CH$_3$CH(CO$_2$H)CH$_2$CO— (65)	57

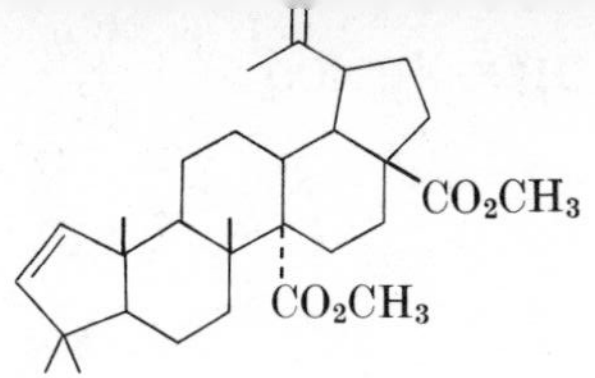

58

(70)

CO_2H CO_2H

8   Reflux   2,4,6-Collidine   LiI

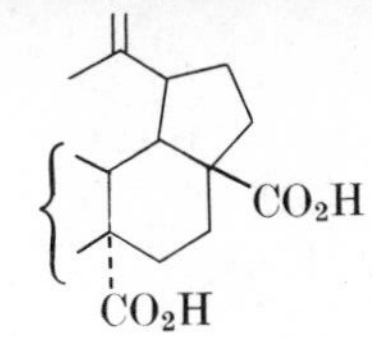

CO_2CH_3 CO_2CH_3

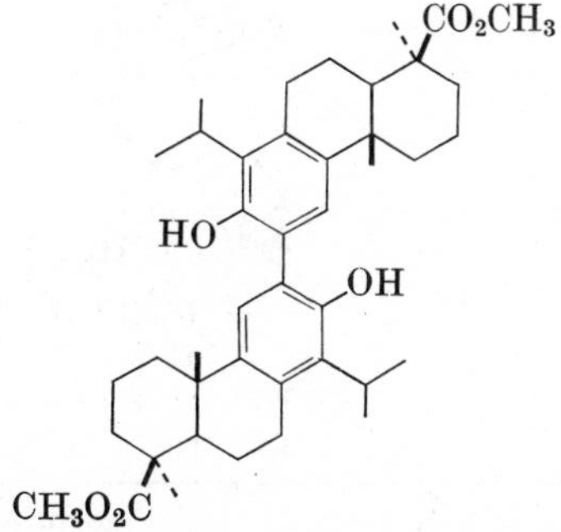

59

(50)

$[\ \ CO_2H\ \ ]_2$ HO

8   Reflux   2,4,6-Collidine   LiI

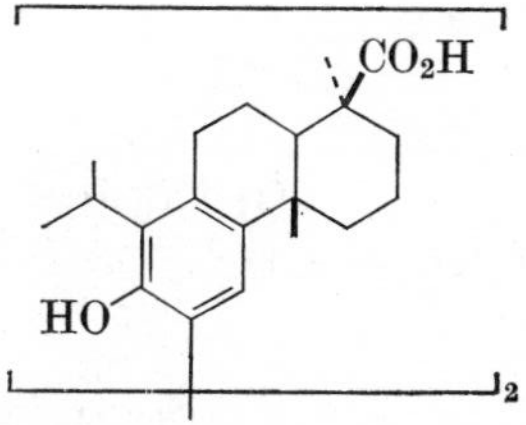

CO_2CH_3 OH HO CH_3O_2C

C_{40}

Note: References 40–88 are on pp. 223–224.

TABLE III. ESTER CLEAVAGE BY THIOLATES

Reactant	Thiolate	Solvent	Temperature	Time (hr)	Product and Yield (%)	Refs.
C7						
C6H5CO2R					C6H5CO2H	
R = CH2C6H5	NaSC6H5	DMF	100°	0.5	(64)	17
R = CH2COC6H5	"	"	25°	"	(87)	17
p-HOC6H4CO2CH3	NaSC2H5	DMF	—	—	p-HOC6H4CO2H (90)	60
C6H5CH2NH					C6H5CH2NH	
R = CH2COC6H4Br-p	NaSC6H5	DMF	25°	0.5	(64)	17
R = CH2COC6H5	"	"	"	0.25	(84)	17
C8						
CO2CH2COC6H5	NaSC6H5	DMF	25°	0.5	CO2H (76)	17

C_{10}	$2,4,6\text{-}(CH_3)_3C_6H_2CO_2CH_3$	$LiSC_3H_7\text{-}n$	HMPA	25°	1	$2,4,6\text{-}(CH_3)_3C_6H_2CO_2H$ (100)	24
C_{11}	$(i\text{-}C_3H_7)_3CCO_2CH_3$	$NaSC_3H_7\text{-}n$	DMF	Reflux	3	(78)	61
		$LiSC_3H_7\text{-}n$	HMPA	25°	1	$(i\text{-}C_3H_7)_3CCO_2H$ (99)	24
C_{17}	(structure: OCH_3 ... CH_3O_2C)	$LiSC_3H_7\text{-}n$	HMPA	25°	1	(structure: HO_2C) (100)	24
C_{20}	(structure: RO ... CO_2CH_3, OH) R = H	$LiSC_3H_7\text{-}n$	HMPA	—	—	(structure: RO ... CO_2H, OH) R = H (High)	25
	R = glucose	"	"	—	—	R = glucose (—)	26

Note: References 40–88 are on pp. 223–224.

211

TABLE IV. ESTER CLEAVAGE BY POTASSIUM t-BUTOXIDE IN DIMETHYL SULFOXIDE

	Reactant	Temperature	Time (hr)	Product and Yield (%)	Refs.
C$_8$	(structure, CO$_2$CH$_3$, OCH$_3$)	75°	—	(structure, CO$_2$H, OCH$_3$) (—)	38
C$_{11}$	$(i$-C$_3$H$_7)_3$CCO$_2$CH$_3$	100°	4	(Mixture of diene isomers) $(i$-C$_3$H$_7)_3$CCO$_2$H (100)	28
C$_{17}$	(structure, OCH$_3$, CH$_3$O$_2$C)	65°	2	(structure, HO$_2$C) (84)	62
	(structure, OCH$_3$, CH$_3$O$_2$C)	50°	2	(structure, HO$_2$C) (97)	28
C$_{20}$	(structure, OCH$_3$, CH$_3$O$_2$C)	100°	24	(structure, HO$_2$C) (90)	63

212

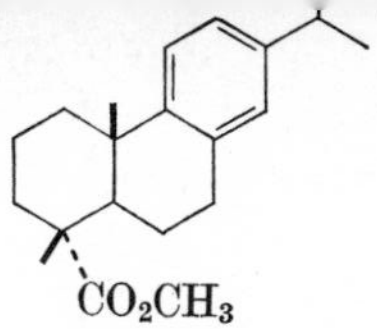

28

(94)

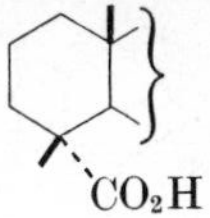

64

(65)

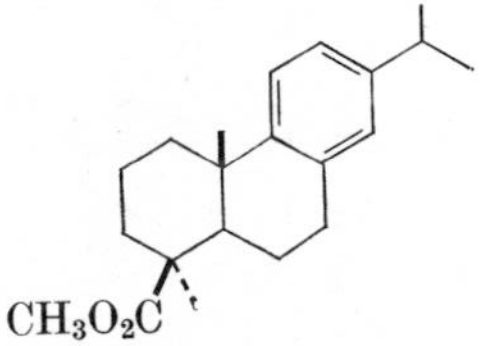

65

(96)

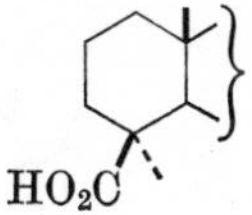

66

(63)

67

(—)

1

6

1

—

—

25°

100°

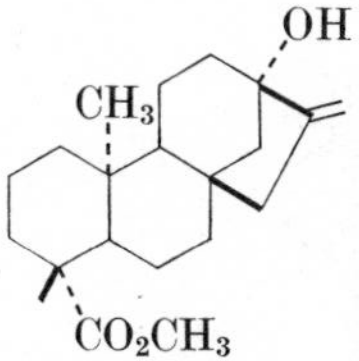

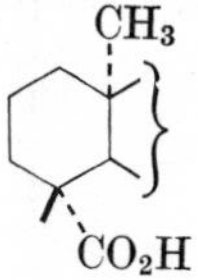

80°

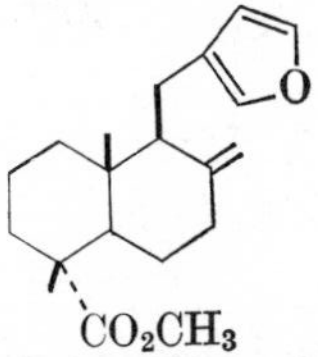

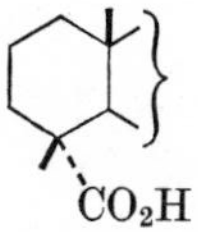

—

—

213

Note: References 40–88 are on pp. 223–224.

TABLE IV. ESTER CLEAVAGE BY POTASSIUM t-BUTOXIDE IN DIMETHYL SULFOXIDE (*Continued*)

Reactant	Temperature	Time (hr)	Product and Yield (%)	Refs.
C$_{24}$	25°	48	CO$_2$H (53)	27
C$_{24}$	25°	48	CO$_2$H (−)	27
C$_{30}$	105°	4	CO$_2$H (−)	68
C$_{30}$	—	—	CO$_2$H (−)	54

214

	Reactant	Temperature	Time (hr)	Product and Yield (%)	Refs.
C$_4$	CH$_3$CH(CO$_2$C$_2$H$_5$)$_2$	160°	4	CH$_3$CH$_2$CO$_2$C$_2$H$_5$ (75)	11
C$_5$	CH$_3$CH$_2$CH(CO$_2$C$_2$H$_5$)$_2$	160°	4	CH$_3$CH$_2$CH$_2$CO$_2$C$_2$H$_5$ (80)	11
C$_6$	(cyclobutane)(CO$_2$C$_2$H$_5$)(CO$_2$C$_2$H$_5$)	160°	4	(cyclobutane)CO$_2$C$_2$H$_5$ (75)	11
	(dioxolane)CH(CO$_2$C$_2$H$_5$)$_2$, H	170°	4	(dioxolane)...CO$_2$C$_2$H$_5$ (50), H	69
C$_7$	(cyclopentane)(CO$_2$C$_2$H$_5$)(CO$_2$C$_2$H$_5$)	160°	4	(cyclopentane)CO$_2$C$_2$H$_5$ (75)	70
	FCH$_2$CH$_2$CH(CH$_3$)CH(CO$_2$C$_2$H$_5$)$_2$	160°	4	FCH$_2$CH$_2$CH(CH$_3$)CH$_2$CO$_2$C$_2$H$_5$ (16)	70
	(THP-O-alkene)CH(CO$_2$C$_2$H$_5$)$_2$	160°	4	(THP-O-alkene)...CO$_2$C$_2$H$_5$ (64)	71
C$_9$	(C$_2$H$_5$O)$_2$CHCH$_2$C(CO$_2$C$_2$H$_5$)$_2$CH$_2$CH(C$_2$H$_5$)CO$_2$CH$_3$	—	—	(C$_2$H$_5$O)$_2$CHCH$_2$CHCH$_2$CH(C$_2$H$_5$)CO$_2$CH$_3$	29
C$_{10}$	(CH$_3$-N-piperidone)(CH$_3$O$_2$C)(CO$_2$CH$_3$), ethyl	160°	4	(CH$_3$-N-piperidone)(CO$_2$CH$_3$), ethyl (70)	12
C$_{13}$	(cyclopentane)(CH$_3$O$_2$C)(CH$_3$O$_2$C), CH$_3$O$_2$C, OTs, OTs	—	—	(cyclopentane)(CH$_3$O$_2$C), CN, CN	72

Note: References 40–88 are on pp. 223–224.

215

TABLE V. ESTER CLEAVAGE AND DECARBOXYLATION BY SODIUM CYANIDE IN DIMETHYL SULFOXIDE (Continued)

Reactant	Temperature	Time (hr)	Product and Yield (%)	Refs.
CH(CO₂CH₃)₂ structure	130°	—	CO₂CH₃ (81)	73
C₂H₅O₂C ... CO₂C₂H₅ structure	170°	4	(80) C₂H₅O₂C structure	74
C₂H₅O₂C ... CO₂C₂H₅ structure	—	—	C₂H₅O₂C structure	75
CH₂C₆H₅ / CO₂CH₃ cyclopentanone structure	75° (HMPA)	1	CH₂C₆H₅ (80) O= structure	16
C₂H₅O₂C ... CO₂C₂H₅ decalin structure (C₁₅)	160°	6	(70) CO₂C₂H₅ structure	76
C₂H₅O₂C ... CO₂C₂H₅ tricyclic structure (C₁₆)	160°	4	(75) H...CO₂C₂H₅ structure	77

	Reactant	Product	%	Temp	Ref.
C$_{19}$	$C_2H_5O_2C(CH_2)_7CH-CH(CH_2)_6CH(CO_2C_2H_5)_2$	$C_2H_5O_2C(CH_2)_7CH-CH(CH_2)_6CH_2CO_2C_2H_5$ (80)	—	160°	78
C$_{20}$		(89)	—	—	79
		(−)	48	80°	80
		(10)	4	140°	81

Note: References 40–88 are on pp. 223–224.

TABLE VI. ESTER CLEAVAGE AND DECARBOXYLATION BY ALKALI HALIDES

Reactant	Halide	Solvent	Temperature	Time (hr)	Product and Yield (%)	Refs.
C_8 CH=CH-CH(CO$_2$C$_2$H$_5$)$_2$	LiI/ NaCN	DMF	130°	—	CO$_2$C$_2$H$_5$ (80)	39
C_9 (succinimide) N-CH$_2$SO$_2$CH$_3$, C(CO$_2$CH$_3$)(CO$_2$CH$_3$)	KI	DMF	110°	0.5	(81) CO$_2$CH$_3$	82
C_{10} (lactone) CO$_2$CH$_3$, CH$_2$N$^+$(CH$_3$)$_3$ I$^-$	none	DMF	80°	—	(—)	23
(lactone) CO$_2$CH$_3$, O-tetrahydropyranyl	LiI	Pyridine	Reflux	3	(37)	83
C_{11} C$_6$H$_5$CH$_2$C(CH$_3$)(CO$_2$C$_2$H$_5$)CON(CH$_3$)$_2$	LiI	2,6-Lutidine	Reflux	2.5	C$_6$H$_5$CH$_2$CH(CH$_3$)CON(CH$_3$)$_2$ (70)	10
R, CO$_2$C$_2$H$_5$, (CH$_2$)$_n$ ketone n = 1; R = (CH$_2$)$_3$CH=CH$_2$	LiI	DMF	—	—	R, (CH$_2$)$_n$ ketone n = 1; R = (CH$_2$)$_3$CH=CH$_2$ (—)	84
n = 1; R = (CH$_2$)$_3$C≡CH	''	''	—	—	n = 1; R = (CH$_2$)$_3$C≡CH (—)	84

Reactant	Reagent	Solvent/Base	Temp.	Time (hr)	Product(s) (% Yield)	Refs.
C_{12} $n = 1$; $R = (CH_2)_4C{\equiv}CH$	"	"	—	—	$n = 1$; $R = (CH_2)_4C{\equiv}CH$ (—)	84
$n = 2$; $R = (CH_2)_3CH{=}CH_2$	"	"	—	—	$n = 2$; $R = (CH_2)_3CH{=}CH_2$ (—)	84
$n = 2$; $R = (CH_2)_3C{\equiv}CH$	"	"	—	—	$n = 2$; $R = (CH_2)_3C{\equiv}CH$ (—)	84
C_{13} $n = 2$; $R = (CH_2)_4CH{=}CH_2$	"	"	—	—	$n = 2$; $R = (CH_2)_4CH{=}CH_2$ (—)	84
$n = 2$; $R = (CH_2)_4C{\equiv}CH$	"	"	—	—	$n = 2$; $R = (CH_2)_4C{\equiv}CH$ (—)	84
(cyclopentanone with $CH_2C_6H_5$, CO_2CH_3)	"	2,4,6-Collidine	Reflux	19	(structure with $CH_2C_6H_5$) (75)	22
	LiCl	HMPA	75°	24	(90)	16
C_{17} (phenanthrenone with CO_2CH_3)	LiI	2,4,6-Collidine	Reflux	10	(structure) (—)	46
	LiI	2,4,6-Collidine	Reflux	10	(90)	46
C_{20} (CH_3O, CH_3O, CH_3O_2C, isopropyl structure)	LiI	2,6-Lutidine	Reflux	10	(CH_3O, OCH_3, OH structure) (100)	85
$(CH_3)_2NCO(CH_2)_7CHCO(CH_2)_8CON(CH_3)_2$, CO_2CH_3	LiI	2,4,6-Collidine	Reflux	14	$(CH_3)_2NCO(CH_2)_7CH_2CO(CH_2)_8CON(CH_3)_2$ (63)	86

Note: References 40–88 are on pp. 223–224.

219

TABLE VI. ESTER CLEAVAGE AND DECARBOXYLATION BY ALKALI HALIDES (*Continued*)

Reactant	Halide	Solvent	Temperature	Time (hr)	Product and Yield (%)	Refs.
C$_{30}$	LiI	2,4,6-Collidine	Reflux	8	(60)	36
C$_{30}$	LiI	2,4,6-Collidine	Reflux	2	(−) + (−)	87

Note: References 40–88 are on pp. 223–224.

TABLE VII. Ester Cleavage by Miscellaneous Reagents

	Reactant	Reagent	Solvent	Temperature	Time (hr)	Product and Yield (%)	Refs.
C$_7$	3,5-(HO)$_2$C$_6$H$_3$CO$_2$CH$_3$ (benzene ring with CO$_2$R)	KSCN	None	300°	—	3,5-(HO)$_2$C$_6$H$_3$CO$_2$H (benzene ring with CO$_2$H) (90)	88
	R = CH$_3$	"	DMF	Reflux	12	" (60)	34b
	R = CH$_2$C$_6$H$_5$	"	"	"	3	" (72)	"
	CH$_3$(CH$_2$)$_5$CO$_2$CH$_3$	"	"	"	12	CH$_3$(CH$_2$)$_5$CO$_2$H (55)	"
C$_8$	(benzene with CO$_2$CH$_3$, CO$_2$CH$_3$)	"	"	"	"	(benzene with CO$_2$H, CO$_2$H) (68)	"
C$_9$	(cinnamyl CO$_2$R)					(cinnamyl CO$_2$H)	
	R = CH$_3$	"	"	"	"	" (74)	"
	R = CH$_2$C$_6$H$_5$	"	"	"	3	" (70)	"

Note: References 40–88 are on pp. 223–224.

221

TABLE VII. Ester Cleavage by Miscellaneous Reagents (*Continued*)

Reactant	Reagent	Solvent	Temperature	Time (hr)	Product and Yield (%)	Refs.
C$_{10}$ $2,4,6\text{-}(CH_3)_3C_6H_2CO_2CH_3$	DBN	Xylene	165°	6	$2,4,6\text{-}(CH_3)_3C_6H_2CO_2H$ (94)	33
	DBU	,,	,,	48	(97)	34a
C$_{11}$ $(i\text{-}C_3H_7)_3CCO_2CH_3$	DBN	Xylene	165°	6	$(i\text{-}C_3H_7)_3CCO_2H$ (94)	33
	DBU	,,	,,	48	(91)	34a
C$_{13}$ [pyridinone structure: OH, C$_6$H$_5$, C$_2$H$_5$O$_2$C, N–H, CH$_3$]	NaCN	DMSO	155°	4	[pyridinone structure: OH, C$_6$H$_5$, HO$_2$C, N–H, CH$_3$, O] (50)	30
C$_{17}$ [tricyclic structure: OCH$_3$, CH$_3$O$_2$C]	DBN	Xylene	165°	6	[tricyclic structure: OCH$_3$, HO$_2$C] (91)	33
	DBU	,,	,,	48	(97)	34a
C$_{24}$ [steroid structure: AcO, CO$_2$CH$_3$]	DBN	Xylene	165°	8	[steroid structure: CO$_2$H] (50)	33
	DBU	,,	,,	3.5	(41)	34a

Note: References 40–88 are on pp. 223–224.

222

REFERENCES TO TABLES II–VII

[40] T. L. Gilchrist and C. W. Rees, *J. Chem. Soc.*, *C*, **1968**, 776.

[41] N. J. Leonard and G. E. Wilson, *J. Amer. Chem. Soc.*, **86**, 5307 (1964).

[42] H. O. House and G. A. Frank, *J. Org. Chem.*, **30**, 2948 (1965).

[43] R. C. Cookson, J. Dance, and J. Hudec, *J. Chem. Soc.*, **1964**, 5416.

[44] W. Schafer and H. Schlude, *Tetrahedron Lett.*, **1968**, 2161.

[45] R. H. Eastman and K. Tamaribuchi, *J. Org. Chem.*, **30**, 1671 (1965).

[46] E. Wenkert, P. Beak, R. W. Carney, J. W. Chamberlin, D. B. R. Johnston, C. D. Roth, and A. Tahara, *Can. J. Chem.*, **41**, 1924 (1963).

[47] C. R. Bennett and R. C. Cambie, *Tetrahedron*, **23**, 927 (1967).

[48] D. K. Black and G. W. Hedrick, *J. Org. Chem.*, **34**, 1940 (1969).

[49] A. Afonso, *J. Org. Chem.*, **35**, 1949 (1970).

[50] W. Nagata, T. Wakabayshi, Y. Hayase, M. Narisada, and S. Kamata, *J. Amer. Chem. Soc.*, **93**, 5740 (1971).

[51] U. Kerb, P. Hocks, and R. Wiechert, *Tetrahedron Lett.*, **1966**, 1387.

[52] U. Kerb, G. Schulz, P. Hocks, R. Wiechert, A. Furlenmeier, A. Furst, A. Langemann, and G. Waldvogel, *Helv. Chim. Acta*, **49**, 1601 (1966).

[53] G. V. Baddeley, J. J. H. Simes, and T. G. Watson, *Aust. J. Chem.*, **24**, 2639 (1971).

[54] S. N. Bose and H. N. Khastgir, *J. Indian Chem. Soc.*, **46**, 860 (1969).

[55] P. Iveson and D. V. Parke, *J. Chem. Soc.*, *C*, **1970** 2038.

[56] S. Rozen, I. Shahak, and E. D. Bergman, *Israel J. Chem.*, **9**, 185 (1971).

[57] A. T. Glen, W. Lawrie, J. McLean, and M. El-Garby Younes, *J. Chem. Soc.*, *C*, **1967**, 510.

[58] P. de Mayo and A. N. Starratt, *Can. J. Chem.*, **40**, 1632 (1962).

[59] A. C. Day, *J. Chem. Soc.*, **1964**, 3001.

[60] G. I. Feutrill and R. N. Mirrington, *Aust. J. Chem.*, **25**, 1731 (1972).

[61] W. R. Vaughan and J. B. Baumann, *J. Org. Chem.*, **27**, 739 (1962).

[62] K. Mori and M. Matsui, *Tetrahedron*, **24**, 3095 (1968).

[63] R. C. Cambie and T. J. Fullerton, *Aust. J. Chem.*, **24**, 2611 (1971).

[64] R. M. Carman and H. C. Deeth, *Aust. J. Chem.*, **20**, 2789 (1967).

[65] K. Mori and M. Matsui, *Tetrahedron*, **24**, 6573 (1968).

[66] K. Mori, Y. Nakahara, and M. Matsui, *Tetrahedron Lett.*, **1970**, 2411.

[67] S. W. Pelletier, L. B. Hawley, and K. W. Gopinath, *Chem. Commun.*, **1967**, 96.

[68] D. R. Misra and H. N. Khastgir, *J. Indian Chem. Soc.*, **46**, 1063 (1969).

[69] O. P. Vig, A. S. Dhindsa, A. K. Vig, and O. P. Chugh, *J. Indian Chem. Soc.*, **49**, 163 (1972).

[70] M. Hudlicky, E. Kraus, J. Korbl, and M. Cech, *Collect. Czech. Chem. Commun.*, **34**, 833 (1969).

[71] E. J. Corey and H. A. Kirst, *J. Amer. Chem. Soc.*, **94**, 667 (1972).

[72] A. P. Krapcho and B. P. Mundy, *Tetrahedron*, **26**, 5437 (1970).

[73] E. E. van Tamelen and R. J. Anderson, *J. Amer. Chem. Soc.*, **94**, 8225 (1972).

[74] O. P. Vig, R. C. Anand, G. L. Kad, and J. M. Sehgal, *J. Indian. Chem. Soc.*, **47**, 999 (1970).

[75] W. S. Johnson, C. A. Harbert, and R. D. Stipanovic, *J. Amer. Chem. Soc.*, **90**, 5279 (1968).

[76] E. E. van Tamelen, M. P. Seiler, and W. Wierenga, *J. Amer. Chem. Soc.*, **94**, 8229 (1972).

[77] H. Cristol, D. Moers, and Y. Pietrasanta, *Bull. Soc. Chim. Fr.*, **1972**, 566.

[78] J. F. McGhie, W. A. Ross, J. W. Spence, and F. J. James, *Chem. Ind.* (London), **1972**, 536.

[79] K. V. Lichman, *J. Chem. Soc.*, *C*, **1971**, 2539.

[80] E. Winterfeldt, A. J. Gaskell, T. Korth, H. Radunz, and M.Walkowiak, *Chem. Ber.*, **102**, 3558 (1969).

[81] F. Texier and R. Carrie, *Bull. Soc. Chim. Fr.*, **1972**, 258.

[82] P. R. Atkins and I. T. Kay, *Chem. Commun.*, **1971**, 430.

[83] E. J. Corey and P. L. Fuchs, *J. Amer. Chem. Soc.*, **94**, 4014 (1972).
[84] G. Mandeville, F. Leyendecker, and J.-M. Conia, *Bull. Soc. Chim. Fr.*, **1973**, 963.
[85] H. Linde, *Helv. Chim. Acta*, **47**, 1234 (1964).
[86] H. Cohen and R. Schubart, *J. Org. Chem.*, **38**, 1424 (1973).
[87] D. H. R. Barton, P. G. Sammes, and M. Silva, *Tetrahedron, Suppl.* 7, 57 (1966).
[88] E. W. Thomas and T. I. Crowell, *J. Org. Chem.*, **37**, 744 (1972).

CHAPTER 3

ARYLATION OF UNSATURATED COMPOUNDS BY DIAZONIUM SALTS (THE MEERWEIN ARYLATION REACTION)

CHRISTIAN S. RONDESTVEDT, JR.*

Jackson Laboratory, E. I. du Pont de Nemours and Company, Inc., Wilmington, Delaware

CONTENTS

* I am greatly indebted to Carolyn Sidor, du Pont Experimental Station, for invaluable assistance in the later stages of the literature survey.

INTRODUCTION

The arylation of unsaturated compounds by diazonium salts with copper salt catalysis was first disclosed by Hans Meerwein.[1, 2] Meerwein arylation proceeds best when the double bond is activated by an electron-attracting group Z, such as carbonyl, cyano, aryl, vinyl, ethynyl, or chloro. The net result is the union of the aryl group with the carbon atom *beta* to the activating group, either by substitution of a β-hydrogen atom or by addition of Ar and Cl to the double bond.

$$ArN_2Cl + RCH{=}CRZ \xrightarrow[\text{salt}]{\text{Copper}} ArCR{=}CRZ \text{ and/or } ArCHRC(R)ClZ$$

The generally accepted mechanism of the reaction involves the aryl radical Ar· from the diazonium salt, though the manner of its formation and its subsequent reaction is still controversial.

The reaction was reviewed in *Organic Reactions* in 1960[3] and again in Russia in 1962.[4] Since the initial review, the Meerwein arylation has found extensive use in synthesis. No really new applications of the general reaction have been described, though the more than 150 papers

---

[1] H. Meerwein, E, Buchner, and K. van Emster, *J. Prakt. Chem.* [2] **152**, 239 (1939).

[2] R. Criegee, *Angew. Chem., Int. Ed. Engl.*, **5**, 333 (1966), presents an obituary of Hans Meerwein.

[3] C. S. Rondestvedt, Jr. *Org. Reactions*, **11**, 189 (1960).

[4] A. V. Dombrovskii, *Reakts. Metody Issled. Org. Soedin.*, **11**, 285 (1962) [*C.A.*, **59**, 7329d (1963)].

published since 1958, mostly in Russian,* have broadened its scope. This chapter follows the presentation of the original one, which should be kept at hand during the reading, and consists of comments describing the advances in the art since 1958. The Tabular Survey includes all significant new work known to the author.

MECHANISM

The Meerwein arylation is one example of the very general group of redox-modulated radical additions to olefins.[5] A cationic mechanism advocated at one time is no longer accepted.[3] A simplified mechanism accounts for the main features of the reaction.

(1) $2CuCl_2 + CH_3COCH_3$ (solvent) $\rightarrow 2CuCl + HCl + ClCH_2COCH_3$

(2) $ArN_2Cl + CuCl \rightarrow ArN_2 \cdot \rightarrow Ar \cdot + N_2 + CuCl_2$

(3) $Ar \cdot + CH_2{=}CHZ \rightarrow ArCH_2CH(Z) \cdot$ †

(4) $ArCH_2CH(Z) \cdot + CuCl_2 \rightarrow ArCH_2CH(Z)Cl + CuCl$
$ {}^{\searrow}ArCH{=}CHZ + CuCl + HCl$

(5) $Ar \cdot + CuCl_2 \rightarrow \underline{ArCl} + CuCl$ (Sandmeyer)

(6) $Ar \cdot + CH_3COCH_3 \rightarrow \underline{ArH} + \cdot CH_2COCH_3 +$ products containing the acetonyl group, including $ArCH_2COCH_3$‡

As noted by the underlining above, chloroacetone, ArCl, and ArH are by-products always encountered in Meerwein arylations. Indeed, the chief challenge in improving yields is minimization of the competitive side reactions leading to these products.

Meerwein's classical conditions involve aqueous acetone solvent and

---

* The majority of papers since 1958 on the Meerwein arylation have appeared in Russian journals. The principal Russian journals were available to the writer in English translation and are cited with their titles in English; the page reference is to the translation. *Chemical Abstracts* references to these are given for the convenience of those who do not have access to the English versions. Where the Russian title of the journal is cited, no translation was available and only the abstract was consulted.

† The radical $ArCH_2CHZ \cdot$ has been detected by electron spin resonance (esr) spectroscopy in mixtures of ArN_2BF_4 and $CH_2{=}CHZ$ after reduction by a one-electron reducing agent.[6]

‡ The reported $ArCH_2COCH_3$ may arise through addition to a low equilibrium concentration of acetone enol.[7, 8]

[5] F. Minisci, *Accts. Chem. Res.*, **8**, 165 (1975), presents a recent review. Many books on free-radical chemistry discuss the general subject briefly. Some newer methods of radical generation are summarized by F. Minisci and O. Porta, *Adv. Heterocycl. Chem.*, **16**, 123 (1974).

[6] A. L. J. Beckwith and M. D. Lawton, *J. Chem. Soc., Perkin II*, **1973**, 2134.

[7] M. Allard and J. Levisalles, *Bull. Soc. Chim. France*, **1972**, 1921.

[8] M. Allard and J. Levisalles, *Bull. Soc. Chim. France*, **1972**, 1926.

cupric chloride catalyst. Reduction of cupric chloride to cuprous chloride by acetone is well established. Most authors therefore ascribe the initiating step to a one-electron reduction of the diazonium salt by *cuprous* chloride. This may be correct in many cases, but it cannot be so in useful solvents that do not reduce cupric chloride, such as water, acetonitrile N-methylpyrrolidone, or sulfolane.[3] Moreover, numerous papers mention that cupric chloride is effective and cuprous chloride is not, for certain systems.

Many of the unsaturated compounds used in the Meerwein arylation are well-known vinyl monomers, yet vinyl polymers are not formed. This feature of the reaction was explained by the extraordinarily potent chain-transfer properties of cupric chloride.[3] Recently, telomers with the general formula $Ar(CH_2CHZ)_nCl$ have been obtained in low yields under conditions of high vinyl monomer concentration.[9, 10]

A serious problem with the simplified mechanism is its requirement that the free aryl radical be created before the monomer becomes involved. Yet the system diazonium salt-copper salt in aqueous organic solvent is normally stable for some time below room temperature. Only when the olefin is added does nitrogen evolution commence. To account for this behavior a ternary complex of the three reagents was proposed.[11] The stable complex bis(acrylonitrile)nickel(0) yields Meerwein products with diazonium salts, and adducts of acrolein, acrylonitrile, and various dienes with cuprous chloride have been characterized as π complexes with the C=C.[12] On the other hand, acrylonitrile forms complexes with *cupric* chloride through the cyano nitrogen.[12] The polarographic and spectroscopic behavior of diazonium solutions containing cupric chloride is interpreted in terms of a complex $(ArN_2)_2CuCl_4$[13] or $(ArN_2CuCl)^+Cl^-$.[10] The latter salt is believed to react with olefin (butadiene) by displacement of the inner-sphere chloride to give $(ArN_2CuC_4H_6)^{2+}2Cl^-$; an internal electron shift expels nitrogen and forms the Ar—C bond within the complex.[10] Only those radicals that become "free" react with the medium to yield ArCl and ArH.

[9] (a) R. Kh. Freidlina, B. V. Kopylova, and L. V. Yashkina, *Dokl. Chem.*, **183**, 1093 (1968) [*C.A.*, **70**, 67834p (1969)]; (b) B. V. Kopylova, L. V. Yashkina, and R. Kh. Freidlina, *Bull. Acad. Sci. USSR, Ser. Chem.*, **1971**, 160 [*C.A.*, **74**, 125009u (1971)]; (c) B. V. Kopylova, V. I. Dostovalova, and R. Kh. Freidlina, *ibid.*, **1971**, 957 [*C.A.*, **76**, 33922z (1972)]; (d) B. V. Kopylova, L. V. Yashkina, and R. Kh. Freidlina, *ibid.*, **1972**, 940 [*C.A.*, **77**, 100616p (1972)].

[10] N. I. Ganushchak, V. D. Golik, and I. V. Migaichuk, *J. Org. Chem. USSR*, **8**, 2403 (1972) [*C.A.*, **78**, 123556d (1973)].

[11] C. S. Rondestvedt, Jr., and O. Vogl., *J. Amer. Chem. Soc.*, **77**, 2313 (1955).

[12] G. N. Schrauzer, *Chem. Ber.*, **94**, 1891 (1961); G. N. Schrauzer and S. Eichler, *ibid.*, **95**, 260 (1962).

[13] A. I. Lopushanskaya, A. V. Dombrovskii, and V. I. Laba, *J. Gen. Chem. USSR*, **30**, 2028 (1960) [*C.A.*, **55**, 6416h (1961)].

The intermediate complex mechanism was criticized on the basis that the ratio of ArCl (Sandmeyer by-product) to ArH (hydrogen abstraction by the radical from the solvent) is not altered by the addition of olefin.[14] Supposedly this result could occur only if the Meerwein, Sandmeyer, and abstraction reactions involve a common intermediate assumed to be the *free* aryl radical. Related results were reported for the decomposition of *p*-nitrobenzenediazonium fluoborate catalyzed by tetrakis(acetonitrile) copper(I) perchlorate in the presence of methyl iodide as a radical trap, and these were interpreted in terms of ArCu(II) and Ar$_2$Cu(III) species.[15] However, if the Sandmeyer and abstraction reactions occur only with aryl radicals which escape from a complex, while the Meerwein reaction occurs chiefly with complexed radicals, the ratio ArCl/ArH would also remain constant.

An alternative to the preceding formulations of the intermediate complex can be built up as follows. The olefin donates an electron to cupric chloride to form a charge-transfer complex of cuprous chloride and a radical cation. This complex transfers an electron to the diazonium cation to give a diazonium radical that promptly loses nitrogen. The resulting aryl radical attacks the nearby radical cation to yield a carbocation that either loses a proton or acquires chloride ion. Details of the

$$ZCH{=\!=}CH_2 + CuCl_2 \rightleftharpoons (Z\overset{+}{C}H\overset{\cdot}{C}H_2CuCl_2{}^-) \xrightarrow{\;ArN_2{}^+\;} (Z\overset{+}{C}H\overset{\cdot}{C}H_2CuCl_2Ar\cdot) + N_2$$

$$\downarrow$$

$$Products \longleftarrow Z\overset{+}{C}HCH_2Ar + CuCl_2$$

bonding and geometry within the complex are unknown at this time. In this formulation it is unnecessary to supply acetone to reduce cupric to cuprous copper formulation the olefin supplies the electron.

This proposal predicts that the more electron-rich olefins would be better electron donors and thus more reactive in the Meerwein arylation (or should yield less of the by-products). Equivalently, the stability of the radical cation would determine the monomer's initiating ability. The proposal shows the diazonium cation being reduced by complexed cuprous copper; in principle, such a function could be assumed by another reducing agent so that copper could be eliminated entirely. This possibility has been realized in the arylation of quinones;[3] here a small amount of hydroquinone generates a semiquinone radical that reduces diazonium

[14] S. C. Dickerman, D. J. DeSouza, and N. Jacobson, *J. Org. Chem.*, **34**, 710 (1969).
[15] T. Cohen, R. J. Lewarchik, and J. Z. Tarino, *J. Amer. Chem. Soc.*, **96**, 7753 (1974).

cation to radical.[16] This radical adds to quinone, and the adduct radical (semiquinone) is aromatized by more quinone; copper is not needed for the last step, in contrast to typical Meerwein adduct radical cations. Yields are normally very high. Since tetraphenylethylene is effective in

$$Q + H_2Q \rightleftharpoons 2QH\cdot$$

$$QH\cdot + ArN_2{}^+ \longrightarrow QH^+ + Ar\cdot + N_2 \longrightarrow ArQH^+ \xrightarrow{Q} ArQ(-H) + QH\cdot + H^+$$

initiating diazoalkane decomposition (in place of the usual copper salt),[8] it may be applicable in certain Meerwein arylations as well.

The foregoing discussion emphasizes the uncertainties in current mechanistic knowledge. The various proposals can serve as working hypotheses for future research. In particular, studies in the absence of acetone are essential, for, in many cases, acetone actually reduces the yield of Meerwein product.[3] In most research, diazonium *halides* have been used; other anions that do not complex copper may sometimes prove beneficial. Some recent work has used acylate or inorganic oxyanions, though the yields were lower than in the presence of chloride.[17, 18] The influence of pH on yield has frequently been noted but never explained; for example, in some work, most diazonium salts give best results at pH 3–4, whereas bisdiazonium salts (*e.g.*, from benzidine) and *p*-nitrobenzenediazonium salts give best results near pH 1.

It seems likely that the Meerwein arylation is really a group of reactions governed by several related mechanisms, not just one.

SCOPE AND LIMITATIONS

The Unsaturated Component

Although many new examples of the Meerwein arylation have been reported since 1960, the list of olefins arylated has not been greatly expanded. Of the olefins studied, the greatest emphasis has been placed on dienes, including the methyl-, dimethyl- and aryl-butadienes and various chlorobutadienes. With dienes and aryldiazonium *chlorides*, the products are usually 1-aryl-4-*chloro*butenes. However, in the absence of chloride ion, 1-aryl-4-hydroxy- or 1-aryl-4-acetoxy-butenes may be prepared.[17, 18] Cycloheptatriene reacts as a conjugated diene, but only

[16] A. A. Matnishyan, G. V. Fomin, E. V. Prut, B. I. Liogon'kii, and A. A. Berlin, *Russ. J. Phys. Chem.*, **45**, 745 (1971); *Bull. Acad. Sci. USSR, Ser. Chem.*, **1972**, 1961 [*C.A.*, **78**, 28742c (1973)]. See also A. N. Grinev, N. V. Arkhangel'skaya, and G. Ya. Uretskaya, *J. Org. Chem. USSR*, 5, 1434 (1969) [*C.A.*, **71**, 112678z (1969)].

[17] N. I. Ganushchak, B. D. Grishchuk, and A. V. Dombrovskii, *J. Org. Chem. USSR*, **9**, 1030 (1973) [*C.A.*, **79**, 52909d (1973)].

[18] N. I. Ganushchak, B. D. Grishchuk, V. A. Baranov, and A. V. Dombrovskii, *J. Org. Chem. USSR*, **9**, 2157 (1973) [*C.A.*, **80**, 36809m (1974)].

monoarylation was observed. Ethylene reacts poorly.[3, 9] It would be instructive to examine bicycloheptene, whose strained double bond is highly reactive in cycloadditions, and to look for transannular interactions in compounds like bicycloheptadiene or 1,5-cyclooctadiene. Acrolein and crotonaldehyde, whose omission was noted previously,[3] have now been arylated. Enol esters and enol ethers react normally; their arylation affords an indirect method of arylating aldehydes and ketones on the α-carbon atom. Unsaturated sulfonic acids, sulfones, and phosphonate

$$RCOCH_3 \longrightarrow R'OCR{=}CH_2 \xrightarrow{ArN_2X} R'OCRXCH_2Ar \longrightarrow RCOCH_2Ar$$

$$R' = alkyl, acyl$$

esters give the expected products. The CH$=$N bond in aldoximes[3] or aldehyde phenylhydrazones[19, 20] is arylated *without copper* by replacement of the aldehyde hydrogen. However, enamines undergo azo coupling at the α position.[21] Heterocyclic N-oxides are ring-arylated.[22, 23] Unsaturated phosphorus compounds, $R_3P{=}CR_2$, have not been evaluated and would make an interesting study.

Reactivities of Unsaturated Compounds

Most of the recent comparisons are drawn from competitive experiments with pairs of olefins against a given diazonium salt.[14, 24] With the olefins styrene, acrylonitrile, methacrylonitrile, methyl acrylate, and methyl methacrylate, different reactivity orders were obtained according to whether the diazonium salt (the aryl radical) bore an electron-attracting (nitro, chloro) or electron-releasing (methyl, methoxyl) group. Thus styrene was more reactive than acrylonitrile toward the p-nitrophenyl radical, but acrylonitrile was more reactive than styrene toward tolyl and p-anisyl radicals. The extremes in the ratio (yield from styrene/ yield from acrylonitrile) are 5.0 and 0.67. "Reactivity" was defined in

[19] W. Reid, K. Sommer, and H. Dickhäuser, *Angew. Chem.*, **67**, 705 (1955).

[20] W. Ried and K. Sommer, *Ann.*, **611**, 108 (1958).

[21] A. G. Cook, Ed., *Enamines*, Dekker, New York, 1969, pp. 158–160; 414–415. See also V. I. Shvedov, L. B. Altukhina, and A. N. Grinev, *J. Org. Chem. USSR*, **1**, 882 (1965) [*C.A.*, **63**, 2928g, 6893h (1965)].

[22] M. Natsume, S. Kumadaki, and R. Tanaba, *Itsuu Kenkyosho Nempo*, **1971**, 25 [*C.A.*, **77**, 61765g (1972)].

[23] M. Colonna, *Atti Accad. Naz. Lincei, Rend., Cl. Sci. Fis., Mat. Nat.*, **26**, 39 (1959)[*C.A.*, **53**, 21929e (1959)].

[24] S. C. Dickerman, D. J. DeSouza, M. Fryd, I. S. Megna, and M. M. Skoultchi, *J. Org. Chem.*, **34**, 714 (1969).

terms of yield; since the yields did not total 100%, these ratios are probably inaccurate. Moreover, the competitive experiments were apparently performed in unbuffered solutions (pH ∼ 1), and yields may vary with pH depending upon the substituent in the aryl radical.

Butadiene and isoprene are chloroarylated satisfactorily at −10 to +5°, whereas 2,3-dimethylbutadiene is chloroarylated only at 30–40°.[25, 26] If the ternary complex mechanism were invoked, one might predict that dienes would react in the *cis* (Z) form; the Z form is not so easily reached with 2,3-dimethylbutadiene because the methyl groups interfere.

Previously the products from isoprene were written as $ArCH_2CH=C$-$(CH_3)CH_2Cl$ on the basis of an ambiguous oxidation procedure.[3] Subsequently this structure was revised to $ArCH_2C(CH_3)=CHCH_2Cl$,[25, 27] which was later supported by nuclear magnetic resonance evidence.[28] The Meerwein adducts from chloroprene were written similarly. No explanation is apparent for attack by the aryl radical on both dienes at the terminus closer to the dissimilar methyl and chloro substituents.

Ene-ynes are arylated exclusively at the ene terminus unless it already bears a substituent. In that case, arylation can occur at either terminus. Though the products have been written as 1,2 adducts, $Ar-C-C(Cl)$-$C≡C$,[29, 30] more detailed study has disclosed the presence of allenic 1,4 adducts, $Ar-C-C=C=C-Cl$, in many cases.[31]

The Diazonium Salt

The earlier discussion[3] is essentially valid today, although negatively substituted diazonium salts do not invariably give higher yields than those bearing electron-releasing groups. Numerous new examples of

[25] A. V. Dombrovskii and N. I. Ganushchak, *J. Gen. Chem. USSR*, **31**, 1191 (1961) [*C.A.*, **55**, 23387d (1961)]. See also Ref. 27.

[26] N. I. Ganushchak, M. M. Yukhomenko, M. D. Stadnichuk, and A. V. Dombrovskii *J. Gen. Chem. USSR*, **34**, 2249 (1964) [*C.A.*, **61**, 10622g (1964)].

[27] A. V. Dombrovskii and N. I. Ganushchak, *Ukr. Khim. Zh.*, **24**, 217 (1958) [*C.A.*, **52**, 18271e (1958)].

[28] N. I. Ganushchak, K. G. Zolotukhina, M. D. Stadnichuk, N. I. Malashchuk, and A. V. Dombrovskii, *J. Org. Chem. USSR*, **4**, 214 (1967) [*C.A.*, **68**, 104940b (1968)].

[29] A. V. Dombrovskii, *J. Gen. Chem. USSR*, **27**, 3080 (1957) [*C.A.*, **52**, 8087d (1958)].

[30] L. G. Grigoryan, F. A. Martirosyan, and V. O. Babayan, *Sb. Nauch. Tr., Erevan. Arm. Gos. Pedagog. Inst., Khim.*, **1970** [1], 23–28 [*C.A.*, **78**, 42923u (1973)].

[31] (a) A. A. Petrov, Kh. V. Bal'yan, Yu. I. Kheruze, and T. V. Yakovleva, *J. Gen. Chem. USSR*, **29**, 2071 (1959) [*C.A.*, **54**, 8677g (1960)]; (b) Yu. I. Keruze and A. A. Petrov, *ibid.*, **30**, 2511 (1960) [*C.A.*, **55**, 21002c (1961)]; (c) Yu. I. Kheruze and A. A. Petrov, *ibid.*, **31**, 708 (1961) [*C.A.*, **55**, 23385g (1961)]; (d) Yu. I. Kheruze and A. A. Petrov, *ibid.*, **31**, 389 (1961) [*C.A.*, **55**, 23330a (1961)].

bisdiazonium salts have been reported.[28, 32-34] The "reactivities" observed in competitive studies have been correlated with the Hammett σ° parameter; however, p-nitrophenyl deviated considerably from the correlation line, perhaps because of the pH factor mentioned above.[24]

Though the Meerwein arylation is now limited to aromatic amines that yield stable diazonium salts, it may be applicable to the rather stable vinylic diazonium salts recently prepared.[35, 36] A general extension of the reaction to "alkylation" will probably use alkyl radicals derived from halides by redox initiation,[5,37a] organopalladium compounds ($RPdXL_n$), or organocopper reagents ($Li^+R_2Cu^-$)[37b] that add conjugatively to activated olefins.

Reagents other than diazonium salts may serve as sources of aryl radicals. Most of the reagents used in arylation of aromatic rings, such as benzoyl peroxide, 1-aryl-3,3-dimethyltriazenes, and N-nitrosoacetanilides, give poor yields in Meerwein arylations,[3] in part because they initiate vinyl polymerization of most substrates. Diaryliodonium salts fail to give Meerwein arylations of olefins,[38] contrary to the earlier report.[1] Certain "stabilized" diazonium salts, such as diazoaminobenzenes or salts with complex anions,[39-41] can be used in place of ordinary diazonium salt solutions. Since diazonium fluoroborates do arylate aromatic rings in dimethyl sulfoxide solution,[42] especially with added sodium nitrite,[39] this technique, which requires no copper catalyst, may be applicable also to Meerwein arylations. Phenylhydrazones may be autoxidized to compounds of the type $PhN=NCR_2OOH$, which decompose to phenyl radicals in the presence of copper or iron salts at -20 to $0°$.

[32] N. I. Ganushchak, K. G. Zolotukhina, and A. V. Dombrovskii, *J. Org. Chem. USSR*, **2**, 1058 (1966) [*C.A.*, **65**, 18510f (1966)].

[33] N. I. Ganushchak, V. A. Vengrzhanovskii, and N. M. Mel'nik, *J. Org. Chem. USSR*, **6**, 787 (1970) [*C.A.*, **73**, 14323b (1970)].

[34] N. I. Ganushchak, B. D. Grishchuk, K. G. Tashchuk, A. Yu. Nemish, and A. V. Dombrovskii, *J. Org. Chem. USSR*, **8**, 2597 (1972) [*C.A.*, **79**, 18249d (1973)].

[35] K. Bott, *Tetrahedron Lett.*, **1971**, 2227.

[36] K. Bott, *Tetrahedron Lett.*, **1968**, 4979; *Chem. Ber.*, **103**, 3850 (1970).

[37a] A. Or, M. Levy, M. Asscher, and D. Vofsi, *J. Chem. Soc., Perkin II*, **1974**, 857. This paper provides entry to Asscher and Vofsi's earlier work. For an extensive review of telomerization with carbon tetrachloride, including the work of Freidlina, see C. M. Starks, *Free Radical Telomerization*, Academic Press, New York, 1974, especially Chapter 5. Radical addition of R-H to activated olefins may be promoted by Mn(III)-Cu(II) couple; for example, see M. G. Vinogradov, T. M. Federova, and G. I. Nikissin, *Bull. Acad. Sci. USSR, Ser. Chem.*, **1974**, 2384 [*C.A.*, **82**, 30921m (1975)].

[37b] G. H. Posner, *Org. Reactions*, **19**, 1 (1972); S. B. Bowlus, *Tetrahedron Lett.*, **1975**, 3591; H. O. House, *Accts. Chem. Res.*, **9**, 59 (1976).

[38] F. M. Beringer and P. Bodlaender, *J. Org. Chem.*, **34**, 1981 (1969).

[39] M. Kobayashi, H. Minato, N. Kobori, and E. Yamada, *Bull. Chem. Soc. Jap.*, **43**, 1131 (1970).

[40] S. Kojima, *Kogyo Kagaku Zasshi*, **64**, 1984 (1961) [*C.A.*, **57**, 2111b (1962)].

[41] S. Kojima, *Kogyo Kagaku Zasshi*, **64**, 2075 (1961) [*C.A.*, **57**, 2111c (1962)].

These radicals phenylate unsaturated compounds.[5, 43] But, since phenyl-hydrazine is prepared by reduction of benzenediazonium salts, the yield by the phenylhydrazine route must be much better than by conventional Meerwein arylation to justify its use.

A reaction synthetically equivalent to the Meerwein arylation, yet independent of aromatic amines, involves oxidative addition of ArHal (and vinylic halides) to palladium(O) complexes. The resulting $ArPdHalL_2$ (L = ligand) will substitute an aryl group for hydrogen in olefins such as propene, styrene, and methyl acrylate to yield the same product obtained in Meerwein arylations.[44] In its present state of development this reaction appears to complement the Meerwein arylation.

$$ArCl + PdL_2 \rightarrow ArPdClL_2$$
$$ArPdClL_2 + CH_2{=}CHZ \rightarrow ArCH{=}CHZ + HCl + PdL_2$$

Factors Influencing Addition vs. Substitution

α-Bromostyrene reacts with arenediazonium *chlorides* to yield α-*chloro*stilbenes and bromide ion.[45] This shows that substitution products

$$C_6H_5CBr{=}CH_2 + ArN_2Cl \rightarrow (C_6H_5CBrClCH_2Ar) \xrightarrow{-HBr} C_6H_5CCl{=}CHAr$$

can be formed from addition products, but it is probably not the general rule.

Side Reactions

A recent study with *p*-chlorobenzenediazonium chloride and five diverse olefins[14, 24] showed that the best yield of Meerwein product was obtained with low cupric ion concentrations. The major side product under these conditions was ArH, while at higher cupric ion concentrations the Sandmeyer product ArCl became more prominent. These conclusions may not apply with other diazonium salts, however. ArH formation is usually ascribed to hydrogen abstraction from a reactive C–H bond (as in acetone),[46] but it has also been ascribed to reduction of diazonium ion by cuprous ion.[25] To phrase it differently, diazonium ion is an oxidizing agent and so ensures that the concentration of *cuprous* ion does not exceed a certain low steady state.

[42] B. L. Kaul and H. Zollinger, *Helv. Chim. Acta*, **51**, 2132 (1968).

[43] F. Minisci and U. Pallini, *Gazz. Chim. Ital.*, **90**, 1318 (1960).

[44] R. F. Heck, *J. Amer. Chem. Soc.*, **96**, 1133 (1974), and P. M. Henry, *Adv. Organomet. Chem.*, **13**, 363 (1975) give leading references. See also M. Watanabe, M. Yamamura, I. Moritani, Y. Fujiwara, and A. Sonoda, *Bull. Chem. Soc. Jap.*, **47**, 1035 (1974).

[45] A. V. Dombrovskii and K. G. Tashchuk, *J. Gen. Chem. USSR*, **34**, 3393 (1964) [*C.A.*, **62**, 3958f (1965)].

[46] T. Cohen and J. U. Tirpak, *Tetrahedron Lett.*, **1975**, 143, and A. H. Lewin and R. J. Michl, *J. Org. Chem.*, **38**, 1126 (1973), provide leading references to recent work on copper catalysis and organocopper compounds. See also A. E. Jukes, *Adv. Organomet. Chem.*, **12**, 215 (1974).

"Diazo resins" frequently accompany the Meerwein product, especially at more alkaline pH. It has been shown repeatedly that most diazonium salts function best (give higher yields) at pH 3–5, whereas nitro-substituted and bisdiazonium salts function best at pH 1–2.[47–49] This effect may be related to foreign anions. It should be noted that the neutralization of the strongly acidic diazotization solution with sodium bicarbonate or calcium oxide usually affords higher yields. This result is in contrast with the sodium acetate of the classical Meerwein conditions.

The well-known ability of an aromatic nitro group to function as a radical trap (e.g., polymerization inhibition) is occasionally manifested with o-nitrobenzenediazonium ion adducts. The intermediate adduct radical may cyclize to the nitro group, and this may account for lower yields of Meerwein adducts from o-nitrobenzenediazonium salts than from the corresponding m- and p-isomers.[6] The cyclic nitroxide radical has been detected by electron spin resonance spectroscopy.[6]

One anticipated side reaction, telomerization, has now been observed. When diazonium salt was added slowly to a large excess of ethylene or acrylonitrile under Meerwein conditions, the yield of the primary adduct $ArCH_2CHZCl$ was considerably reduced, and telomers were formed in low yields (0.7 % with ethylene, 16 % with acrylonitrile and o-toluenediazonium chloride).[9] Under standard Meerwein conditions, with diazonium salt and olefin present in equimolar quantities, telomerization seems to be insignificant.

$$ArN_2Cl + CH_2{=}CHZ(excess) \rightarrow ArCH_2CHZCl + Ar(CH_2CHZ)_nCl$$

Copper-promoted diazonium salts may initiate acrylonitrile polymerization, especially in the absence of chloride ion.[50]

[47] A. V. Dombrovskii and N. I. Ganushchak, *J. Gen. Chem. USSR*, **31**, 1774 (1961) [*C.A.*, **55**, 24675e (1961)].

[48] A. V. Dombrovskii, Ya. G. Bal'on, and K. G. Tashchuk, *J. Gen. Chem. USSR*, **32**, 592 (1962) [*C.A.*, **58**, 1383e (1963)].

[49] K. G. Tashchuk and A. V. Dombrovskii, *J. Org. Chem. USSR*, **1**, 2034 (1965) [*C.A.*, **64**, 9617a (1966)].

[50] S. C. Chiang, K. C. Liu, H. Y. Hung, and Y. H. Chang, *Ko Fen Tzu T'ung Hsun*, **7**, 79 (1965) [*C.A.*, **64**, 3691c (1966)].

SYNTHETIC APPLICATIONS OF THE MEERWEIN ARYLATION

Since 1958, Meerwein adducts $ArCH_2CHClZ$ and related compounds have been converted to a wide variety of useful materials. The allylic halides $ArCH_2CH=CHCH_2Cl$ from dienes can be treated with a great variety of nucleophiles to yield numerous derivatives such as $ArCH_2$-$CH=CHCH_2NR_2$ and $ArCH_2CH=CHCH_2CH(CO_2Et)_2$. When phosphines are alkylated, the derivatives can be carried through the Wittig reaction to yield new conjugated dienes.[28, 32, 51-64] The adducts with vinyl chloride ($ArCH_2CHCl_2$) yield arylacetaldehydes on hydrolysis;[65-66] the vinyl acetate adducts $ArCH_2CHClOAc$ are more readily hydrolyzed to the same aldehydes.[67] Similarly, the adducts from vinylidene chloride ($ArCH_2CCl_3$) yield arylacetic acids.[68]

The adducts from acrylic acid derivatives ($ArCH_2CHClZ$) have been aminated to an extensive series of ring-substituted phenylalanines

[51] A. V. Dombrovskii and A. P. Terent'ev, J. Gen. Chem. USSR, 26, 3091 (1956) [C.A., 51, 7337c (1957)].

[52] N. I. Ganushchak, K. G. Zolotukhina, and A. V. Dombrovskii, J. Org. Chem. USSR, 5, 301 (1969) [C.A., 70, 106080m (1969)].

[53] N. I. Ganushchak, M. M. Yukhomenko, and A. V. Dombrovskii, Dopov. Akad. Nauk Ukr. RSR, 1962 [2], 211 [C.A., 58, 2400h (1963)].

[54] K. G. Zolotukhina, N. I. Ganushchak, M. M. Yukhomenko, A. V. Dombrovskii, J. Gen. Chem. USSR, 33, 1197 (1963) [C.A., 59, 9978e (1963)].

[55] M. M. Yukhomenko, N. I. Ganushchak, and A. V. Dombrovskii, J. Gen. Chem. USSR, 33, 2464 (1963) [C.A., 60, 521g (1964)].

[56] N. O. Pastushak, N. F. Stadniichuk, and A. V. Dombrovskii, J. Gen. Chem. USSR, 33, 2877 (1963) [C.A., 60, 1639g (1964)].

[57] M. M. Yukhomenko, N. I. Ganushchak, and A. V. Dombrovskii, Ukr. Khim. Zh., 32, 61 (1966) [C.A., 64, 19401b (1966)].

[58] N. I. Ganushchak, M. M. Yukhomenko, M. D. Stadnichuk, and M. I. Shevchuk, J. Gen. Chem. USSR, 36, 1164 (1966) [C.A., 65, 10613f (1966)].

[59] M. M. Yukhomenko and A. V. Dombrovskii, Ukr. Khim. Zh., 33, 76 (1967); [C.A., 66, 115373d (1967)].

[60] A. F. Tolochko, N. I. Ganushchak, and A. V. Dombrovskii, J. Gen. Chem. USSR, 38, 1068 (1968) [C.A., 69, 106825n (1968)].

[61] N. I. Ganushchak and A. V. Dombrovskii, USSR Pat. 255,255 [C.A., 72, 121158c (1970)].

[62] A. V. Dombrovskii, L. G. Pribytkova, N. I. Ganushchak, and V. A. Vengrzhanovskii, J. Org. Chem. USSR, 6, 969 (1970) [C.A., 73, 34990v (1970)].

[63] N. I. Ganushchak, B. D. Grishchuk, and A. V. Dombrovskii, USSR Pat. 363,734; [C.A., 78, 159169u (1973)].

[64] V. O. Babayan, L. G. Grigoryan, and S. V. Toganyan, Armen. Khim. Zh., 22, 805 (1969) [C.A., 72, 31390t (1970)].

[65] V. M. Naidan and A. V. Dombrovskii, J. Gen. Chem. USSR, 34, 3391 (1964) [C.A., 62, 3958e (1965)].

[66] V. M. Naidan, N. V. Dzumedzei, and A. V. Dombrovskii, J. Org. Chem. USSR, 1, 1395 (1965) [C.A., 64, 721a (1966)].

[67] V. M. Naidan and G. D. Naidan, J. Org. Chem. USSR, 8, 2216 (1972) [C.A., 78, 42962f (1973)].

[68] A. V. Dombrovskii and V. M. Naidan, J. Gen. Chem. USSR, 32, 1256 (1962) [C.A., 58, 1383g (1963)]; see also ibid., 34, 1474 (1964) [C.A., 61, 5554g (1964) and 62, 10353f (1965)].

$ArCH_2CH(NH_2)CO_2H$ after transformation of group Z to a carboxyl group.[69-77]

EXPERIMENTAL CONDITIONS

No extensive research has been directed toward greater convenience and yields. Perhaps the most rapid progress will come from an understanding of the effect of pH and of buffer constituents on the reaction. Acetate ion, commonly used in the earlier studies,[3] may function as a ligand to copper and thus modify its properties.

The concentration of chloride or other anion is another parameter. Meerwein products have been obtained from diazonium acetates, fluoroborates, sulfates, and nitrates, in the absence of halides, though the yields are lower than for the corresponding reactions in the presence of halide.[3, 18, 67] On the other hand, addition of extra chloride ion may improve the yield; p-$CH_3C_6H_4CH_2CHClCN$ was prepared in 56% yield by the standard Meerwein conditions, but in 89% yield on addition of an extra mole of chloride ion.[78]

Low yields are usually attributable to copper-catalyzed side reactions that consume aryl radicals. Reduced copper concentrations tend to favor Meerwein arylation over the Sandmeyer reaction and hydrogen abstraction.[14, 24] Since ArH is probably formed by the reaction of Ar· with a readily abstracted hydrogen atom, solvents like ethers and alcohols with the grouping H—C—O are unsuitable in the Meerwein arylation. Despite the ease of hydrogen abstraction from acetone, most workers persist in using it because of availability, convenience, cost, and tradition. To the list of alternative solvents such as acetonitrile, N-methylpyrrolidone,[71] dimethyl sulfoxide, and sulfolane,[3] one might add formamide, whose use has not yet been reported. γ-Butyrolactone was suggested previously[3] but proved not to be useful.[71] Acetic acid improves yields, according to one report.[79]

Substantially all research on the Meerwein arylation has employed deliberately homogeneous solutions. A proposed two-phase system has

[69] A. M. Yurkevich, A. V. Dombrovskii, and A. P. Terent'ev, *J. Gen. Chem. USSR*, **28**, 226 (1958) [*C.A.*, **52**, 12797d (1958)].

[70] R. Filler and H. Novar, *Chem. Ind.* (London), **1960**, 468.

[71] R. Filler, L. Gorelic, and B. Taqui-Khan, *Proc. Chem. Soc.*, **1962**, 117.

[72] R. Filler and H. Novar, *J. Org. Chem.*, **26**, 2707 (1961).

[73] R. Filler, B. Taqui-Khan, and C. W. McMullen, *J. Org. Chem.*, **27**, 4660 (1962).

[74] R. Filler and W. Gustowski, *Nature*, **205**, 1105 (1965).

[75] R. Filler, A. B. White, B. Taqui-Khan, and L. Gorelic, *Can. J. Chem.*, **45**, 329 (1967).

[76] G. H. Cleland, *J. Org. Chem.*, **26**, 3362 (1961).

[77] G. H. Cleland, *J. Org. Chem.*, **34**, 744 (1969).

[78] Y. Mori and J. Tsuji, *Jap. Kokai*, **73**, 67,236 [*C.A.*, **80**, 3280d (1974)].

[79] J. R. Brepoels and J. M. Vaneghen, Ger. Offen. 2,016, 809 (Oct. 29, 1970) [*C.A.*, **74**, 42117d (1971)]; Belg. Pat. 741,640 [*C.A.*, **74**, 76154b (1971)].

not been tried.[3] A new field of study is suggested by the recent under-standing of reactions in micelles and reactions with phase-transfer catalysts.[80] Readers are reminded that emulsion polymerization is an extremely important (and successful) reaction in which radicals generated in an aqueous phase react in high yields in a micelle. Polymerization during a Meerwein arylation is controllable by the chain-transfer abilities of copper derivatives.

No alternative to copper salts has been demonstrated to be widely applicable, though ferrous iron is sometimes effective.[3, 67] Yet important side reactions are also associated with copper ions. Perhaps the desirable properties of copper could be separated from the undesirable by addition of a suitable ligand (compare Ref. 37), but those which complex too tightly to copper destroy the catalytic activity.[3, 10, 25] Recent work with organocopper reagents and the influence of copper on other organic reactions[15, 46] may suggest appropriate candidates. Alternative to copper may be nonmetallic reagents which reduce diazonium salts; tetraphenyl-ethylene and hydroquinone are two possibilities already mentioned.

CAVEAT. It is highly probable that the various experimental con-ditions are not truly independent, but rather are linked. Thus a change in pH will probably change the ionic/covalent proportion of the diazonium salt. Adjustment of pH is frequently accomplished with foreign anions, *e.g.*, acetate, which may function as ligands for copper. A change in solvent will change the dielectric constant and solvating power of the medium and the activities of ions. Thus arylation does not proceed in *dry* acetone, but it commences when water is added.[25] The experimenter will therefore save time by applying the principles of statistical design of experiments.* The approach in which all variables but one are held constant almost certainly will not lead to optimization of a procedure.

CONCLUSIONS

The Meerwein arylation is a valuable tool which will continue to be applied to the synthesis of compounds with an aryl group linked to aliphatic carbon. The availability and low cost of aromatic amines and activated olefins (including enol derivatives) readily offset the frequently

---

* O. L. Davies and P. L. Goldsmith, *Statistical Methods in Research and Production*, Hafner Publ. Co., New York, 1972, and O. L. Davies, *Design and Analysis of Industrial Experiments*, 2nd. Ed., Hafner, New York, 1967, provide excellent treatments of the statistical design of experiments. See also the important papers by G. E. P. Box and D. W. Behnken, *Techno-metrics*, **2**, 455–475 (1960). and by R. L. Plackett and J. P. Burman, *Biometrika*, **33**, 305–325 (1946).

[80] J. Dokcx, *Synthesis*, **1973**, 441; E. V. Dehmlow, *Angew. Chem., Int. Ed. Engl.*, **13**, 170 (1974); E. H. Cordes and R. B. Dunlap, *Accts. Chem. Res.*, **2**, 329 (1969), present reviews of phase-transfer catalysis.

low yields. The mechanism is not yet fully understood. Better understanding, and a systematic study of reaction variables, should result in greatly improved yields and wider applicability. Most importantly, the Meerwein products are frequently valuable synthons for further molecule building.

EXPERIMENTAL PROCEDURES

The following collection of procedures supplements those given earlier.[3] The present examples were chosen to illustrate new techniques or new classes of compounds.

α-Chloro-β-phenylpropionaldehyde from Acrolein.[81]

Aniline (37.2 g, 0.40 mol) in 92 ml of concentrated hydrochloric acid was diazotized in the usual way with 28 g (0.40 mol) of sodium nitrite. The solution was neutralized to pH 4 with sodium bicarbonate and diluted to 350 ml with ice water. A 1-l three-necked flask fitted with a stirrer, thermometer, dropping funnel, and an exit tube leading to a bubble counter was charged with 22.5 g (0.40 mol) of acrolein, 200 ml of acetone, 10 g of cupric chloride dihydrate, and 4 g of calcium oxide. The mixture was maintained at 0–2° while the diazonium solution was added from the dropping funnel at such a rate that nitrogen was evolved at 2–3 bubbles/second. The pH was monitored during the reaction, either continuously with a pH meter or intermittently with pH paper; sodium bicarbonate was added as required to maintain a pH of 5–6. Nitrogen evolution was complete in 2 hours. The mixture was extracted with ether, and the ether layer was washed with calcium chloride solution. The product was then vacuum-distilled; yield 35.0 g (52.8%); bp 104–106° (12 mm). The aldehyde readily formed a solid hydrate on exposure to water vapor.

1,1-Dichloro-2-phenylethane from Vinyl Chloride.[66] *CAUTION.*

Because vinyl chloride is a carcinogen, all operations with it should be carried out in a good hood, and any breathing of the gas should be avoided.

A four-necked flask equipped with stirrer, dropping funnel, thermometer, and gas inlet and outlet tubes was charged with 3 g of calcium hydroxide and 200 ml of acetone. The mixture was then saturated with vinyl chloride at − 20°. Benzenediazonium chloride solution, prepared as above from 0.2 mol of aniline, was neutralized to pH 4 with sodium bicarbonate, mixed with 8.4 g of cupric chloride dihydrate, and diluted to 160 ml with ice water. While vinyl chloride was passed in continuously at −5 to −7°, the diazonium solution was added to the flask during 30 minutes. The vinyl chloride stream was continued for an hour and then

[81] A. V. Dombrovskii, A. M. Yurkevich, and A. P. Terent'ev, *J. Gen. Chem. USSR*, **27**, 3077 (1957) [*C.A.*, **52**, 8087b (1958)].

stopped. The mixture was stirred for 3 hours while the temperature was allowed to rise slowly to 18–20°. The upper layer was extracted with ether, and the ether layers were washed three times with water and dried with calcium chloride. The product was vacuum-distilled; weight 18.4 g (52%); bp 77–78° (5 mm).

The reaction was conducted similarly with m-nitroaniline, but the reaction mixture was steam-distilled to remove solvent and by-products. The solid m-nitro derivative was crystallized from ethanol; yield 52%; mp 77°.

Procedures for converting the dichloro compounds to the α-chlorostyrenes with alcoholic potassium hydroxide and to the arylacetylenes with molten potassium hydroxide-sodium hydroxide eutectic are given in the reference.

The dichloro compounds can be converted to the arylacetaldehyde cyclic acetals be heating with potassium hydroxide in ethylene glycol. Acid hydrolysis yields the unstable arylacetaldehydes which are trapped as dinitrophenylhydrazones or as oximes.

Arylacetaldehydes may also be prepared by chloroarylation of vinyl acetate and hydrolysis. Yields are comparable.[67]

1,1,1-Trichloro-2-p-tolylethane from Vinylidene Chloride.[68] The foregoing procedure was followed, using 2 mol of vinylidene chloride (bp 30°) per mol of diazonium salt to compensate for the loss of the low-boiling olefin with the escaping nitrogen. The product was vacuum-distilled to give a yield of 67% of the title compound; bp 122° (11 mm).

Conversion to p-tolylacetic acid was accomplished by refluxing the trihalide with 4 equivalents of lead nitrate in 50% aqueous acetic acid for 8 hours. The acid was isolated by acidification, extraction with benzene, evaporation, and recrystallization from water; yield 69%; mp 94°. An alternative procedure uses mercuric oxide in acetic acid. The conventional method for converting a trichloromethylarene to carboxylic acid with concentrated sulfuric acid fails with most 1,1,1-trichloro-2-arylethanes because of ring sulfonation.

1-p-Nitrophenyl-2-chloroethane from Ethylene. Isolation of Telomers. (Illustration of an Autoclave Reaction).[9d] *SAFETY NOTE. High-pressure reactions must be conducted behind an appropriate barricade with remotely actuated gas inlet and vent valves. The critical temperature of ethylene is 9.7°, hence chilling the autoclave before admitting ethylene permits overcharging unless the ethylene cylinder can be weighed during charging. Refer to the autoclave manufacturer's operating instructions for proper techniques of closing the vessel.*

A 1-l agitated (rocker, shaker, or stirrer) stainless steel autoclave equipped with valved gas inlet and outlet tubes, a pressure gauge, and a *suitable rupture disk* was charged with 0.3 mol of p-nitrobenzenediazonium chloride solution, 8 g of cupric chloride dihydrate, 40 g of sodium acetate, and 250 ml of acetone. The autoclave was sealed and ethylene added above 10° to a gauge pressure of 70–75 atm. The autoclave was then heated to 42° and held for 3 hours. The autoclave was vented, and the reaction mixture taken up in ether, washed with dilute hydrochloric acid to remove copper salts, then with water, sodium carbonate solution, and water, and dried over calcium chloride. The residue after removal of the ether weighed 47.5 g; analysis by gas chromatography[76] with an internal standard gave 1-p-nitrophenyl-2-chloroethane in a yield of 45–50%.

The residue was fractionated under reduced pressure. The first fraction, bp 160–162° (3 mm), was recrystallized from ethanol to give 18 g (33%) of the chloroethane, mp 49–50°. The second fraction (*CAUTION: It may decompose on strong heating*), bp 175–185° (3 mm), was analyzed by gas chromatography*; it contained 4-p-nitrophenyl-1-chlorobutane (yield 4–5%) and 6-nitro-1,2,3,4-tetrahydronaphthalene.

This autoclave technique should also be applicable to vinyl chloride and vinylidene chloride arylations.

1-(5-Nitro-2-thiazolyl)-2-(2-pyridyl)ethene from 2-Vinylpyridine. (Low-Temperature Diazotization of a Heterocyclic Amine).[82]

A slurry of 145 g (1.0 mol) of 5-nitro-2-aminothiazole in 450 ml of 12 N hydrochloric acid and 100 ml of water was stirred at about −70° while 69 g (1.0 mol) of sodium nitrite in 100 ml of water was added during a half hour. The pale-green mixture was stirred 10 minutes longer, and 160 g (1.52 mol) of 2-vinylpyridine in 600 ml of acetone was added rapidly below −30°. Then 28 g of cupric chloride dihydrate was added, and after 10 minutes the mixture was allowed to warm. At −10° the green mixture became reddish and vigorous nitrogen evolution ensued. When nitrogen evolution ceased, the mixture was diluted with 500 ml of water, neutralized with sodium bicarbonate, mixed with dichloromethane, filtered, and separated. The aqueous phase was extracted again with dichloromethane, and the combined organic layers were dried over magnesium sulfate and evaporated. The viscous residue yielded 25.3 g of product on trituration with methanol, mp 179–182°. An additional 6 g

---

* Gas chromatographic conditions were not specified in detail. Three columns were mentioned: (1) silicone oil E-301 on Celite 545, 1 m, 185°; (2) poly(ethylglycol)adipate on brick, 1 m, 185°; 3) 15% Apiezon L on Chromosorb W, 2 m, 210°. The internal standard was 1-(2,4-dinitrophenyl)-4-chlorobutane.

82 G. Asato, *J. Org. Chem.*, **33**, 2544 (1968).

of less pure material was obtained by concentrating the filtrate. Crystallization from chloroform (activated carbon) yielded 25.4 g (10.5%) of the title compound, mp 180–183°.

An alternative synthesis from 2-methyl-5-nitrothiazole and pyridine-2-carboxaldehyde gave the same product in 65% yield.

6-β-p-Chlorophenyl-4-cholesten-3-one. (Arylation of an Enol Acetate).[8] A solution of 3-acetoxy-3,5-cholestadiene (from 4-cholestenone) (500 mg, 1.25 mmol) in 75 ml of acetone and 5 ml of dichloromethane was treated with the diazonium salt prepared from 2 g (15.7 mmol) of p-chloroaniline in 20 ml of water and 3 equivalents of hydrochloric acid (buffered with sodium acetate to pH 3). Then 20 ml of a catalyst solution, prepared under nitrogen from 5 g of cuprous chloride, 160 ml of acetone, 80 ml of water, and 4 ml of 12 N hydrochloric acid, was added. One hour after nitrogen evolution ceased, the acetone was evaporated and the crude product washed with water. It was chromatographed on 50 g of silica gel, using petroleum ether (30–60°)—ether (84:16) for elution. The crude eluate (340 mg) was crystallized from methanol; the yield of arylcholestenone was 400 mg (63%), mp 155°.

α-Chloro-β-p-methoxyphenylpropionamide from Acrylamide.[83] p-Anisidine (0.1 mol) was diazotized in 25 ml of cold 12 N hydrochloric acid with 8 g (0.114 mol) of sodium nitrite in 15 ml of water; the solution (65 ml) was adjusted to pH 3–4 with sodium bicarbonate. This cold solution was added slowly from a dropping funnel to a vigorously stirred mixture of 300 ml of acetone, 2.5 g of cupric chloride dihydrate, and 0.1 mol of aqueous 8% acrylamide solution. (Crystalline acrylamide gave very poor results). Nitrogen evolution continued for 8–10 hours at 18–22°. The mixture was diluted with 900 ml of water, and the oily product was extracted into benzene, washed with water, and dried with sodium sulfate. Removal of the solvent and crystallization from 50% aqueous alcohol yielded 7.9 g (37%) of the α-chloroamide; mp 103–104°.

4,4'-Bis(4-chloro-2-butenyl)biphenyl from Benzidine and Butadiene. (Use of a Diamine).[32] *CAUTION. Benzidine is a potent carcinogen. All operations with benzidine should be conducted in a good hood. Rubber gloves should be worn whenever there is a possibility of skin contact with this amine.*

Benzidine (18.4 g, 0.1 mol) in 40 ml of 12 N hydrochloric acid and 50 ml of water, was heated until dissolved. The solution was chilled rapidly to 0° before diazotization with 14 g (0.20 mol) of sodium nitrite in 30 ml of water. The product solution was added dropwise to a solution, at 15°, of

[83] Ya. Sh. Shkolnik, A. V. Dombrovskii, and B. M. Perepletchik, *J. Org. Chem. USSR*, **4**, 220 (1968) [*C.A.*, **68**, 104680s (1968)].

0.25 mol of butadiene in 150 ml of acetone containing 0.025 mol of cupric chloride. Nitrogen evolution began at 15° but continued energetically at 0–5°. When it stopped, the organic layer was taken up in ether, washed with water, dried with calcium chloride, and evaporated. The dark, oily residue in 250 ml of benzene was passed through a 15 × 3 cm column of granulated active carbon. The first 20–30 ml of eluate contained 4,4'-dichlorobiphenyl and was discarded. The subsequent eluate was evaporated to dryness and held for several days in a vacuum desiccator over paraffin wax. The bis(chlorobutenyl)biphenyl, a light-yellow viscous liquid, weight 21.1 g (64%), n^{20}D 1.6020, was sufficiently pure for most purposes, as shown by elementary analysis and conversion to a series of diamines.

1-Phenyl-2-buten-4-ol and 1-Phenyl-3-buten-2-ol from Butadiene in the Absence of Chloride Ion.[17] The benzenediazonium sulfate solution from 18.6 g (0.20 mol) of aniline, 15.9 ml (0.3 mol) of concentrated sulfuric acid, 60 ml of water, and 15.8 g (0.23 mol) of sodium nitrite was neutralized to pH 6–7 with sodium bicarbonate and added to an ice-cold mixture of 14.0 g (0.25 mol) of butadiene, 10.0 g (0.04 mol) of cupric sulfate, 22.2 g (0.08 mol) of ferrous sulfate, and 80 ml of acetone. Evolution of nitrogen began at 0° and continued for 6–8 hours near 0°. As the reaction proceeded, an additional 4 g of sodium bicarbonate was added in small portions to maintain the pH at 5–7. The organic product was taken into ether, washed with 2–3 portions of water, dried with anhydrous copper sulfate, and distilled. The first fraction was 8.3 g (28%) of 1-phenyl-2-butene-4-ol, bp 114–115° (3 mm); the second fraction was 2.0 g (7%) of the isomeric 1-phenyl-3-butene-2-ol, bp 142–145° (3 mm).

TABULAR SURVEY OF THE MEERWEIN ARYLATION REACTION

The following tables contain examples of the Meerwein arylation found in the literature from 1958 to the end of 1974. The search was conducted with *Chemical Abstracts* Subject Indexes and Author Indexes through Vol. **81** (1974). A supplementary search with *Science Citation Index* through 1974 uncovered several additional examples. No new preparations were seen in the 1975 literature during the preparation of this review.

In each table the unsaturated components are arranged in the following order: the parent compound of the series; its halogen derivatives; its alkyl derivatives in the order of increasing number of carbon atoms; its phenyl derivatives and its nuclear-substituted phenyl derivatives; and finally heterocyclic derivatives of the parent compound.

Under each unsaturated component the diazonium salts used are arranged in the following sequence: benzenediazonium chloride, then nuclear substitution products [in the order F, Cl, Br, I, NO_2, OH, OCH_3, NH_2, $NHCOCH_3$, SO_3H, SO_2NH_2, alkyl (in the order of increasing size), aryl (including condensed aryl as in naphthalenediazonium chloride), CHO, CO_2H, CO_2R, COR, CN], and finally heterocyclic diazonium salts.

The individual diazonium salts are not entered in the tables, except in rare instances, since they are adequately identified by inspection of the products.

The abbreviations Ac (acetyl), Bz (benzoyl), Pr (propyl), and Bu (butyl) are used.

Considerable space has been saved by modifying the customary *Organic Reactions* format for yield citation. Since most authors publish the results of arylating one olefin with several different diazonium salts, the table entries are presented as a type formula, X designating a ring substituent, Y a side-chain halogen, Z the "activating" group, and R a side-chain alkyl group. The yields for each substituent are then collected in linear form *for each individual paper*, so that a given result can be unambiguously identified with a given reference. This format also permits comparison of the results of different workers. The symbol (—) indicates that no yield was reported. Unsuccessful experiments have been included in the tables. Table entries are amplified, where necessary, by footnotes.

TABLE I.   NONCONJUGATED OLEFINS AND ACETYLENES

Unsaturated Compound	Product(s) and Yield(s) (%)	Refs.
CH_2=CH_2	$ClCH_2CH_2C_6H_5$ (7–10)	9b
	$ClCH_2CH_2C_6H_4NO_2$-p (45–50)	9d
CH_2=CHCl	$Cl_2CHCH_2C_6H_4X$	
	$\quad$ X = H (52); Cl-p (74);	
	$\qquad$ Br-p (70); NO_2-m (52);	66
	$\qquad$ OCH_3-p (36); CH_3-p (47)	
	$\quad$ X = NO_2-p (62)	65
	$\quad$ X = CH_2CHCl_2-p (70)^a	84
CH_2=CCl_2	$Cl_3CCH_2C_6H_4X$	
	$\quad$ X = H (79); Cl-p (76);	68,85
	$\qquad$ Br-p (66); NO_2-m (55);	
	$\qquad$ NO_2-p (70);	
	$\qquad$ 3-NO_2-3-CH_3-4 (32);	
	$\qquad$ OCH_3-p (48); CH_3-p (67)	

Note: References 84–133 are on pp. 258–259.

^{a} Prepared from $Cl_2CHCH_2C_6H_4NH_2$-p.

TABLE I. NONCONJUGATED OLEFINS AND ACETYLENES (*Continued*)

Unsaturated Compound	Product(s) and Yield(s) (%)	Refs.
ClCH=CHCl (*cis*)	Cl$_2$CHCHClC$_6$H$_5$ (13)	86
	Cl$_2$CHCHClC$_6$H$_4$Cl-*p* (18)	86
ClCH=CHCl (*trans*)	Cl$_2$CHCHClC$_6$H$_5$ (26)	86
	Cl$_2$CHCHClC$_6$H$_4$Cl-*p* (36)	86
CCl$_2$=CHCl	Cl$_3$CCHClC$_6$H$_4$X-*p*	
	X = H (45); Cl (51);	87
	Br(51); NO$_2$(49);	
	CH$_3$ (52)	
CH$_2$=CHOAc	AcOCHClCH$_2$C$_6$H$_4$X-*p*	
	X = H (51); Cl (53);	67
	NO$_2$ (41); OCH$_3$ (30);	
	CH$_3$ (44)	
CH$_2$=C(CH$_3$)OAc	CH$_3$COCH$_2$C$_6$H$_4$Cl-*p* (10)[b]	8
1–Acetoxycyclohexene	2-*p*-ClC$_6$H$_4$-cyclohexanone (10)[b]	8
CH$_2$=C(C$_6$H$_5$)OAc	C$_6$H$_5$COCH$_2$C$_6$H$_4$Cl-*p* (53)[b]	8
C$_6$H$_5$CH=C(C$_6$H$_5$)OAc	C$_6$H$_5$COCH(C$_6$H$_5$)C$_6$H$_4$Cl-*p* (—)[b]	8
CH$_2$=C(OAc)CH=CHC$_6$H$_5$	C$_6$H$_5$CH=CHCOCH$_2$C$_6$H$_4$Cl-*p* (35)[b]	8
	C$_6$H$_5$CHClCHArCOCH$_2$C$_6$H$_4$Cl-*p* (20)[b]	8
CH$_2$=C(CH$_3$)- ⎤[c] CH=C(CH$_3$)OAc	*p*-ClC$_6$H$_4$CH$_2$C(CH$_3$)=CHCOCH$_3$ (53)⎤[b]	
(CH$_3$)$_2$- C=CHC(OAc)=CH$_2$ ⎦	*p*-ClC$_6$H$_4$CH$_2$COCH=C(CH$_3$)$_2$ (14) ⎦	8

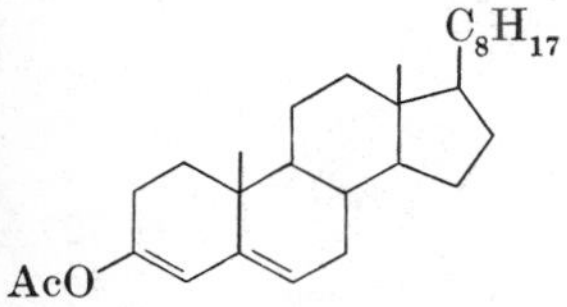

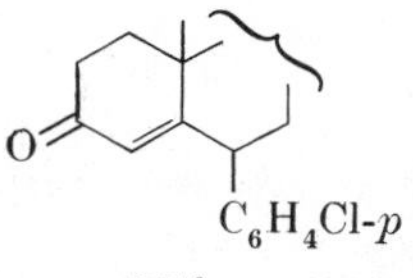

(63)[b]

8

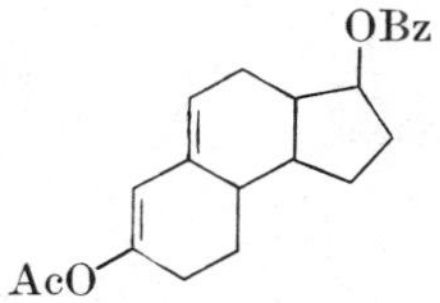

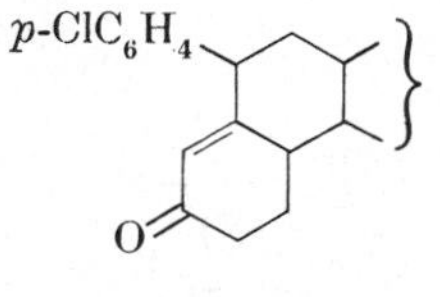

(25)[b]   (22)[b]

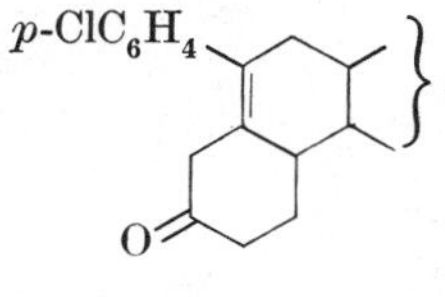

8

Note: References 84–133 are on pp. 258–259.

[b] The *gem*-chloroacetates were not isolated but were hydrolyzed to the ketones shown.

[c] This mixture of enol acetates was prepared from mesityl oxide.

TABLE II. CONJUGATED DIENES AND ENE-YNES

Unsaturated Compound	Product(s) and Yield(s) (%)	Refs.
CH_2=CHCH=CH_2	$ClCH_2CH$=$CHCH_2C_6H_4X$-p	
	X = Cl (70)[a]	25
	X = F (66); SO_2NH_2 (70);	
	SO_2NH-thiazolyl-2 (45);	
	$CO_2CH_2CH_2N(C_2H_5)_2$ (60)	52
	From diazonium salt p-$XC_6H_4N_2^+A^-$:	
	$HOCH_2CH$=$CHCH_2C_6H_5$[b]	
	A = SO_4^{2-} (28); NO_3^- (36);	17
	BF_4^- (40)	
	$HOCH_2CH$=$CHCH_2C_6H_4X$-p	
	(A = SO_4^{2-})	
	X = NO_2 (18); OCH_3 (25);	17
	CH_3 (26)	
	From diazonium salt p-$XC_6H_4N_2^+OAc^-$:	
	$AcOCH_2CH$=$CHCH_2C_6H_4X$-p	
	X = H (30); Cl (25); OCH_3 (—);	18
	CH_3 (25)	
ClCH=CHCH=CH_2	Cl_2CHCH=$CHCH_2C_6H_4X$-p	
	X = H (53); Cl (72); CH_3 (54)	88
CH_2=CClCH=CH_2	$ClCH_2CH$=$CClCH_2C_6H_4X$-p	
	X = H (60); Cl (50); NO_2 (40);	89
	CH_3 (54)	
	$AcOCH_2CH$=$CClCH_2C_6H_5$ (35)[c]	18
ClCH=CClCH=CH_2	$ClCH$=$CClCHClCH_2C_6H_5$ (55)[d]	90
CH_2=CClCCl=CH_2	$ClCH_2CCl$=$CClCH_2C_6H_4X$	
	X = H (81); Br-o (80);	91
	NO_2-p (41); NO_2-2-CH_3-4 (38);	
	p-OCH_3-p (42); CH_3-o (25)	
	X = 2-Cl-2-CH_3-6 (72)	79
	X = C_2H_5-o (44); i-Pr-o (24)	64
	$ClCH_2CCl$=$CClCH_2C_{10}H_7$-1 (22)	64
	$ClCH_2CCl$=$CClCH_2C_{10}H_7$-2 (17)	64
CH_3CH=CHCH=CH_2	$CH_3CHClCH$=$CHCH_2C_6H_4X$	
	X = H (70); Cl-p (57);	92,93
	NO_2-o (40); NO_2-m (42);	
	OCH_3-p (45); CH_3-p (60)	
	CH_3CH=$CHCH$=CHC_6H_4X-p	
	X = Br (38); I (32); NO_2 (50)	92,93

Note: References 84–133 are on pp. 258–259.

[a] Ref. 25 contains an extensive study of the effect of reaction conditions on yields.

[b] The product contains a few percent of $XC_6H_4CH_2CHOHCH$=CH_2.

[c] From $C_6H_5N_2^+OAc^-$.

[d] Note 1,2 addition of ArCl. Eight other compounds are mentioned in the abstract but are not described.

TABLE II. CONJUGATED DIENES AND ENE-YNES (*Continued*)

Unsaturated Compound	Product(s) and Yield(s) (%)	Refs.
$CH_2=CHC(CH_3)=CH_2$	$ClCH_2CH=C(CH_3)CH_2C_6H_4X$-$p$	
	X = H (—); NO_2 (—);	27
	OCH_3 (—); CH_3 (—)	
	X = Cl (70)	25,27
	$HOCH_2CH=C(CH_3)CH_2C_6H_4X$-$p^e$	
	X = H (—); Cl (—); NO_2 (—);	94
	OCH_3 (—)	
	$RCO_2CH_2CH=C(CH_3)CH_2C_6H_4X$-$p^e$	
	X = H (34,32,8); Cl (40,14);	18
	OCH_3 (36); CH_3 (36,12)^f	
$CH_2=C(CH_3)C(CH_3)=CH_2$	$ClCH_2C(CH_3)=C(CH_3)CH_2C_6H_4X$-$p$	
	X = H (58); Cl (55); OCH_3 (50);	26
	CH_3 (56)	
	X = F (65); Br (50); I (45);	52
	SO_2NH_2 (70);	
	SO_2NH-thiazolyl-2 (40);	
	$CO_2C_2H_5$ (65);	
	$CO_2CH_2CH_2N(C_2H_5)_2$ (60)	
Cycloheptatriene	$C_7H_7-C_6H_4X$-p	
	X = H (16); Cl (29); NO_2 (7)	95
$C_6H_5CH=CHCH=CH_2$	$C_6H_5CH=CHCH=CHC_6H_5$ (90)	47
	$C_6H_5CH=CHCH=CHC_6H_4CH_3$-$p$ (75)	47
$C_6H_5CH=CClCH=CH_2$	$C_6H_5CH=CClCH=CHC_6H_4X$-p	
	X = H (37); NO_2 (45); CH_3 (45)	47,89
$C_6H_5CH=CHCH=CHCH_3$	$C_6H_5CH=CHCH=C(CH_3)C_6H_4Cl$-$p$ (32)	96
	$C_6H_5CH=CHCH=C(CH_3)C_6H_4NO_2$-$p$	96
	(32)	
p-$ClC_6H_4CH=CHCH=$ $CHCH_3$	p-$ClC_6H_4CH=CHCH=C(CH_3)C_6H_4Cl$-$p$ (28)	96
	p-$ClC_6H_4CH=CHCH=C(CH_3)C_6H_4NO_2$-$p$ (30)	96
$C_6H_5CH=C(CH_3)CH=CH_2$	$C_6H_5CH=C(CH_3)CH=CHC_6H_4X$-$p$	
	X = H (58); NO_2 (36); CH_3 (50)	47
p-$CH_3C_6H_4CH=CHCH=$ $CHCH_3$	p-$CH_3C_6H_4CH=CHCH=C(CH_3)C_6H_4X$-$p$	
	X = Cl (34); NO_2 (29)	96
p-$CH_3C_6H_4CH=C(CH_3)$- $CH=CH_2$	p-$CH_3C_6H_4CH=C(CH_3)$- $CH=CHC_6H_4CH_3$-p (54)	47
$C_6H_5CH=C(CH_3)$- $C(CH_3)=CH_2$	$C_6H_5CH=C(CH_3)C(CH_3)=CHC_6H_5$ (46)	47

Note: References 84–133 are pp. 258–259.

^{e} The reaction was run under chloride-free conditions.
^{f} The yield figures refer to R = CH_3, R = CF_3, and R = CCl_3, in that order.

TABLE II. CONJUGATED DIENES AND ENE-YNES (*Continued*)

Unsaturated Compound	Product(s) and Yield(s) (%)	Refs.
CH_2=CHC≡CH	$XC_6H_4CH_2CHClC$≡CH $XC_6H_4CH_2CH$=C=$CHCl$ ⎤	
	X = H (46)	31a, 29[g]
	X = m-Cl (30); p-Cl (27); p-CH_3 (38)	31d
	X = p-CH_3O (20)	29[g]
CH_3CH=CHC≡CH	$CH_3CHClCH$=C=CHC_6H_5 $C_6H_5CH(CH_3)CH$=C=$CHCl$ ⎤ (34)	33c
CH_2=$C(CH_3)C$≡CH	p-$XC_6H_4CH_2C(CH_3)$=C=$CHCl$ p-$XC_6H_4CH_2C(CH_3)ClC$≡CH ⎤	
	X = H (63); Cl (33)	29[g]
	X = H (48)	31c
CH_2=CHC≡CCH_3	$C_6H_5CH_2CHClC$≡CCH_3 (41)	31e
CH_2=CHC≡CC_2H_5	p-$XC_6H_4CH_2CHClC$≡CC_2H_5	
	X = H (42); Cl (30); CH_3O (32); CH_3 (54)	31b
CH_2=CHC≡CCR_2OH	$C_6H_5CH_2CHClC$≡CCR_2OH (35–52, various R)	30
CH_2CHC≡$C(CH_3)_2OR$	$C_6H_5CH_2CHClC$≡$CC(CH_3)_2OR$ (51–68, various R)	30
CH_2=CHC≡$CC(CH_3)$=CH_2	$C_6H_5CH_2CHClC$≡$CC(CH_3)$=CH_2 (Low)	31b

[g] The allenic isomers were not reported in Ref. 29.

TABLE III. STYRENES AND ARYLACETYLENES[a]

Unsaturated Compound	Product(s) and Yield(s) (%)	Refs.
C_6H_5CH=CH_2	$C_6H_5CHClCH_2C_6H_4X$	
	X = H (44); Cl-p (68); Br-p (74); NO_2-o (46); NO_2-m(56); NO_2-p (80); OCH_3-p (42); CH_3-p (52)	49
	X = Cl_2-2,4 (—)	24
	X = SO_2NH_2-p (78)	97
	X = $CO_2C_2H_5$-p (79); $CO_2CH_2CH_2N(C_2H_5)_2$ (68)	98
p-ClC_6H_4CH=CH_2	p-$ClC_6H_4CHClCH_2C_6H_4Cl$-p (56)	97
p-$CH_3C_6H_4CH$=CH_2	p-$CH_3C_6H_4CHClCH_2C_6H_4X$-p	
	X = H (49); Cl (54); NO_2 (68)	97
	X = SO_2NH_2 (64); $CO_2C_2H_5$ (76)	98
2-Pyridyl-CH=CH_2	2-Pyridyl-$CHClCH_2C_6H_4X$	
	X = H; Cl-p; Br-p; NO_2-m; NO_2-p OCH_3-p; CH_3-p (—)	99
	2-Pyridyl-CH=CH-thiazolyl-2 (11)	82

Note: References 84–133 are on pp. 258–259.

TABLE III. Styrenes and Arylacetylenes (Continued)

Unsaturated Compound	Product(s) and Yields(s) (%)	Refs.
$C_6H_5CCl{=}CH_2$	$C_6H_5CCl{=}CHC_6H_4X$ X = H (66); Cl-p (84); NO_2-o (0); NO_2-m (68); NO_2-p (82); OCH_3-p (52); CH_3-p (64)	100
p-$ClC_6H_4CCl{=}CH_2$	p-$ClC_6H_4CCl{=}CHC_6H_4NO_2$-$p$ (28)	97
p-$CH_3C_6H_4CCl{=}CH_2$	p-$CH_3C_6H_4CCl{=}CHC_6H_4X$-$p$ X = NO_2 (70); CH_3 (78)	100
$2,4$-$(CH_3)_2C_6H_3CCl{=}CH_2$	$2,4$-$(CH_3)_2C_6H_3CCl{=}CHC_6H_4NO_2$-$p$ (77)	100
p-$C_2H_5C_6H_4CCl{=}CH_2$	p-$C_2H_5C_6H_4CCl{=}CHC_6H_4X$-$p$ X = NO_2 (60); CH_3 (62)	97
$C_6H_5CBr{=}CH_2$	$C_6H_5CCl{=}CHC_6H_4X$-p^b X = H (45); NO_2 (75); CH_3 (42)	45
$C_6H_5C(CH_3){=}CH_2$	$C_6H_5C(CH_3){=}CHC_6H_4X$ X = H (50); Cl-p (47); NO_2-o (63); NO_2-m (53); NO_2-p (72); CH_3-p (64) OCH_3-p (56)	48 97
p-$CH_3OC_6H_4C(CH_3){=}CH_2$	p-$CH_3OC_6H_4C(CH_3){=}CHC_6H_4NO_2$-$p$ (50)	97
p-$CH_3C_6H_4C(CH_3){=}CH_2$	p-$CH_3C_6H_4C(CH_3){=}CHC_6H_4X$-$p$ X = H (58); NO_2 (78); CH_3 (70)	48
$C_6H_5C(C_2H_5){=}CH_2$	$C_6H_5C(C_2H_5){=}CHC_6H_4X$-$p$ X = H (48); Cl (46); NO_2 (42)	97
$(C_6H_5)_2C{=}CH_2$	$(C_6H_5)_2C{=}CHC_6H_4X$ X = H; NO_2-o; NO_2-m; NO_2-p; CH_3-p (—)	101, 102
$C_6H_5C{\equiv}CH$	$C_6H_5CCl{=}CHC_6H_4X$-p X = H (46); NO_2 (36)	29

[a] In addition to the compounds tabulated, 22 stilbenes p-$RC_6H_4CR'{=}CHC_6H_4R''$ were prepared from combinations of p-$RC_6H_4CR'{=}CH_2$ and p-$R''C_6H_4N_2Cl$ in yields of 58–96%; R = H, CH_3; R' = Cl, CH_3, C_6H_5; R'' = SO_2NH_2, $CO_2C_2H_5$, $CO_2CH_2CH_2N(C_2H_5)_2$.[98] The intermediate chloroethanes were not isolated.

[b] Note loss of hydrogen bromide from the assumed intermediate C_6H_5CBr-$ClCH_2Ar$.

TABLE IV. α,β-Unsaturated Aldehydes and Ketones[a]

Unsaturated Compound	Product(s) and Yield(s) (%)	Refs.
$CH_2{=}CHCHO$	$C_6H_5CH_2CHClCHO$ (53)	81
	$C_6H_5CH_2CHBrCHO$ (22)	81
$CH_3CH{=}CHCHO$	$C_6H_5CH(CH_3)CHClCHO$ (41)	81
$trans$-$C_6H_5CH{=}CHCOCH_3$	$C_6H_5CHClCH(COCH_3)C_6H_4Cl$-$p$ (45)	7
cis-$C_6H_5CH{=}CHCOCH_3$	$C_6H_5CHClCH(COCH_3)C_6H_4Cl$-$p$ (12)	7

Note: References 84–133 are on pp. 258–259.

[a] Two compounds were prepared from methacrolein and methyl vinyl ketone but were not described in detail in Ref. 78.

TABLE V. Aliphatic Monobasic α,β-Unsaturated Acid Derivatives

Unsaturated Compound	Product(s) and Yield(s) (%)	Refs.

A. Acids

$CH_2{=}CHCO_2H$	$C_6F_5CH_2CHBrCO_2H$ (10)	74, 76
	$ArCH_2CHClCO_2H$ (0–71)[a]	71–73, 75, 76
	$ArCH_2CHBrCO_2H$ (32–97)[a]	76,77
	5-Nitro-2-thiazolyl-CH${=}$CHCO$_2$H (15)	103
$CH_2{=}C(CH_3)CO_2H$	$C_6H_5CH_2C(CH_3)ClCO_2H$ (50)	104

[a] The emphasis in these reports was placed upon ammonolysis of these halides to ring-substituted phenylalanines $ArCH_2CH(NH_2)CO_2H$. The Meerwein arylations, using a variety of ArN_2X, were not always described in detail.

B. Esters

$CH_2{=}CHCO_2CH_3$	m-$XC_6H_4CH_2CHClCO_2CH_3$[a] X = H (24); Cl (39); O$_2$N (56); CH$_3$O (21)	24
$CH_2{=}CClCO_2CH_3$	$XC_6H_4CCl_2CO_2CH_3$ X = H (53); p-Cl (65) m-O$_2$N (52); p-O$_2$N (77); p-CH$_3$O (32); p-CH$_3$ (61)	105
$CH_2{=}CBrCO_2CH_3$	p-$XC_6H_4CH_2CBrClCO_2CH_3$ X = H (49); Cl (61); NO$_2$ (70); CH$_3$ (53)	106
$CH_2{=}C(CH_3)CO_2CH_3$	$XC_6H_4CH_2CCl(CH_3)CO_2CH_3$ X = m-F (35); p-F (36); m-CH$_3$O (44) m-CH$_3$ (31)	24

[a] See also Ref. 75 for numerous other examples not described in detail.

C. Nitriles

$CH_2{=}CHCN$	$XC_6H_4CH_2CHClCN$ X = m-F (26); p-F (28); m-Cl (45); m-Br (15); m-CH$_3$O (43); m-CH$_3$ (27)	24
	X = p-Br (75)	104
	X = p-CH$_3$ (89[a], 56) 78[b]	

[a] The higher yield was obtained by adding an extra mole of chloride ion.
[b] Several other examples are given, but are not described in detail.

Note: References 84–133 are on pp. 258–259.

Unsaturated Compound	Product(s) and Yield(s) (%)	Refs.

$CH_2{=}CHCN$

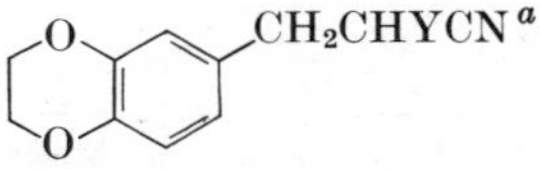

CH_2CHYCN [a]

	$Y = Cl$ (60); Br (35)	107
	$X_2C_6H_3CH_2CHClCN$	
	$X_2 = 3,4\text{-}(CO_2CH_3)_2$ (42)	108
	$X_2 = 3,5\text{-}(CO_2CH_3)_2$ (—)	109
	$p\text{-}XC_6H_4CH{=}CHCN$	
	$X = CH_3S$ (4); CH_3SO (25);	110
	NCS (87); CF_3 (50)	
$CH_2{=}CClCN$	$XC_6H_4CH_2CCl_2CN$	
	$X = H$ (53); $p\text{-}Cl$ (68)	111
	$p\text{-}Br$ (68); $o\text{-}O_2N$ (62);	
	$p\text{-}O_2N$ (65); $p\text{-}CH_3O$	
	(53); $p\text{-}CH_3$ (52)	
	$m\text{-}O_2NC_6H_4CH{=}CClCN$ (36)	111
$CH_2{=}C(CH_3)CN$	$XC_6H_4CH_2C(CH_3)ClCN$	
	$X = H(46)$; $p\text{-}Cl$ (78);	112
	$p\text{-}Br$ (49); $o\text{-}O_2N$ (75);	
	$m\text{-}O_2N$ (55); $p\text{-}O_2N$	
	(60); $p\text{-}CH_3$ (65)	
	$X = m\text{-}Cl$ (19); $m\text{-}CH_3O$	24
	(57)	

[a] See also Ref. 78 for several other examples not described in detail.

D. *Amides*

$CH_2{=}CHCONH_2$	$XC_6H_4CH_2CHClCONH_2$	
	$X = H$ (31); $p\text{-}Cl$ (58);	83
	$o\text{-}O_2N$ (43); $m\text{-}O_2N$	
	(56); $p\text{-}O_2N$ (64);	
	$p\text{-}CH_3O$ (37); $p\text{-}CH_3$	
	(42)	
	$X = H$ (55)	76

TABLE VI. Aromatic α,β-Unsaturated Acids

Unsaturated Compound	Product(s) and Yield(s) (%)	Refs.
$C_6H_5CH{=}CHCO_2H$	$C_6H_5CH{=}CHC_6H_4C_6H_5$-$o$, -$m$, and -$p$ -o (6); -m (38); -p (35)	113
CH=CHCO$_2$H	CH=CHC$_6$H$_4$NO$_2$-p (25)	107

TABLE VII. α,β-Unsaturated Keto Acids and Esters

Unsaturated Compound	Product(s) and Yield(s) (%)	Refs.
$CH_3COCH{=}CHCO_2H$	$CH_3COCH{=}CHC_6H_4NO_2$-p (17) $CH_3COCH{=}C(CO_2H)C_6H_4NO_2$-$p$ (17)	114[a]
$C_6H_5COCH{=}CHCO_2H$	$C_6H_5COCH{=}CHC_6H_4OCH_3$-$o$, -$m$, and -$p$ -o (Trace); -m (3); -p (5)	115
	$C_6H_5COCH{=}CHC_6H_2(OCH_3)_3$-3,4,5 (10)	115
p-$ClC_6H_4COCH{=}CHCO_2H$	p-$ClC_6H_4COCH{=}CHC_6H_4X$ X = H (10); Cl-o (7); Cl-p (30); Br-p (22); OCH$_3$-p (10); p-CH$_3$ (7)	116
2-CH_3O-5-$CH_3C_6H_3COCH{=}CHCO_2H$	2-CH_3O-5-$CH_3C_6H_3COCH{=}CHC_6H_4Cl$-$p$ (14)	114
	(12)	117

[a] No yields were reported when the substituent was p-Cl, p-Br, or o-NO$_2$.

TABLE VIII. Quinones

Unsaturated Compound	Product(s) and Yield(s) (%)	Refs.
2-$ClC_6H_3O_2$	Ar-2-$ClC_6H_2O_2$ $(-)^a$	118
$2,5$-$Cl_2C_6H_2O_2$	 I $\qquad$ II	

Note: References 84–133 are on pp. 258–259.

TABLE VIII. Quinones   (*Continued*)

Unsaturated Compound	Product(s) and Yield(s)(%)	Refs.
	Ar = 3,4-$(CH_3O)_2C_6H_3$ (I, 26; II, 11)	119
	Ar = 3-OCH_3-4-$CH_3C_6H_3$ (—)	119
	Ar = 3-CH_3-4-$OCH_3C_6H_3$ (—)	119
2,6-$Cl_2C_6H_2O_2$	Results similar to 2,5-$Cl_2C_6H_2O_2$	119

[a] For various aryl groups, mixtures of isomers were obtained with Ar at position 3, 5, or 6; they were separated by chromatography.

TABLE IX.   Coumarins

Unsaturated Compound	Product(s) and Yield(s) (%)	Refs.
4-HO-coumarin[a]	4-HO-3-Ar-coumarin[a] (—)	120[b]
	7-HO-4-CH_3-3-XC_6H_4-coumarin (2–12) X = H; o-Cl; m-Cl; o-NO_2; m-NO_2; p-NO_2; o-OCH_3; m-OCH_3; o-CH_3; m-CH_3; p-CH_3	121
7-CH_3O-coumarin	7-CH_3O-3-p-X-C_6H_4-coumarin X = Cl, SO_2NH_2, SO_2CH_3, CH_3, CO_2H, $CONH_2$ (—)	122[b]
Various coumarins	Various 3-p-XC_6H_4-coumarins	123[b]

Note: References 84–133 are on pp. 258–259.

[a] The 7-CH_3, 8-CH_3, and 5,6-benzo derivatives were also studied. In addition to the 3-aryl derivative, the 3-azo compound and the substituted salicylic acid were formed.

[b] Yield details are not given in the abstract.

TABLE X. MISCELLANEOUS REACTIONS

A. Bisdiazonium Salts

	Unsaturated Compound	Product(s) and Yield(s) (%)	Refs.
Benzidine	$CH_2=CCl_2$	$(Cl_3CCH_2C_6H_4)_2$ (78)	34
	$CH_2=CHCH=CH_2$	$(ClCH_2CH=CHCH_2C_6H_4)_2$ (64)	32
	$CH_2=CHCCl=CH_2$	$(ClCH_2CH=CClCH_2C_6H_4)_2$ (60)	28
	$CH_2=CHC(CH_3)=CH_2$	$[ClCH_2CH=C(CH_3)CH_2C_6H_4)_2$ (60)	32
	$CH_3CH=CHCH=CH_2$	$\left[\begin{array}{l}(CH_3CHClCH=CHCH_2C_6H_4)_2\\(CH_3CH=CHCHClCH_2C_6H_4)_2\end{array}\right]$ (60)	32
	$XC_6H_4CH=CHCH=CH_2$ ᵃ	$XC_6H_4CH=CHCH=CHC_6H_4C_6H_4Cl\text{-}p$	33,124
		X = H (38); o-Cl (38); p-Cl (36);	
		o-Br (32); p-CH₃ (34)	
	$XC_6H_4CH=C(CH_3)CH=CH_2$ ᵃ	$p\text{-}XC_6H_4CH=C(CH_3)CH=CHC_6H_4C_6H_4Cl\text{-}p$	33,124
		X = H (33); Cl(35); CH₃ (32)	34,125
	$C_6H_5CH=CH_2$ ᵃ	$C_6H_5CHClCH_2C_6H_4C_6H_4Cl\text{-}p$ (57)	34
	$C_6H_5CCl=CH_2$ ᵃ	$C_6H_5CCl=CHC_6H_4C_6H_4Cl\text{-}p$ (54)	34
	$p\text{-}CH_3C_6H_4CCl=CH_2$ ᵃ	$p\text{-}CH_3C_6H_4CCl=CHC_6H_4C_6H_4Cl\text{-}p$ (52)	34
	$(C_6H_5)_2C=CH_2$ ᵃ	$(C_6H_5)_2C=CHC_6H_4C_6H_4Cl\text{-}p$ (26)	34
	$CH_2=CHCN$	$(NCCHClCH_2C_6H_4)_2$ (78)	34
3,3'-(CH₃O)₂benzidine	$CH_2=CHCH=CH_2$	$[ClCH_2CH=CHCH_2C_6H_3(OCH_3)]_2$ (45)	28
3,3'-(CH₃)₂benzidine	$CH_2=CHC(CH_3)=CH_2$	$[ClCH_2CH=C(CH_3)CH_2C_6H_3(OCH_3)]_2$ (40)	28
	$CH_2=CHCR=CH_2$	$[ClCH_2CH=CRCH_2C_6H_3(CH_3)]_2$	28
		R = H (55); Cl (50); CH₃ (50);	
$(p\text{-}H_2NC_6H_4)_2S$	$C_6H_5CH=CH_2$	$(C_6H_5CHClCH_2C_6H_4)_2S$ (40)	34
	$C_6H_5CCl=CH_2$	$(C_6H_5CCl=CHC_6H_4)_2S$ (33)	34
	$CH_2=CHCN$	$(NCCHClCH_2C_6H_4)_2S$ (44)	34

CH$_2$=CCl$_2$	(Cl$_3$CCH$_2$C$_6$H$_4$)$_2$CH$_2$ (40)	34
C$_6$H$_5$CH=CH$_2$	(C$_6$H$_5$CHClCH$_2$C$_6$H$_4$)$_2$CH$_2$ (67)	34
C$_6$H$_5$CCl=CH$_2$	(C$_6$H$_5$CCl=CHC$_6$H$_4$)$_2$CH$_2$ (54)	34
(C$_6$H$_5$)$_2$C=CH$_2$	[(C$_6$H$_5$)$_2$C=CHC$_6$H$_4$]$_2$CH$_2$ (49)	34
CH$_2$=CHCN	(NCCHClCH$_2$C$_6$H$_4$)$_2$CH$_2$ (66)	34

(p-H$_2$NC$_6$H$_4$)$_2$CH$_2$

B. Vinyl Derivatives of Oxygen, Sulfur, and Phosphorus

1-Methoxycyclohexene	2-p-Chlorophenylcyclohexanone (20)	8
2-Butoxy-2-cyclohexenone	(fused ring structure) X = H (27); X$_2$ = OCH$_2$O (26)	126 127, 128
CH$_3$SO$_2$CH=CH$_2$	CH$_3$SO$_2$CHClCH$_2$C$_6$H$_4$X X = H (6); NO$_2$-m (45); NO$_2$-p (48); SO$_2$NH$_2$-p (35); CO$_2$H-p (45)	127, 128
C$_2$H$_5$SO$_2$CH=CH$_2$	C$_2$H$_5$SO$_2$CHClCH$_2$C$_6$H$_4$X-p X = Cl (31); NO$_2$ (28)	129
p-CH$_3$C$_6$H$_4$SO$_2$CH=CH$_2$	p-CH$_3$C$_6$H$_4$SO$_2$CHYCH$_2$C$_6$H$_4$X-p X = H (20); NO$_2$ (26) for Y = Cl; X = NO$_2$ (24) for Y = Br	127

Note: References 84–133 are on pp. 258–259.

[a] When the unsaturated compound holds an aryl group, the bis adduct cannot be formed; instead the second diazonium group is replaced by chlorine.[33,34,124]

[b] The intermediate Meerwein adduct from (structure with N$_2^+$, CO$_2$CH$_3$, X substituents) cyclizes spontaneously.

TABLE X. MISCELLANEOUS (Continued)

Unsaturated Compound	Product(s) and Yield(s) (%)	Refs.
B. Vinyl Derivatives of Oxygen, Sulfur, and Phosphorus (Continued)		
$(CH_2=CH)_2SO_2$	$SO_2(CHClCH_2C_6H_4X\text{-}p)_2$ X = H (15); Cl (23)	130
	$CH_2=CHSO_2CHClCH_2C_6H_4NO_2\text{-}p$ (Low)	130
$CH_2=CHSO_3K$	$KO_3SCHClCH_2C_6H_4NO_2\text{-}p$ (Low)	130
$CH_2=CHP(O)(OC_2H_5)_2$	$p\text{-}XC_6H_4CH_2CHClP(O)(OC_2H_5)_2$ X = H (47); Cl (45); O_2N (17)	131
C. Oximes and Other C=N Compounds		
$CH_2=NOH$	$3,4,5\text{-}(CH_3O)_3C_6H_2CHO$ (20)	115
$CH_3CH=NOH$	$3,5\text{-}(CH_3O_2C)_2C_6H_3COCH_3$ (17)	108
$RC(=NNHC_6H_5)CH=NNHC_6H_5$	$RC(=NNHC_6H_5)C(=NNHC_6H_5)Ar$ (7–31) (Ten examples with various R and Ar combinations)	19, 20

Pyridine-N-oxide	2-Ar-pyridine-N-oxide (—)	22, 23
Quinoline-N-oxide	2-Ar-quinoline-N-oxide (16–42)	22
Isoquinoline-N-oxide	1-Ar-isoquinoline-N-oxide (16–42)	22

D. Thiophene and Furan Derivatives

$2\text{-}C_4H_3SCHO$	$5\text{-}XC_6H_4\text{-}2\text{-}C_4H_2SCHO$ $X = p\text{-}Br$ (25); $o\text{-}O_2N$ (11); $m\text{-}O_2N$ (16); $p\text{-}O_2N$ (24); $p\text{-}CH_3O$ (11); $p\text{-}CH_3$ (7)	132
$2\text{-}C_4H_3OCO_2H$	$5\text{-}XC_6H_4\text{-}2\text{-}C_4H_3OCO_2H$ $X = o\text{-}Cl$ (41); $m\text{-}Cl$ (13); $p\text{-}Cl$ (—); $o\text{-}O_2N$ (46)	133
$2\text{-}C_4H_3OCO_2CH_3$	$5\text{-}XC_6H_4\text{-}2\text{-}C_4H_3OCO_2CH_3$ $X = o\text{-}Cl$ (38); $m\text{-}Cl$ (10); $p\text{-}Cl$ (11); $o\text{-}O_2N$ (8); $m\text{-}O_2N$ (4)	133

Note: References 84–133 are on pp. 258–559.

257

REFERENCES TO TABLES I-X

[84] V. M. Naidan, A. K. Grabovoi, and A. V. Dombrovskii, *Ukr. Khim. Zh.* (*Russ. Ed.*), **39**, 805 (1973) [*C.A.*, **79**, 115476h (1973)].

[85] V. M. Naidan, *Nauk. Zap. Chernivetsk. Derzh. Univ., Ser. Prirodn. Nauk*, **51**, 40 (1961) [*C.A.*, **62**, 10353f (1965)].

[86] V. M. Naidan and A. V. Dombrovskii, *J. Org. Chem. USSR*, **2**, 883 (1966) [*C.A.*, **65**, 13580g (1966)].

[87] V. M. Naidan and A. V. Dombrovskii, *J. Org. Chem. USSR*, **1**, 2037 (1965) [*C.A.*, **64**, 9617c (1966)]. See also Ref. 68.

[88] N. I. Ganushchak, N. F. Stadniichuk, and A. V. Dombrovskii, *J. Org. Chem. USSR*, **5**, 678 (1969) [*C.A.*, **71**, 21764h (1969)].

[89] N. I. Ganushchak and F. V. Kvasnyuk-Mudrii, *Nauk. Zap. Chernivets'k. Univ.*, **53**, 77 (1961) [*C.A.*, **60**, 14410f (1964)].

[90] S. V. Toganyan, L. G. Grigoryan, and V. O. Babayan, *Armen. Khim. Zh.*, **24**, 421 (1971) [*C.A.*, **76**, 24803j (1972)].

[91] V. O. Babayan, L. G. Grigoryan, and S. V. Toganyan, *J. Org. Chem. USSR*, **5**, 306 (1969) [*C.A.*, **70**, 106081n (1969)].

[92] A. V. Dombrovskii and N. I. Ganushchak, *J. Gen. Chem. USSR*, **32**, 1867 (1962) [*C.A.*, **58**, 4463g (1963), **62**, 7664g (1965)].

[93] A. V. Dombrovskii and N. I. Ganushchak, *Sintez Svoistva Monomerov, Akad. Nauk SSSR, Inst. Neftekhim. Sinteza, Sb. Rabot. 12-oi Konf. po Vysokomolekul. Soedin.*, **1962**, 51–58 (Publ. 1964) [*C.A.*, **62**, 7664g (1965)]. Compare Ref. 92.

[94] N. I. Ganushchak and B. D. Grishchuk, *Zh Vses. Khim. Obshchest.*, **18**, 357 (1973) [*C.A.*, **79**, 78276v (1973)].

[95] K. Weiss and M. Lalande, *J. Amer. Chem. Soc.*, **82**, 3117 (1960).

[96] N. I. Ganushchak, V. A. Vengrzhanovskii, and A. V. Dombrovskii, *J. Org. Chem. USSR*, **5**, 111 (1969) [*C.A.*, **70**, 87181b (1969)].

[97] K. G. Tashchuk and A. V. Dombrovskii, *J. Org. Chem. USSR*, **5**, 479 (1969) [*C.A.*, **71**, 12680a (1969)].

[98] K. G. Tashchuk, A. A. Yatsishin, and A. V. Dombrovskii, *J. Org. Chem. USSR*, **9**, 1511 (1973) [*C.A.*, **79**, 91692x (1973)].

[99] K. G. Tashchuk, A. V. Dombrovskii, and V. S. Federov, *Ukr. Khim. Zh.*, **30**, 496 (1964) [*C.A.*, **61**, 5606f (1964)].

[100] A. V. Dombrovskii and K. G. Tashchuk, *J. Gen. Chem. USSR*, **33**, 158 (1963) [*C.A.*, **59**, 491c (1963)]; see *ibid.*, **34**, 1201 (1964).

[101] A. V. Dombrovskii and N. D. Bodnarchuk, *Ukr. Khim. Zh.*, **25**, 477 (1959) [*C.A.*, **54**, 9843d (1960)].

[102] A. V. Dombrovskii and N. D. Bodnarchuk, *Ukr. Khim. Zh.*, **27**, 369 (1961) [*C.A.*, **56**, 4692g (1962)].

[103] Y. Lin, P. B. Hulbert, E. Bueding, and C. H. Robinson, *J. Med. Chem.*, **17**, 835 (1974).

[104] A. V. Dombrovskii, A. M. Yurkevich, and A. P. Terent'ev, *J. Gen. Chem. USSR*, **27**, 3381 (1957) [*C.A.*, **52**, 9019i (1958)].

[105] N. O. Pastushak, A. V. Dombrovskii, and A. N. Mukhova, *J. Org. Chem. USSR*, **1**, 566 (1965) [*C.A.*, **63**, 1727c (1965)].

[106] N. O. Pastushak, A. V. Dombrovskii, and A. N. Mukhova, *J. Org. Chem. USSR*, **1**, 1907 (1965) [*C.A.*, **64**, 3403e (1966)].

[107] B. S. Federov, L. G. Pribytkova, M. I. Kanishchev, and A. V. Dombrovskii, *J. Org. Chem. USSR*, **9**, 1517 (1973) [*C.A.*, **79**, 92124a (1973)].

[108] R. A. Clendenning and W. H. Rauscher, *J. Org. Chem.*, **26**, 2963 (1961).

[109] J. J. Glynn, *Diss. Abstr.*, **28**(12)B, 4973 (1968) [*C.A.*, **69**, 87535z (1968)].

[110] J. A. Claisse, M. W. Foxton, G. I. Gregory, A. H. Sheppard, E. P. Tiley, W. K. Warburton, and M. J. Wilson, *J. Chem. Soc., Perkin I*, **1973**, 2241.

[111] N. O. Pastushak, A. V. Dombrovskii, and L. I. Rogovik, *J. Gen. Chem. USSR*, **34**, 2254 (1964) [*C.A.*, **61**, 10623a (1964)].

[112] N. O. Pastushak and A. V. Dombrovskii, *J. Gen. Chem. USSR*, **34**, 3150 (1964) [*C.A.*, **62**, 1594c (1965)].

[113] S. C. Dickerman and I. Zimmerman, *J. Org. Chem.*, **39**, 3429 (1974).

[114] K. B. L. Mathur, H. S. Mehra, D. R. Sharma, and V. P. Chachra, *Indian J. Chem.*, **1**, 388 (1963) [*C.A.*, **60**, 1633d (1964)].

[115] K. P. Sarbhai and K. B. L. Mathur, *Indian J. Chem.*, **1**, 482 (1963) [*C.A.*, **60**, 4047d (1964)].

[116] H. S. Mehra, *J. Indian Chem. Soc.*, **45**, 178 (1968) [*C.A.*, **69**, 76849d (1968)].

[117] V. P. Bhatia and K. B. L. Mathur, *Tetrahedron Lett.*, **1971**, 2371.

[118] J. F. Bagli and P. L'Ecuyer, *Can. J. Chem.*, **39**, 1037 (1961).

[119] O. C. Musgrave and C. J. Webster, *J. Chem. Soc., Perkin I*, **1974**, 2260, 2263.

[120] V. V. Bhat and J. L. Bose, *Symp. Syn. Heterocycl. Compounds Physiol. Interest*, Hyderabad, India, **1964** (publ. 1966), 12–16 [*C.A.*, **68**, 95624x (1968)].

[121] J. N. Gadre and R. A. Kulkarni, *Curr. Sci.*, **38**, 340 (1969) [*C.A.*, **71**, 70449t (1969)].

[122] T. Sakane, I. Tsuda, E. Mihara, and M. Ichida, *Kogyo Kagaku Zasshi*, **74**, 1174 (1971) [*C.A.*, **75**, 130761h (1971)].

[123] P. C. Taunk, S. K. Jain, and R. L. Mital, *Ann. Soc. Sci. Bruxelles, Ser. 1*, **84**, 383 (1971) [*C.A.*, **74**, 141445p (1971)].

[124] N. I. Ganushchak, V. A. Vengrzhanovskii, A. M. Dumanski, and A. V. Dombrovskii, *Dopov. Akad. Nauk Ukr. RSR, Ser. B*, **31**, 517 (1969) [*C.A.*, **71**, 70201f (1969)].

[125] F. Bell and C. J. Olivier, *Chem. Ind. (London)*, **1965**, 1558.

[126] A. Mondon, K. Schattka, and K. Krohn, *Chem. Ber.*, **105**, 3748 (1972).

[127] W. E. Truce, J. J. Breiter, and J. E. Tracy, *J. Org. Chem.*, **29**, 3009 (1964).

[128] E. Siegel and S. Petersen, *Angew. Chem.*, **74**, 873 (1962). The structures assigned here were corrected by Truce et al.[127]

[129] C. Nakashima, S. Tanimoto, and R. Oda, *Kogyo Kagaku Zasshi*, **67**, 1705 (1964) [*C.A.*, **62**, 10357f (1965)].

[130] C. Nakashima, S. Tanimoto, and R. Oda, *Nippon Kagaku Zasshi*, **86**, 442 (1965) [*C.A.*, **63**, 8239g (1965)].

[131] Y. Wada and R. Oda, *Kogyo Kagaku Zasshi*, **67**, 2093 (1964); [*C.A.*, **62**, 13177e (1965)].

[132] R. Frimm, L. Fišera, and J. Kováč, *Collect. Czech. Chem. Commun.*, **38**, 1809 (1973).

[133] M. A. Khan and J. B. Polya, *Austr. J. Chem.*, **26**, 1147 (1973).

CHAPTER 4

SELENIUM DIOXIDE OXIDATION

NORMAN RABJOHN

University of Missouri, Columbia

CONTENTS

INTRODUCTION

Over twenty-six years have passed since selenium dioxide oxidation was reviewed in this series.[1] During the intervening years the reagent has developed considerable stature, and its use has become so common that it should be classed with the standard oxidizing agents. A search for examples of oxidation by selenium dioxide or selenious acid is difficult because the terms seldom are indexed in abstract journals unless included in the title of a publication, and are buried in the discussion and

---

[1] N. Rabjohn, *Org. Reactions*, **5**, 331 (1949).

experimental parts of the literature.[2] A number of the references cited were obtained from other reviews,[3-14] and from abstracts. Many examples of oxidations by selenium dioxide probably have not been located, but it is hoped that enough have been included to cover all phases of the use of this most versatile reagent.

Frequently it is difficult to decide whether selenium dioxide or selenious acid was the oxidizing agent employed in a reaction because it is impossible to determine if water was present. The determination is complex because the free acid or anhydride may react reversibly or irreversibly with various solvents, such as alcohols and acids, to give intermediate selenium-containing compounds of unknown structure which might be the real oxidizing agents. The oxidizing agent is referred in here as selenium dioxide even though water or selenious acid was used to a reaction system.

Reactions brought about by other oxidizing agents in combination with catalytic amounts of selenium dioxide, e.g., the ring contraction of cycloalkanones to cycloalkanecarboxylic acids by hydrogen peroxide and selenium dioxide* are not considered here.

In this review the practical aspects of oxidation with selenium dioxide are emphasized, and much of the information is summarized in the Tabular Survey. Even though much knowledge has been gained in recent years on the mechanisms of selenium dioxide oxidations, these mechanisms are not treated in detail. These phases of the chemistry of selenium dioxide have been covered very adequately in recent excellent reviews by Jerussi[7] and by Trachtenberg.[13] The latter also presents interesting discussions and views of a number of other aspects of selenium dioxide oxidations.

*See Ref. 6 for leading references.

[2] *C.A.* registration number of selenium dioxide 7446-08-4; see Ref. 39.

[3] T. W. Campbell, H. G. Walker, and G. M. Coppinger, *Chem. Rev.*, **50**, 279 (1952).

[4] J. R. Carver, *Diss. Abstr. B*, **30**, 3092 (1970).

[5] L. F. Fieser and M. Fieser, *Reagents for Organic Synthesis*, Vols. I–V, Wiley-Interscience, New York, 1967–1975.

[6] H. O. House, *Modern Synthetic Reactions*, 2nd ed. W. A. Benjamin, New York, 1972.

[7] R. A. Jerussi in *Selective Organic Transformations*, B. S. Thyagarajan, Ed., Vol. I, Wiley-Interscience, New York, 1970, p. 301.

[8] H. P. Klein, *Diss. Abstr.*, *B*, **28**, 4942 (1968).

[9] C. H. Nelson, *Diss. Abstr.*, **26**, 1920 (1965).

[10] P. J. Neustaedter in *Steroid Reactions*, *C.* Djerassi, Ed., Holden-Day, San Francisco, Ca., 1963, Chap. 2.

[11] Y. Ogata and I. Tabushi, *Kagaku no Ryoiki*, **12**, 489 (1958) [*C.A.*, **53**, 1093c (1959)].

[12] R. Owyang in *Steroid Reactions*, C. Djerassi, Ed., Holden-Day, San Francisco, Ca., 1963, Chap. 5.

[13] E. N. Trachtenberg in *Oxidation, Techniques and Applications in Organic Synthesis*, R. L. Augustine, Ed., Marcel Dekker, New York, 1969, Chap. 3.

[14] J. W. White, *Diss. Abstr.*, *B*, **28**, 853 (1967).

NATURE OF THE REACTION

Selenium dioxide has been recognized as a selective oxidizing agent for organic compounds since 1932, when a systematic study of its oxidizing properties was first undertaken.[15] It effects the oxidation of carbon-hydrogen bonds which are attached to activating groups such as carbonyl, carboxyl and related derivatives, olefinic, acetylenic, and other unsaturated systems. Reactions of these types may be illustrated by the generalized structures in the accompanying summary in which reference is made to specific examples in the Tabular Survey.

Substrate	Product	Table(s)
O ‖ RCH_2CY Y = H, R, OH, OR, OCOR, NH_2	OO ‖ ‖ RCCY	II, IV, VIII, XI, XIII
RCH=$CHCH_2$—	RCH=CHCH(Y)— Y = OH, OR, OCOR	II, VI, VII, XIII
RC≡CCH_2—	RC≡CCH(OH)—	VI
$ArCH_2$—	O ‖ ArC—	VII
O ‖ $RCCH_2CH_2$—	O ‖ RCCH=CH—	II, XI, XIII

In addition to the indicated products of oxidation, which are normally carbonyl compounds, both alcohols and acids are encountered. The best known perhaps are allylic alcohols from olefins and heterocyclic carboxylic acids obtained by oxidation of alkyl-substituted heterocyclic compounds.

Selenium dioxide also accomplishes dehydrogenation of certain carbonyl compounds, many steroids, acids, and hydroaromatic substances.

$$-CCH_2CH_2- + SeO_2 \rightarrow -CCH=CH- + Se + H_2O$$
$$\quad\; \overset{\|}{O} \qquad\qquad\qquad\qquad \overset{\|}{O}$$

$$HO_2CCHCHCO_2H + SeO_2 \rightarrow HO_2CC=CCO_2H + Se + H_2O$$

$$+ SeO_2 \longrightarrow \qquad + Se + H_2O$$

[15] H. L. Riley, J. F. Morley, and N. A. C. Friend, *J. Chem. Soc.*, **1932**, 1875.

Less well-known reactions of selenium dioxide, such as the conversion of olefins to diols and the cleavage of allyl and propargyl ethers, or those which possess limited utility, such as oxidation of alcohols, alkanes, amines, mercaptans, nitriles, phenols, and several miscellaneous types of compounds, are discussed later and may be found by consulting the tabular survey tables.

Numerous publications on the mechanisms of selenium dioxide oxidations of organic compounds have appeared in recent years (Refs. 4, 7, 8, 9, 13, 16–47). Much of the information in them has been discussed in the two reviews referred to earlier (p. 263). This chapter summarizes the

[16] I. Alkonyi, *Chem. Ber.*, **94**, 2486 (1961); **95**, 279 (1962).

[17] D. Arigoni, A. Vasella, K. B. Sharpless, and H. P. Jensen, *J. Amer. Chem. Soc.*, **95**, 7917 (1973).

[18] E. J. Corey and J. P. Schaefer, *J. Amer. Chem. Soc.*, **82**, 918 (1960).

[19] V. Dovinola and L. Mangoni, *Gazz. Chim. Ital.*, **99**, 206 (1969).

[20] F. R. Duke, *J. Amer. Chem. Soc.*, **70**, 419 (1948).

[21] L. F. Fieser and G. Ourisson, *J. Amer. Chem. Soc.*, **75**, 4404 (1953).

[22] B. Horvath, *Diss. Abstr.*, **25**, 5560 (1965).

[23] J. L. Huguet, *Advan. Chem. Ser.*, **76**, 345 (1968).

[24] N. Iordanov and L. Futekov, *Izv. Inst. Obshta Neorg. Khim.*, *Bulg. Akad. Nauk*, **4**, 25 (1966) [*C.A.*, **67**, 21259q (1967)].

[25] H. H. Jaffé, *Chem. Rev.*, **53**, 191 (1953).

[26] K. A. Javaid, N. Sonoda, and S. Tsutsumi, *Tetrahedron Lett.*, **1969**, 4439.

[27] K. A. Javaid, N. Sonoda, and S. Tsutsumi, *Ind. Eng. Chem.*, *Prod. Res. Develop.*, **9**, 87 (1970).

[28] D. Jerchel and H. E. Heck, *Ann. Chem.*, **613**, 180 (1958).

[29] R. A. Jerussi and D. Speyer, *J. Org. Chem.*, **31**, 3199 (1966).

[30] G. Langbein, *J. Prakt. Chem.*, [4] **18**, 244 (1962).

[31] N. N. Mel'nikov and Yu. A. Baskakov, *Zhr. Obshch. Khim.*, **21**, 694 (1951) [*C.A.*, **45**, 9019i (1951)].

[32] N. N. Mel'nikov and Yu. A. Baskakov, *J. Gen. Chem. USSR*, **21**, 763 (1951) [*C.A.*, **46**, 10145a (1952)].

[33] N. N. Mel'nikov and Yu. A. Baskakov, *Doklady Akad. Nauk SSSR*, **85**, 337 (1952) [*C.A.*, **47**, 7461f (1953)].

[34] N. N. Mel'nikov and M. S. Rokitskaya, *J. Gen. Chem. USSR*, **14**, 1054 (1944) [*C.A.*, **41**, 5780f (1947)].

[35] H. L. Riley, *Nature*, **159**, 571 (1947).

[36] Y. Sakuda, *Bull. Chem. Soc. Jap.*, **42**, 3348 (1969) [*C.A.*, **72**, 43894y (1970)].

[37] J. P. Schaefer, *Diss. Abst.*, **19**, 2773 (1959).

[38] J. P. Schaefer and E. J. Corey, *J. Org. Chem.*, **24**, 1825 (1959).

[39] J. P. Schaefer, B. Horvath, and H. P. Klein, *J. Org. Chem.* **33**, 2647 (1968).

[40] K. B. Sharpless and R. F. Lauer, *J. Amer. Chem. Soc.*, **94**, 7154 (1972); **95**, 2697 (1973); *J. Org. Chem.*, **37**, 3973 (1972); and unpublished results.

[41] K. B. Sharpless, M. W. Young, and R. F. Lauer, *Tetrahedron Lett.*, **1973**, 1979; K. B. Sharpless, R. F. Lauer, and A. Y. Teranishi, *J. Amer. Chem. Soc.*, **95**, 6137 (1973).

[42] S. K. Talapatra, S. Sengupta, and B. Talapatra, *Tetrahedron Lett.*, **1968**, 5963.

[43] A. F. Thomas and W. Bucher, *Helv. Chim. Acta*, **53**, 770 (1970).

[44] E. N. Trachtenberg and J. R. Carver, *J. Org. Chem.*, **35**, 1646 (1970).

[45] E. N. Trachtenberg, C. H. Nelson, and J. R. Carver, *J. Org. Chem.*, **35**, 1653 (1970).

[46] W. A. Waters, *Mechanism of Oxidation of Organic Compounds*, Methuen and Co., Ltd., London, 1964, p. 94.

[47] K. B. Wiberg and S. D. Nielson, *J. Org. Chem.*, **29**, 3353 (1964).

more detailed accounts and adds recent views and findings. A discussion of the best mechanistic studies of selenium dioxide reactions currently available follows.

Allylic Oxidations

Depending upon the solvent, temperature, and stoichiometry, the oxidation of olefins at allylic positions with selenium dioxide yields compounds such as allylic alcohols and their dehydration products, esters, ethers, and α, β-unsaturated carbonyls. In addition, rearrangements of the double bond may occur with some olefins.

$$
\underset{\substack{| \\ H}}{-C}-C=C\diagup \xrightarrow[\substack{H_2O, \\ ROH, \\ RCO_2H}]{SeO_2}
\begin{cases}
\underset{\substack{| \\ OH}}{-C}-C=C\diagup \\[2ex]
\underset{\substack{| \\ OR}}{-C}-C=C\diagup \\[2ex]
\underset{\substack{| \\ OCOR}}{-C}-C=C\diagup \\[2ex]
\underset{\substack{\| \\ O}}{-C}-C=C\diagup
\end{cases}
$$

Oxidation may occur at the double bond under atypical conditions of temperature or acidity. Certain olefins such as ethylene, propylene, and stilbene give 1,2-dicarbonyl compounds when treated with selenium dioxide at higher temperatures. Also, it has been shown more recently that

$$2\ H_2C{=}CH_2 + 3\ SeO_2 \xrightarrow{50\text{-}300°} \underset{\substack{\|\ \| \\ O\ O}}{HCCH} + 3\ Se + 2\ H_2O$$

$$2\ CH_3CH{=}CH_2 + 3\ SeO_2 \xrightarrow{220\text{-}240°} 2\ CH_3COCHO + 3\ Se + 2H_2O$$

$$2\ C_6H_5CH{=}CHC_6H_5 + 3\ SeO_2 \xrightarrow{190\text{-}200°} 2\ C_6H_5COCOC_6H_5 + 3\ Se + 2\ H_2O$$

olefins may be oxidized to glycols or their derivatives in the presence of mineral acids (p. 287).[26, 27, 48–56]

[48] K. A. Javaid, N. Sonoda, and S. Tsutsumi, *Bull. Chem. Soc. Jap.*, **42**, 2056 (1969).

[49] K. A. Javaid, N. Sonoda, and S. Tsutsumi, *Bull. Chem. Soc. Jap.*, **43**, 3475 (1970).

[50] D. H. Olson, *Tetrahedron Lett.*, **1966**, 2053.

[51] D. H. Olson, U.S. Pat. 3,427,348 [*C.A.*, **71**, 12587a (1969)].

[52] D. H. Olson, U.S. Pat. 3,632,776 [*C.A.*, **76**, 85401y (1972)].

[53] N. Sonoda, *Yuki Gosei Kagaku Kyokai Shi*, **30**, 739 (1972) [*C.A.*, **78**, 3317e (1973)].

[54] N. Sonoda, S. Furui, K. A. Javaid, and S. Tsutsumi, *Ann. N.Y. Acad. Sci.*, **192**, 49 (1972).

[55] N. Sonoda and S. Tsutsumi, *Bull. Chem. Soc. Jap.*, **38**, 958 (1965).

[56] N. Sonoda, Y. Yamamoto, S. Murai, and S. Tsutsumi, *Chem. Lett.*, **1972**, 229 [*C.A.*, **76**, 126492z (1972)].

$$CH_2{=}CH_2 + SeO_2 \xrightarrow[HCl]{AcOH} AcOCH_2CH_2OAc + (AcOCH_2CH_2)_2Se +$$

$$(AcOCH_2CH_2)_2Se_2$$

The correlation of the structure of an olefin with the position of attack by selenium dioxide has been summarized by a set of rules.[57] Although exceptions to them have been noted (p. 271), they are useful, and their applicability has been discussed in detail.[13] Allylic hydroxylation of olefins by selenium dioxide may be correlated empirically as follows.

1. In trisubstituted olefins, oxidation occurs on the more highly substituted side of the double bond.

$$(CH_3)_2C{=}CHCH_3 \xrightarrow{SeO_2} HOCH_2C(CH_3){=}CHCH_3$$

2. The order of ease of oxidation of groups in trisubstituted olefins is $CH_2 > CH_3 > CH$.

$$CH_3CH_2C(CH_3){=}CHCH_3 \xrightarrow{SeO_2} CH_3CHOHC(CH_3){=}CHCH_3$$

$$(CH_3)_2CHC(CH_3){=}CHCH_3 \xrightarrow{SeO_2} (CH_3)_2CHC(CH_2OH){=}CHCH_3$$

3. For cyclic hydrocarbons, oxidation occurs in the ring, if possible, and α to the more substituted carbon atom. If the favored position for attack is tertiary, dienes may result.

4. Disubstituted olefins are oxidized in the α position, and CH_2 is more reactive than CH_3. If a methylene group is present on each side of the

$$CH_3CH{=}CHCH_2CH_3 \xrightarrow{SeO_2} CH_3CH{=}CHCHOHCH_3$$

[57] A. Guillemonat, *Ann. Chim.* (Paris), **11**, 143 (1939).

ethylenic carbon, both are oxidized and a mixture of alcohols is formed.

$$CH_3CH_2CH_2CH=CHCH_2C_3H_7\text{-}n \xrightarrow{SeO_2}$$

$$CH_3CH_2CHOHCH=CHCH_2C_3H_7\text{-}n + CH_3CH_2CH_2CH=CHCHOHC_3H_7\text{-}n$$

5. For cyclic olefins unsubstituted on the ethylenic carbons, attack is α and methylene is more active than methinyl. Small amounts of 4-methyl-cyclohexen-3-ol, as well as toluene and 4-methylcyclohexene, were obtained in this reaction, showing that allylic rearrangement products can arise.

6. Terminal olefins yield primary alcohols with allylic rearrangement of the double bond.

$$CH_3(CH_2)_3CH=CH_2 \xrightarrow{SeO_2} CH_3(CH_2)_2CH=CHCH_2OH$$

Guillemonat postulated that the organoselenium intermediates in allylic oxidations are selenides. More recently the selenium dioxide oxidation of (+)-p-menth-1-ene (1) was studied and the involvement of an allylseleninic acid intermediate 2 was proposed.[47] It was believed that the latter underwent displacement by the solvent to give the observed products;[47] however, others found that they are racemic in acetic acid-acetic anhydride.[13]

An investigation has been made of the oxidation of 1,3-diphenylpropene (3) by selenium dioxide in acetic acid at 115°.[39] An isotope effect (k_H/k_D) of 3.2 observed for oxidation at the benzylic position indicated that an allylic carbon-hydrogen bond was being broken in the rate-determining step. Also, oxidation was most rapid for electron-rich-olefins and slower for sterically hindered olefins. The mechanism proposed involves the formation of an allylic selenium(II) ester **4** that decomposes to product **5** through a solvolysis reaction. Also, the intermediate

$$C_6H_5\overset{H}{\underset{\underset{C_6H_5}{|}}{\overset{|}{C}H}} + \underset{O}{\overset{RO\quad OH}{\underset{\underset{O}{\|}}{Se}}} \xrightarrow{\text{Slow}} C_6H_5CH{=}CH\underset{4}{\overset{O}{\overset{\overset{SeOH}{\diagup}}{CH}}}C_6H_5 + ROH \longrightarrow$$

3

$$C_6H_5CH{=}CH\overset{+}{C}HC_6H_5 + HSeO_2^- \xrightarrow{\text{AcOH}} C_6H_5\underset{\underset{OAc}{|}}{\overset{}{C}HCH}{=}CHC_6H_5$$

5

isolated was thought to have a selenoxide structure, but it decomposed too slowly to **5** to be the major source of the product.

Selenoxides have been reported to be formed as intermediates in the oxidation of cyclohexene in acetic acid-acetic anhydride;[47] however, it is now believed that these intermediates are selenides, **6**.[40]

$$\text{cyclohexene} + SeO_2 \xrightarrow{\text{AcOH}} \text{product}$$

6

Although arguments have been raised against the involvement of allylseleninic acids (**2**) in the oxidation of olefins because of the inertness of benzylseleninic acid to solvolysis,[39, 45] an alternative mechanism has been suggested. It involves initial ene addition of an $>Se^+{-}O^-$ moiety (step *a*) followed by dehydration or its equivalent (step *b*), and a [2, 3]-sigmatropic shift (step *c*) of the resulting allylseleninic acid, such as **7** from 2-methyl-2-heptene.[40] The observed products arise by solvolysis or decomposition of the selenium(II) ester **8**. Support for the [2, 3] shift was provided by the finding that the appropriate intermediates could be prepared and oxidized to the same products as those obtained from the oxidation reaction mixtures. Additionally, the sigmatropic rearrangement of the allylseleninic acid **7** was shown to lead stereoselectively

(X, Y = OH, OAc, or OR)

to the anticipated (E) -isomer **9**. Evidence in favor of the initial ene reaction has been provided by demonstrating that with appropriate substrates the expected intermediates can be trapped.[17]

Oxidation of β-myrcene (**10**) gave the (E) -alcohol **11** and (E) -aldehyde **12**.[58]

[58] G. Büchi and H. Wüest, *Helv. Chim. Acta*, **50**, 2440 (1967).

Detailed investigation of this exclusive formation of (E) -oxidation products from *gem*-dimethyl olefins has confirmed that the (E) -alcohols and -aldehydes are formed stereoselectively.[59-61] In the series of tri-substituted olefins studied the sequence of reactivity was $CH_2 > CH_3 >$ CH, in agreement with findings of Guillemonat;[57] however, isopropenyl-cyclohexane was found to follow the reactivity sequence $CH > CH_2 > CH_3$, corroborating some of the results obtained by other investigators.[43-45]

(+)-Carvone (13) was oxidized with selenium dioxide in alcohol solution to give the optically active aldehyde 14, the optically inactive tertiary alcohol 15, and dehydrocarvacrol (16).[62] The major volatile product was neither primary alcohol 17, nor the corresponding aldehyde 14, but the tertiary alcohol 15, which was transformed in to the

phenol 16 by brief exposure to sodium hydroxide. The formation of the tertiary alcohol also is contrary to the rule of Guillemonat.[57]

Since early reports of the reaction of limonene with selenium dioxide were not substantiated by spectral data, a reinvestigation of the reaction was undertaken.[43] In ethanol, (+)-limonene (18) gave racemic *p*-mentha-1, 8-dien-4-ol (20) as the main product. Others reported that they obtained optically active 20, [44, 63] but a more recent detailed study showed these

[59] U. T. Bhalerao, J. J. Plattner, and H. Rapaport, *J. Amer. Chem. Soc.*, **92**, 3429 (1970); J. J. Plattner, U. T. Bhalerao, and H. Rapaport, *ibid.*, **91**, 4933 (1969).
[60] U. T. Bhalerao and H. Rapaport, *J. Amer. Chem. Soc.*, **93**, 4835 (1971).
[61] U. T. Bhalerao and H. Rapaport, *J. Amer. Chem. Soc.*, **93**, 5311 (1971).
[62] G. Büchi and H. Wüest, *J. Org. Chem.*, **34**, 857 (1969).
[63] C. W. Wilson and P. E. Shaw, *J. Org. Chem.*, **38**, 1684 (1973).

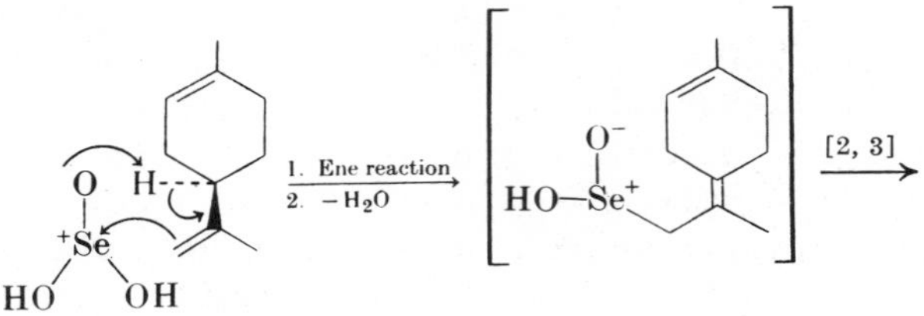

results to be in error.[64] On the basis of the mechanistic scheme proposed for the selenium dioxide oxidation of olefins (p. 270),[40] the reaction proceeds through the symmetric seleninic acid **19**, and **20** would be produced as the racemate.

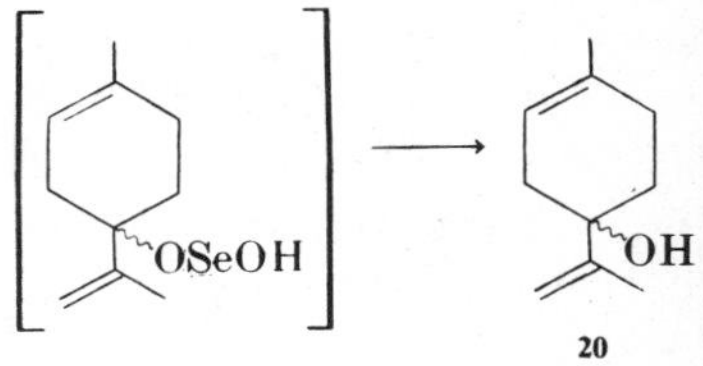

Oxidation of Aliphatic Carbonyl Compounds

The oxidation of a methylenic group α to a carbonyl group is one of the oldest and more exploited reactions of selenium dioxide, since it frequently affords high yields of α-diketones and glyoxals. A kinetic investigation was made of the oxidation of desoxybenzoin and various substituted derivatives **21** by selenium dioxide in 70% acetic acid.[18] It was concluded that an enol selenite ester **22** is formed directly from the ketone by attack

$$\text{ArCOCH}_2\text{Ar}' + \text{SeO}_2 \xrightarrow{\text{AcOH}} \text{ArCOCOAr}' + \text{Se} + \text{H}_2\text{O}$$

21

of an electrophile-nucleophile pair, $H_3SeO_3^+$ and H_2O, in the acid-catalyzed process. The enol selenite ester was presumed to rearrange to an α-substituted selenium(II) ester **23** which decomposed rapidly to a diketone, selenium, and water.

$$\text{ArCOCH}_2\text{Ar}' + \text{H}_3\text{SeO}_3^+ + \text{B} \underset{\text{Slow}}{\overset{}{\rightleftarrows}} \overset{\overset{\text{OSeO}_2\text{H}}{|}}{\text{Ar}\overset{}{\text{C}}{=}\text{CHAr}'} + \text{H}_2\text{O} + \text{BH}^+$$

21   22

$$\xrightarrow{\text{Fast}} \overset{\overset{\text{OSeOH}}{|}}{\text{ArCO}\overset{}{\text{C}}\text{HAr}'} \xrightarrow{\text{Fast}} \text{ArCOCOAr}' + \text{Se} + \text{H}_2\text{O}$$

23

---

[64] H. P. Jensen and K. B. Sharpless, *J. Org. Chem.*, **40**, 264 (1975).

It also has been suggested that the oxidation involves a rapid, concerted reaction of an enol to give a hypothetical selenium(II) ester of the ketone, which decomposes to products.[46, 65]

$$
\underset{Ar'}{\overset{H-O}{\underset{}{\bigvee}}}\underset{H}{\overset{Ar}{\bigvee}} \text{—} O\overset{\frown}{=}Se\overset{OH}{\underset{OH}{}} \rightleftharpoons \underset{Ar'}{\overset{O}{\underset{H}{\bigvee}}}\underset{}{\overset{Ar}{\bigvee}}\text{—}O\text{—}Se\underset{OH}{\overset{}{}} \longrightarrow \underset{Ar'}{\overset{O}{\bigvee}}\underset{O}{\overset{Ar}{\bigvee}} + Se + H_2O
$$

The closely related dehydrogenation of carbonyl compounds has been recognized since the original studies on selenium diooide,[15] but it was not exploited to any great extent until it was reported (in 1947) that 12-keto steroids are converted to 9(11)-unsaturated ketones, instead of the expected 11,12-diketones, by this agent;[66] e.g., 3-acetoxy-12-oxocholanic acid affords 3-hydroxy-12-oxochol-9(11)-enic acid after oxidation and saponification (Eq. 1). Later investigations showed that un-

$$
\begin{array}{ccc}
& \text{1. SeO}_2\text{, AcOH} & \\
& \text{8 hr, reflux} & \\
& \xrightarrow{} & \text{(Eq. 1)} \\
& \text{2. OH}^-\text{, 3. H}^+ &
\end{array}
$$

saturation may be introduced also in the 1, 2 position of 5α-3-keto- or Δ⁴-3-keto-steroids by means of selenium dioxide.[67, 68] Cortisone 21-acetate is converted to prednisone 21-acetate in high yields (Eq. 2).

$$
\xrightarrow[\text{reflux}]{\substack{\text{SeO}_2, \, t\text{-C}_4\text{H}_9\text{OH} \\ \text{AcOH, 4 hr.}}}
$$

(90)%

(Eq.2)

[65] P. A. Best, J. S. Littler, and W. A. Waters, J. Chem. Soc., 1962, 822.

[66] E. Schwenk and E. Stahl, Arch. Biochem., 14, 125 (1947).

[67] C. Meystre, H. Frey, W. Voser, and A. Wettstein, Helv. Chim. Acta, 39, 734 (1956).

[68] S. Szpilfogel, T. Posthumus, M. DeWinter, and D. A. van Dorp, Rec. Trav. Chim., 75, 475 (1956).

The mechanism of the dehydrogenation reaction has been studied fairly extensively.[29, 30, 69] The accompanying scheme has been outlined for the dehydrogenation of a monoketone.[7] It involves either 1,4 elimination from a selenite ester 24, or 1, 2 elimination from an α-ketoselenium(II) ester 25.

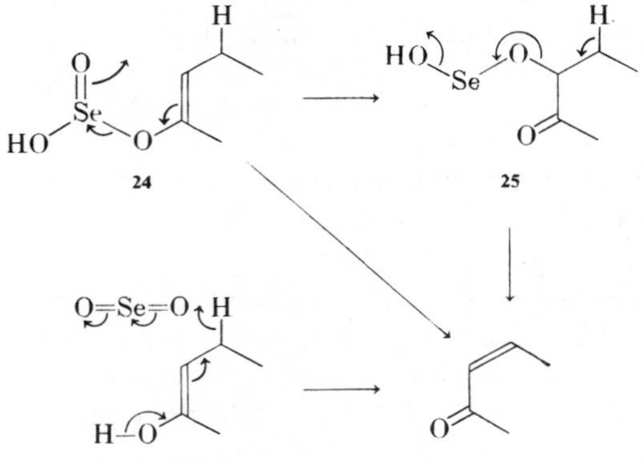

The possibility of a direct attack on the allylic position of an enol 26 by selenium dioxide to remove a hydride ion was also suggested.

The oxidation of 4, 4-dimethyl-2-cyclohexen-1-one (27a) gave, in addition to the dienone 28, 14% of the diselenide 29a.[70] Similar results were obtained with 4-ethyl-4-methyl-2-cyclohexen-1-one (27b). Oxidation of a 1:1 mixture of 27a and 27b produced a diselenide fraction con-

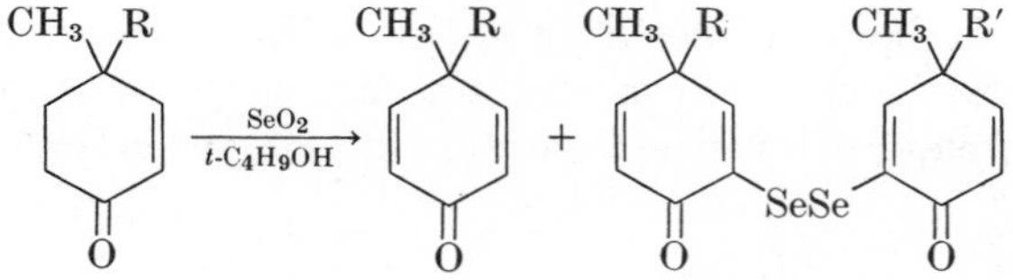

27a (R = CH₃)
27b (R = C₂H₅)

28

29a (R = R′ = CH₃)
29b (R = R′ = C₂H₅)
29c (R = CH₃, R′ = C₂H₅)

taining 29a, 29b, and 29c in a ratio of about 1:1:2. The formation of the crossover product 29c in statistical amount must involve dimerization of a dienone selenium radical as the final step. It is thought that the diselenides must represent an alternative pathway to the normal course of oxidation since the ratio of dienone 28 to diselenide 29 is independent of the amount of selenium dioxide used.[70]

[69] J. P. Schaefer, J. Amer. Chem. Soc., 84, 713, 717 (1962).
[70] J. N. Marx and L. R. Norman, Tetrahedron Lett., 1973, 2867.

It has been suggested that the mechanism of the selenium dioxide oxidation of aliphatic carbonyl compounds is based on the common intermediate **30** that accounts for the types of products isolated.[71] The course of a specific oxidation undoubtedly is influenced considerably by the nature

of the solvent and also by steric factors. The α-dicarbonyl compounds usually are prepared in dioxane or ethanol, whereas dehydrogenation reactions often are accomplished in tertiary alcohols, either with or without added acetic acid, as well as in aromatics and other solvents.

Other Oxidations

Oxidation of the α-methylenic, or benzylic, position of alkyl groups attached to aromatic systems by selenium dioxide leads to carbonyl compounds or acids. The reaction has been used mostly on heterocyclic substrates, particularly methyl-substituted nitrogen heterocycles.[72] Frequently, excellent yields of aldehydes or acids are obtained. Mechanistic proposals for benzilic-type oxidations have been made by a number of investigators.[18, 28, 46, 73, 74]

The oxidation of alcohols to carbonyl compounds or acids by selenium dioxide has been well-documented from the original studies on the reagent. Some of the earliest mechanistic investigations of selenium dioxide were performed on alcohols. They afforded dialkyl selenites which

[71] K. B. Sharpless and K. M. Gordon, *J. Amer. Chem. Soc.* **98**, 300 (1976).
[72] R. C. Elderfield in *Heterocyclic Compounds*, R. C. Elderfield, Ed., Vol. 4, Wiley, New York, 1953, Chap. 1.
[73] D. Jerchel, J. Heider, and H. Wagner, *Ann. Chem.*, **613**, 153 (1958).
[74] D. Jerchel, E. Bauer, and H. Hippchen, *Chem. Ber.*, **88**, 156 (1955).

CH₃ + SeO₂ $\xrightarrow{\text{C}_5\text{H}_5\text{N}}{115°}$ CO₂H + Se   (Ref. 73)

(87%)

CH₂OH + SeO₂ $\xrightarrow{\text{C}_5\text{H}_5\text{N}}{90°}$ CHO + Se   (Ref. 28)

(100%)

CH₃ + SeO₂ $\xrightarrow{\text{C}_5\text{H}_5\text{N}}{115°}$ CO₂H + Se   (Ref. 73)

(75%)

were found to decompose thermally to aldehydes, selenium, and water (Eq. 3).*

$$2\ RCH_2OH + SeO_2 \rightarrow (RCH_2O)_2SeO + H_2O \xrightarrow{\text{Heat}}$$
$$2\ RCHO + Se + H_2O$$ (Eq. 3)

Organoselenium Compounds

The reactions of selenium dioxide with substrates may involve intermediate selenium-containing compositions that vary in their stabilities. Some of these unstable intermediates decompose into oxidation products, water, and selenium during the course of the reaction, but others can be isolated and purified. The more stable selenium compounds, such as the selenites which arise from the reactions of alcohols with selenium dioxide, then may be decomposed under more vigorous conditions to afford aldehydes (Eq. 3). For example, $o\text{-}C_6H_4(CH_2O)_2SeO$ obtained by heating the corresponding diol with selenium dioxide at 130–140° gives o-phthalaldehyde in 68 % yield when heated under reduced pressure over a free flame.[75] Other quite stable selenium compounds obtained by reactions with selenium dioxide are tabulated in a review.[76]

Yields of products isolated from selenium dioxide oxidations are often low, suggesting that residual selenium-containing compounds were present. Monoselenium derivatives of $\Delta^{1,4}$-3-ketosteroids were isolated on

* See Ref. 1, p. 333.
[75] F. Weygand, K. G. Kinkel, and D. Tietjen, Chem. Ber., 83, 394 (1950); H. P. Kaufmann and D. B. Spannuth, ibid., 91, 2127 (1958).
[76] G. R. Waitkins and C. W. Clark, Chem. Rev., 36, 235 (1945).

oxidation of Δ^4-3-ketosteroids with selenium dioxide[77, 78] and the structure **31** was proposed for seleno-1-dehydrotestosterone.[79] Other selenium-containing products from the reactions of steroids and enones (p. 274) with selenium dioxide have been documented.[70, 80, 81]

31

Selenium-containing compounds have been isolated as by-products from the oxidations of alkenes and dienes. Ethylene and other olefins through 1-octene have been converted to selenides, diselenides, and sele-

$$CH_2{=}CH_2 + SeO_2 \xrightarrow[100-200°]{AcOH} AcOCH_2CH_2OAc +$$

$$(AcOCH_2CH_2)_2Se + (AcOCH_2CH_2)_2Se_2$$

$$(CH_3)_2C{=}C(CH_3)_2 + SeO_2 \xrightarrow{500°}$$

32

nophenes such as **32** (Refs. 23, 49, 50–52, 54, 82–84). The high-temperature reactions of alkenes and alkanes with selenium dioxide that yield selenophenes perhaps should not be considered simple oxidation processes.

A study has been made of the organoselenium intermediate from the selenium dioxide oxidation of cyclohexene in the presence of sulfuric acid.[49] Other cyclic olefins such as 6,7-dihydro-5H-benzocycloheptene have been shown to produce di- and tetra-selenides when refluxed with

[77] K. G. Florey, U.S. Pat. 2,917,507 [*C.A.*, **54**, 6821i (1960)].
[78] K. Florey and A. R. Restivo, *J. Org. Chem.*, **22**, 406 (1957).
[79] J. S. Baran, *J. Amer. Chem. Soc.*, **80**, 1687 (1958).
[80] E. Merck A.-G., Belg. Pat. 623,277 [*C. A.*, **60**, 10758c (1964)].
[81] M. Uskokovic̀, M. Gut, and R. I. Dorfman, *J. Amer. Chem. Soc.*, **81**, 4561 (1959).
[82] A. Horeau and J. Jacques, *Bull. Soc. Chim. Fr.*, **1956**, 1467.
[83] Y. Nakanishi, N. Kurata, and Y. Okuda, Jap. Pat. 72 13,018 [*C.A.*, **77**, 19179b (1972)].
[84] Yu.K. Yur'ev and L. I. Khmel'nitskiǐ, *Dokl. Akad. Nauk SSSR*, **94**, 265 (1954) [*C.A.*, **49**, 3121i (1955)].

selenium dioxide in pyridine.[85] 1,3-Butadiene,[27] isoprene,[86] and 2,3-dimethyl-1,3-butadiene[84, 87] have been found to give selenium compounds in addition to the expected oxidation products when allowed to react with selenium dioxide. The product from the latter diene has been shown to have structure 33 instead of 34 as previously proposed.[87]

33   34

The reaction of 2-styrylpyridine with selenium dioxide was found to give 2-(α-pyridyl)selenonaphthene (in 20% yield) in addition to phenyl-2-pyridylglyoxal (in 31% yield); 4-styrylpyridine afforded only the corresponding selenonaphthene.[88] These results contrast with those with stilbene which is oxidized to benzil (in 86% yield).

Primary and secondary amines react readily with selenium dioxide to produce compounds with nitrogen-selenium bonds.[89-95] Other easily oxidized nitrogen compounds such as oximes,[89, 94-98] semicarbazones,[99-102] and phenylhydrazones,[98] yield products with selenium and nitrogen or oxygen bonds.[103]

As might be expected, certain sulfur compounds react with selenium dioxide, and some form sulfur-selenium bonds in addition to undergoing

[85] P. Rona, J. Chem. Soc., 1962, 3629.
[86] J. Tanaka, T. Suzuki, K. Takabe, and T. Katagiri, J. Chem. Soc. Jap., Chem. Ind. Chem., 1973, 292 [C.A., 78, 123896q (1973)].
[87] W. L. Mock and J. H. McCausland, Tetrahedron Lett., 1968, 391.
[88] C. A. Buehler, J. O. Harris, and W. F. Arendale, J. Amer. Chem. Soc., 72, 4953 (1950).
[89] V. Bertini, Gazz. Chim. Ital., 97, 1870 (1967).
[90] N. P. Buu-Hoi, J. Chem. Soc., 1949, 2882.
[91] P. Cukor and P. F. Lott, J. Phys. Chem., 69, 3232 (1965).
[92] A. V. El'tsov, V. S. Kuznetsov, and L. S. Efros, Zh. Obshch. Khim., 33, 3965 (1963) [C.A., 60, 9263e (1964)].
[93] M. V. Gorelik, Khim. Geterosikl. Soedin., 1967, 541 [C.A., 68, 21730g (1968)].
[94] R. Paetzold and E. Roensch, Angew. Chem., 76, 992 (1964).
[95] T. F. Stepanova, D. P. Sevbo, and O. F. Ginzburg, Zh. Org. Khim. 7, 1921 (1971) [C.A., 76, 3509c (1972)].
[96] F. E. King and D. G. I. Felton, J. Chem. Soc., 1949, 274.
[97] D. Paquer, M. Perrier, and J. Vialle, Bull. Soc. Chim. Fr., 1970, 4517.
[98] M. Perrier and J. Vialle, Bull. Soc. Chim. Fr., 1971, 4591.
[99] I. Lalezari, A. Shafiee, and M. Yalpani, Tetrahedron Lett., 1969, 5105.
[100] I. Lalezari, A. Shafiee, and M. Yalpani, J. Heterocycl. Chem., 9, 1411 (1972).
[101] H. Meier and E. Voight, Tetrahedron, 1972, 187.
[102] M. Yalpani, I. Lalezari, and A. Shafiee, J. Org. Chem., 36, 2836 (1971).
[103] R. C. Pal, R. D. Sharma, and K. C. Malhotra, Indian J. Chem., 10, 428 (1972).

oxidation. Cysteine is oxidized to cystine and selenium dicysteine,[104] and glutathione affords selenium diglutathione plus oxidation products.[105] *meso*-2, 3,-Dimercaptosuccinic acid and related compounds are reported to yield cyclic selenium tetrasulfides when caused to react with selenium dioxide in methanol.[106]

SCOPE OF THE REACTION

The general types of reactions effected by selenium dioxide have been outlined on p. 264 and many of them have been discussed previously.[1] The following discourse expands upon this material and summarizes reactions listed in the tables that demonstrate new or interesting variations of the application of selenium dioxide.

Numerous additional examples of the oxidation by selenium dioxide of methylene or methyl groups, activated by carbonyl, have been recorded in recent years. The products of such reactions are usually 1, 2-diketones or α-ketoaldehydes (glyoxals), as shown in the accompanying equations.

$$
\underset{\text{reflux}}{\overset{\text{SeO}_2}{\underset{\text{C}_2\text{H}_5\text{OH, 6 hr}}{\longrightarrow}}} \qquad \text{(Ref. 107)}
$$

(90%)

$$
p\text{-RC}_6\text{H}_4\text{COCH}_3 \xrightarrow[\substack{\text{Dioxane, H}_2\text{O}\\ \text{4 hr, reflux}\\ \text{R}=\text{C}_2\text{H}_5\text{- to } n\text{-C}_{10}\text{H}_{21}\text{-}}]{\text{SeO}_2} p\text{-RC}_6\text{H}_4\text{COCHO (Ref. 108)}
$$
(20–82 %)

α,β-Unsaturated ketones, such as the *o*-hydroxychalcones, undergo oxidation and cyclization to flavones (Eq. 4).[109-112]

$$
\xrightarrow[(20-80\%)]{\text{SeO}_2} \qquad \text{(Eq. 4)}
$$

[104] H. L. Klug and D. F. Petersen, *Proc. S. Dakota Acad. Sci.*, **28,** 87 (1949) [*C.A.*, **46,** 900e (1952)].

[105] D. F. Petersen, *Proc. S. Dakota Acad. Sci.*, **30,** 53 (1951) [*C.A.*, **48,** 11507e (1954)].

[106] E. A. H. Friedheim, U.S. Pat. 3,544,593 [*C.A.*,**74,** 87991y (1971)].

[107] W. VanderHaar, R. C. Voter, and C. V. Banks, *J. Org. Chem.*, **14,** 836 (1949).

[108] H. Schubert, I. Eissfeldt, R. Lange, and F. Trefflich, *J. Prakt. Chem.*, **33,** 265 (1966).

[109] W. Baker and F. Glockling, *J. Chem. Soc.*, **1950,** 2759,

[110] A. Corvaisier, *Bull. Soc. Chim. Fr.*, **1962,** 528.

[111] C. L. Huang, T. Weng, and F. C. Chen, *J. Heterocycl. Chem.*, **7,** 1189 (1970).

[112] S. S. Kumari, K. S. R. K. M. Rao, A. V. S. Rao, and N. V. S. Rao, *Curr. Sci.*, **36,** 430 (1967) [*C.A.*, **68,** 59399u (1968)].

o-Diacetylbenzene and related compounds also lead to cyclized products.

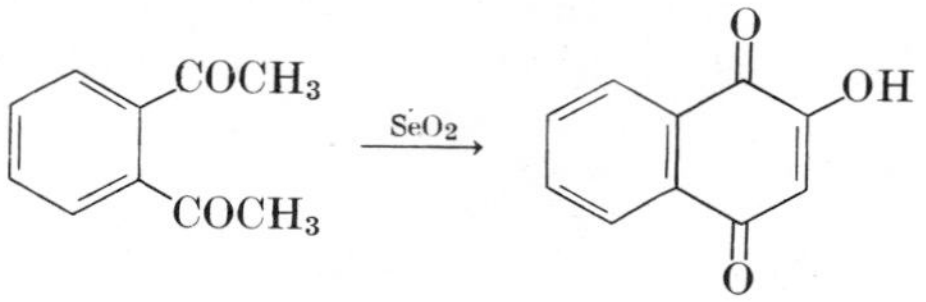

(Ref. 113)

α-Keto esters can be prepared by the reaction of selenium dioxide with an α-bromo ketone in an anhydrous alcohol.[38] Instead of glyoxals, α-keto

$$C_6H_5COCH_2Br \xrightarrow[\text{CH}_3\text{OH}]{\text{SeO}_2} C_6H_5COCO_2CH_3$$

(80%)

acids have been obtained by oxidation of substituted acetophenones in pyridine with excess selenium dioxide.[114]

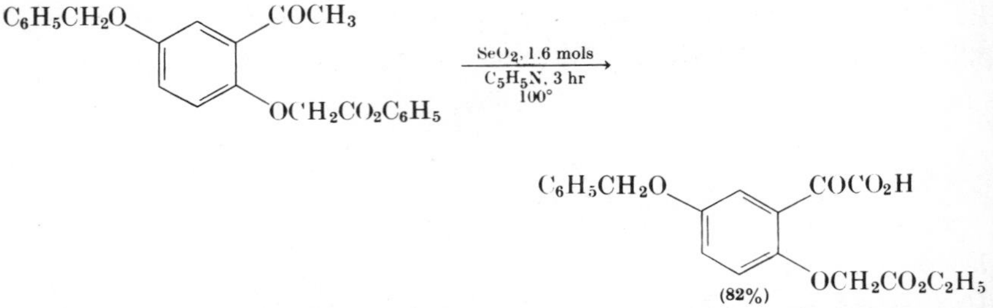

(82%)

Imides and amides were oxidized to carbonyl compounds such as 35 and 36.

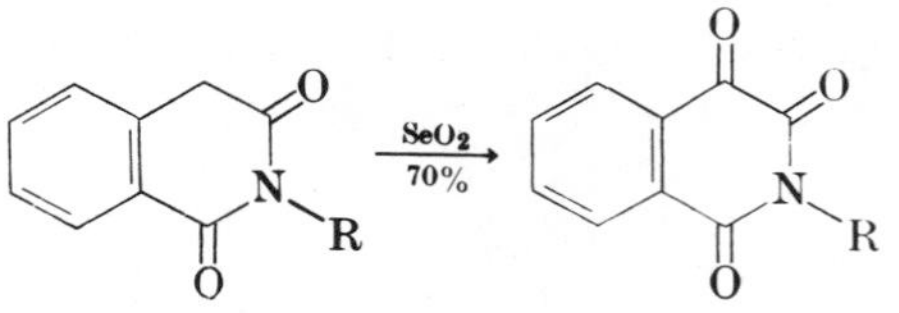

(Ref. 115)

35

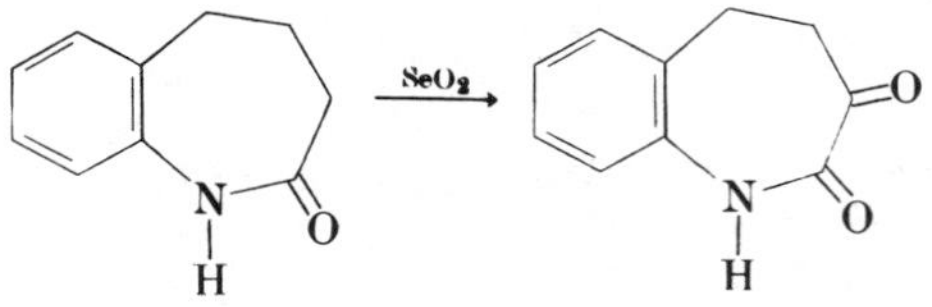

(Ref. 116)

36

[113] F. Weygand, H. Weber, and E. Maekawa, Chem. Ber., 90, 1879 (1957).
[114] G. Hallmann and K. Hägele, Ann. Chem., 662, 147 (1963).
[115] N. P. Buu-Hoi, G. Saint-Ruf, and J. C. Bourgeade, J. Heterocycl. Chem., 5, 545 (1958).
[116] C. G. Hughes and A. H. Rees, Chem. Ind. 1971, 1439.

It was observed recently that α-substituted β-diketones undergo oxidative fission when caused to react with selenium dioxide.[117] 3-Acetyl-1, 2-dihydro-4-hydroxy-1-isoquinolone was cleaved to 1, 2, 3, 4-tetrahydro-1, 3, 4-isoquinolinetrione (Eq. 5), while 1, 2, 3-triphenyl-1, 3-propanedione was oxidized similarly to benzil and benzoic acid. It is believed that this

$$\text{(Eq. 5)}$$

(89%)

β-diketone fission may be general, but not always recognized because of further oxidation of the products.

$$C_6H_5COCH(C_6H_5)COC_6H_5 \xrightarrow[100°]{SeO_2} C_6H_5COCOC_6H_5 + C_6H_5CO_2H$$

The selenium dioxide oxidation of olefins, as described earlier (p. 266) leads to allylic derivatives. Selenium dioxide oxidation has been extended to a series of terminally unsaturated alkyl acetates, the acetates of other types of alcohols, acyclic and cyclic, a number of olefins with aryl and alkoxyaryl substituents, and several aliphatic unsaturated acids.[118–121] In many instances, oxidation was accompanied by allylic rearrangements. A terminal olefin such as 37 afforded isomeric oxidation products.[118, 121]

$$H_2C{=}CHCH_2CH(OAc)CH_2CH_3 \xrightarrow[Ac_2O]{SeO_2} H_2C{=}CHCH(OAc)CH(OAc)CH_2CH_3$$
37

$$+ \text{AcOCH}_2CH{=}CHCH(OAc)CH_2CH_3$$

The oxidation of the acetates of alcohols with terminal isopropylidene groups yielded only glycol diacetates in which oxidation had taken place on the less highly substituted α-carbon atom, e.g., 4-acetoxy-2-methyl-2-hexene (38) gave the diacetate 39.[119] Terminal olefins of the type

[117] R. Howe and D. Johnson, J. Chem. Soc., Perkin Trans. I, 1972, 977.
[118] J. Colonge and N. Reymermier, Bull. Soc. Chim. Fr., 1955, 1531.
[119] J. Colonge and M. Reymermier, Bull. Soc. Chim. Fr., 1956, 188.
[120] J. Colonge and M. Reymermier, Bull. Soc. Chim. Fr., 1956, 195.
[121] J. Colonge and M. Reymermier, C. R. Acad. Sci., 237, 266 (1953).

$CH_2=C(CH_3)CH_2$- produced two products, as shown for the hexene **40**.[119]

$$(CH_3)_2C=CHCH(OAc)CH_2CH_3 \xrightarrow[\text{AcOH,Ac}_2\text{O}]{\text{SeO}_2}$$
$$\underset{38}{}$$

$$AcOCH_2C(CH_3)=CHCH(OAc)CH_2CH_3$$
$$\underset{39}{}$$

$$\underset{40}{CH_2=C(CH_3)CH_2CH(OAc)CH_3} \xrightarrow[\text{AcOH, Ac}_2\text{O}]{\text{SeO}_2}$$

$$\underset{(17\%)}{CH_2=C(CH_3)CH(OAc)CH(OAc)CH_3} + \underset{(16\%)}{CH_2=C(CH_2OAc)CH_2CH(OAc)CH_3}$$

The stereospecific oxidation by selenium dioxide of a select class of trisubstituted olefins with the general formulas **41** and **42** was investigated in detail. (E)-alcohols and (E)-aldehydes were formed stereoselectively (p. 270).[60] The oxidation of other terminal dimethyl-substituted olefins,

$$\underset{41}{}\qquad\qquad\underset{42}{}$$

$$\mathbf{41}\left(\mathbf{a},\ R = CH_3;\ \mathbf{b},\ R = C_2H_5\right.\qquad \mathbf{42}\ (\mathbf{e},\ R = C_4H_9\text{-}t)$$

$$\mathbf{c},\ R = C_3H_7\text{-}i;\ \mathbf{d},\ R = \left.\left\langle\!\!\!\bigcirc\!\!\!\right\rangle\right)$$

such as methyl 5-methyl-4-hexenoate, **(43)**,[122] geranyl esters, **44**,[123, 124] and related compounds,[40, 60, 61, 125, 126] to (E)-α, β-unsaturated aldehydes is of value.

$$\underset{43}{}\qquad\qquad\qquad\qquad (41\%)$$

$$\underset{44}{}\qquad\qquad\qquad\qquad (54\%)$$

[122] E. J. Corey and B. B. Snider, *J. Amer. Chem. Soc.*, **94**, 2549 (1972).

[123] J. Meinwald and K. Opheim, *Tetrahedron Lett.*, **1973**, 281.

[124] C. H. Miller, J. A. Katzenellenbogen, and S. B. Bowlus, *Tetrahedron Lett.*, **1973**, 285.

[125] M. Gates, *J. Amer. Chem. Soc.*, **70**, 617 (1948).

[126] S. Wakayama, S. Namba, K. Hosoi, and M. Ohno, *Bull. Chem. Soc. Jap.*, **44**, 875 (1971) [*C.A.*, **75**, 6118q (1971)].

It has been observed that aryl-substituted olefins such as propenyl-benzene are oxidized to allyl acetates, while anethole (45), gives a di-acetate; however, the isomeric ether 46 led to the rearranged ester.[120]

$$C_6H_5CH{=}CHCH_3 \xrightarrow[Ac_2O]{SeO_2} C_6H_5CH{=}CHCH_2OAc$$

$$p{-}CH_3OC_6H_4CH{=}CHCH_3 \xrightarrow[Ac_2O]{SeO_2} p{-}CH_3OC_6H_4CH(OAc)CH(OAc)CH_3$$
45

$$p{-}CH_3OC_6H_4CH_2CH{=}CH_2 \xrightarrow[Ac_2O]{SeO_2} p{-}CH_3OC_6H_4CH{=}CHCH_2OAc$$
46

Oxidation of several unsaturated aliphatic acids by selenium dioxide has been investigated.[120] Crotonic and 3-methyl-2-butenoic acid gave the expected products, while undecylenic acid produced an undistillable

$$CH_3CH{=}CHCO_2H \xrightarrow[Ac_2O]{SeO_2} AcOCH_2CH{=}CHCO_2H$$

$$(CH_3)_2C{=}CHCO_2H \xrightarrow[Ac_2O]{SeO_2} AcOCH_2C(CH_3){=}CHCO_2H + \text{(lactone)}$$

resin, but ethyl undecylenate (47) afforded a mixture of two isomeric oxidation products.

$$CH_2{=}CH(CH_2)_8CO_2C_2H_5 \xrightarrow[Ac_2O]{SeO_2} CH_2{=}CHCH(OAc)(CH_2)_7CO_2C_2H_5$$
47

$$+ \; AcOCH_2CH{=}CH(CH_2)_7CO_2C_2H_5$$

The direct oxidation of 1, 3, 5-cycloheptatriene gives tropone (48) in a simple one-step process.[127, 128]

48

The selenium dioxide oxidation of α, β-unsaturated esters to γ-lactones has been applied to terpenes and steroids.[129-134] The ester 49 was oxidized

[127] P. Radlick, *J. Org. Chem.*, **29**, 960 (1964).
[128] D. I. Schuster, J. M. Palmer, and S. C. Dickerman, *J. Org. Chem.*, **31**, 4281 (1966).
[129] N. Danieli, Y. Mazur, and F. Sondheimer, *Tetrahedron Lett.*, **1961**, 310.
[130] N. Danieli, Y. Mazur, and F. Sondheimer, *Tetrahedron Lett.*, **1962**, 1281.
[131] N. Danieli, Y. Mazur, and F. Sondheimer, *J. Amer. Chem. Soc.*, **84**, 875 (1962).
[132] N. Danieli, Y. Mazur, and F. Sondheimer, *Tetrahedron*, **22**, 3189 (1966).
[133] N. Danieli, Y. Mazur, and F. Sondheimer, *Tetrahedron*, **23**, 509 (1967).
[134] J. N. Marx and F. Sondheimer, *Tetrahedron Suppl.* **8**, Part I, 1 (1966).

to the γ-lactone, which was an important intermediate in the synthesis of the polycyclic triterpene, α-onocerin.[133] The synthesis of the cardenolide agylcone, digitoxigenin **(50)**, has been accomplished by the aid of a similar oxidation.[131–132]

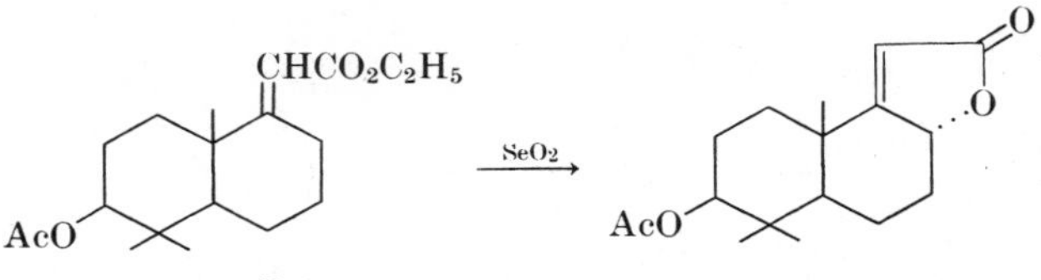

49

Oxidation of α-cyclodihydrocostunolide has been reported to take place

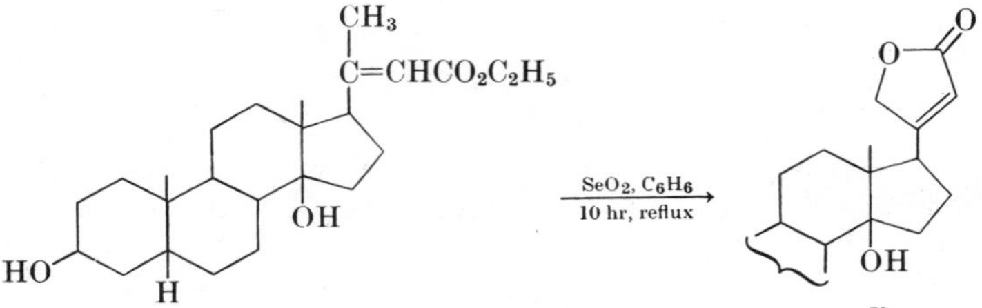

50

at the methinyl allylic position, while its isomer, β-cyclodihydrocostunolide, is oxidized at the methylene group allylic to the double bond (Eqs. 6 and 6a).[135–136]

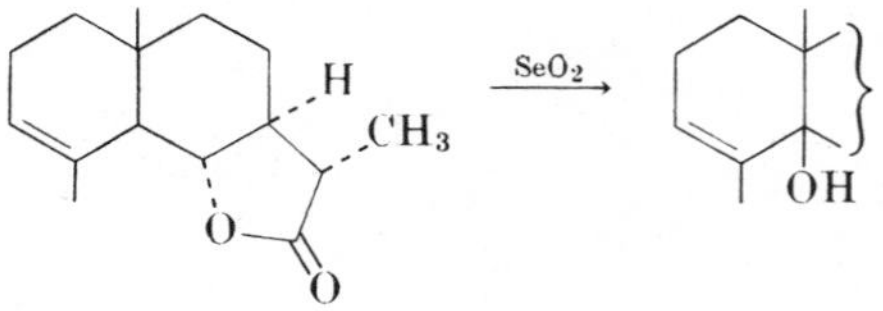

(Eq. 6)

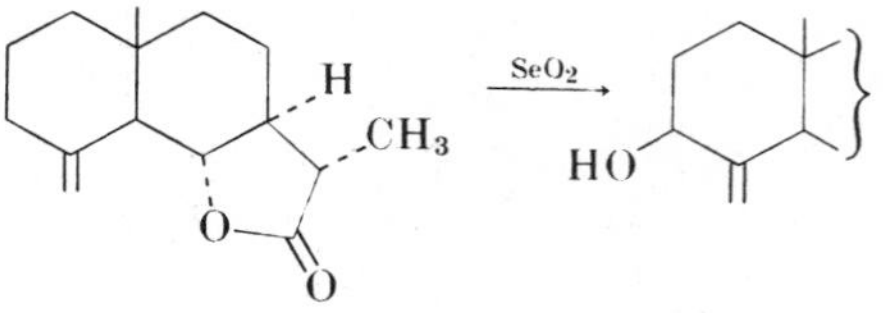

(Eq. 6a)

[135] S. P. Pathak and G. H. Kulkarni, *Chem. Ind. (London)*, **1968**, 913.
[136] S. P. Pathak and G. H. Kulkarni, *Chem. Ind. (London)*, **1968**, 1566.

The closely related benzylic oxidation affords carbonyl compounds as shown in the accompanying equations.

$$\alpha\text{-}C_{10}H_7CH_2CN \xrightarrow{\text{SeO}_2} \alpha\text{-}C_{10}H_7COCN \qquad \text{(Refs. 31, 32)}$$
$$(47\%)$$

$$\alpha\text{-}C_{10}H_7CH_2CO_2CH_3 \xrightarrow{\text{SeO}_2} \alpha\text{-}C_{10}H_7COCO_2CH_3 \qquad \text{(Refs. 31, 32)}$$
$$(68\%)$$

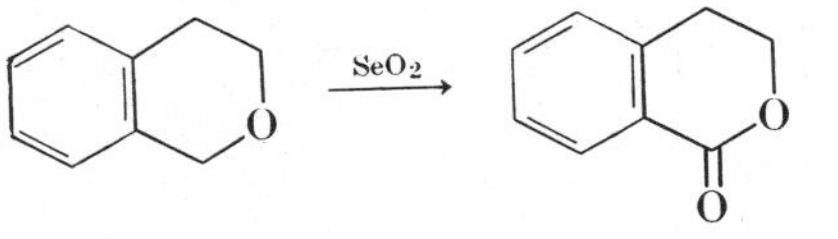

(Refs. 137–138)

(85%)

Methyl-substituted aromatic or heterocyclic systems are oxidized to aldehydes or acids, depending upon the solvent system and the amount of selenium dioxide used. The reaction has found more utility with nitrogen heterocycles than with carbocycles; in the accompanying formulations two examples are shown for naphthalene derivatives.

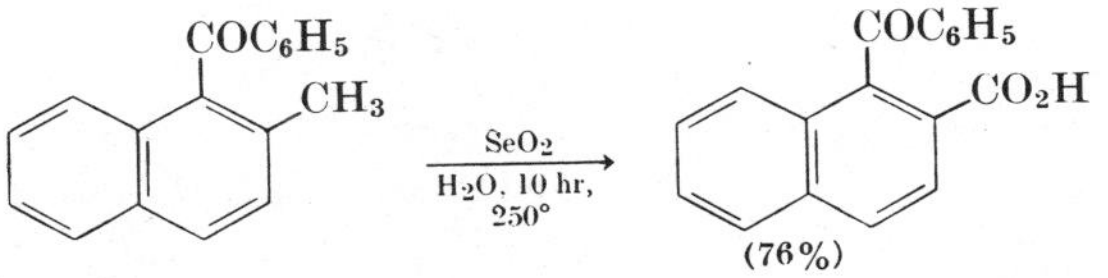

(Ref. 139)

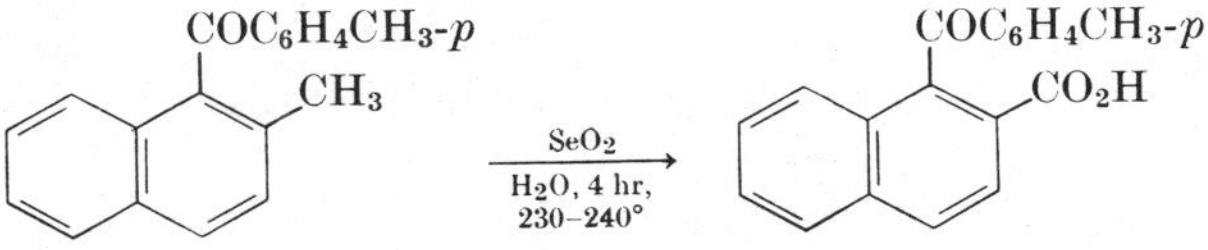

(Ref. 140)

The ring nitrogen activates methyl groups in the *ortho* and *para* positions of N-heterocycles so that they are attacked more readily than those in carbocycles.[72] On pyridines and quinolines, 4-methyl (alkyl) groups are more reactive than 2-methyl groups and 3-methyl groups are not affected by selenium dioxide.[28, 73, 74] The oxidations of 2-methylquinoline (quinaldine) and 4-methylquinoline (lepidine) give the products shown in Eqs. 7 and 8. The aldehydes were obtained with freshly prepared selenium

[137] J. Colonge and P. Boisde, *Bull. Soc. Chim. Fr.*, **1956**, 1337.
[138] P. Maitte, *Ann. Chim.*, **9**, 473 (1954).
[139] E. Clar and D. G. Stewart, *J. Chem. Soc.*, **1951**, 687.
[140] G. M. Badger, *J. Chem. Soc.*, **1947**, 764.

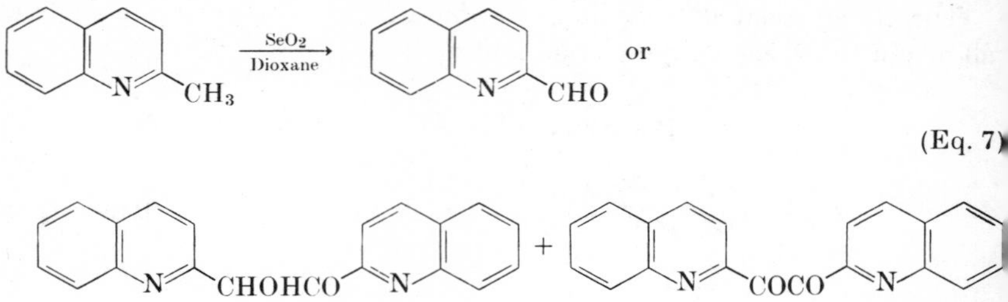

(Eq. 7)

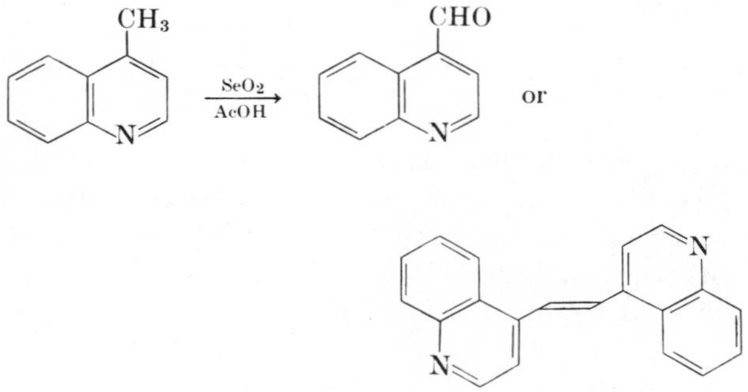

or

(Eq. 8)

dioxide, the bimolecular products with aged selenium dioxide.* Oxidation of lepidine in acetic acid has been reported to afford the same yields of quinoline-4-carboxaldehyde with freshly prepared and one-year-old selenium dioxide.[141] It has been found that 9-methylphenanthridine is oxidized by selenium dioxide in dioxane solution to the corresponding aldehyde (in 60% yield) and the ethylene derivative (7.5% in yield).[142] It is possible that the solvent has an important effect in these systems (see discussion on p. 295).

Other N-heterocycles such as 1,4-dimethylcarbostyril (51) have been oxidized in quite good yields to the corresponding aldehydes.

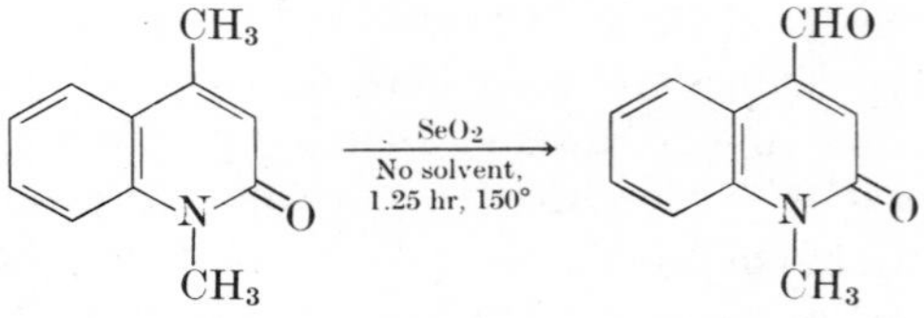

(Ref. 143)

51   (70%)

*Ref. 1, pp. 367, 369.
[141] S. F. MacDonald, J. Amer. Chem. Soc., **69,** 1219 (1947).
[142] A. G. Caldwell, J. Chem. Soc., **1952,** 2035.
[143] D. J. Cook and M. Stamper, J. Amer. Chem. Soc., **69,** 1467 (1947).

Carbocyclic benzyl alcohols are converted to aldehydes, as demonstrated for 4,5-methylenedioxy-o-phthalyl alcohol (Eq. 9). The intermediate selenium ester is isolated and thermally decomposed to the dialdehyde.[144]

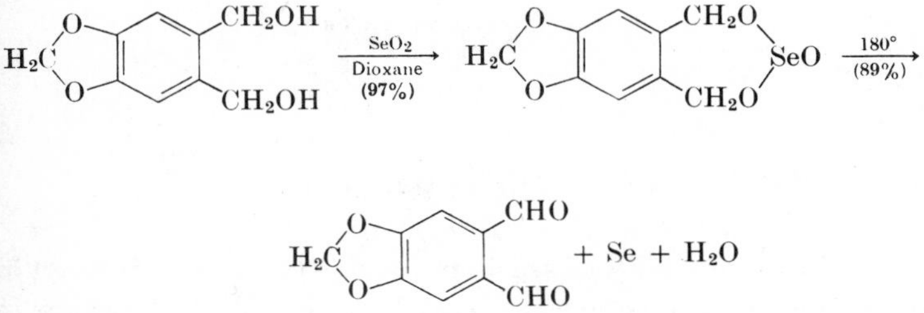

(Eq. 9)

Mention has been made (p. 266) that olefins and dienes are oxidized to glycols and their esters by selenium dioxide in the presence of mineral acids.[26, 27, 48-56] Representative examples are the oxidation of 1-hexene, cyclohexene, and 1, 3-butadiene.

$$CH_3(CH_2)_3CH{=}CH_2 \xrightarrow[\substack{AcOH,\ H_2SO_4 \\ 10\ hr,\ 115°}]{SeO_2} CH_3(CH_2)_3CHOAcCH_2OAc \quad (35\%) +$$

$$CH_3(CH_2)_2CHOAcCH{=}CH_2 \quad (12\%) \quad + \quad CH_3(CH_2)_2CH{=}CHCH_2OAc \quad (5\%)$$
(Ref. 26)

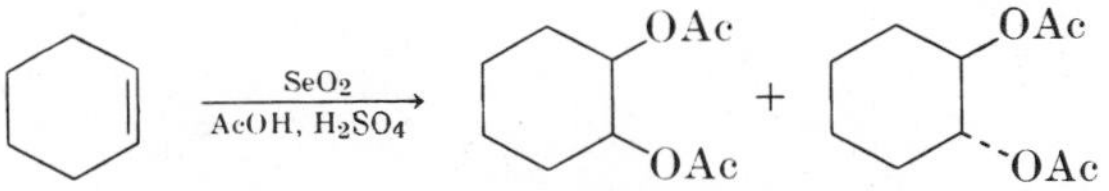

(32% ; cis:trans, 55:45)   (Ref. 26)

$$CH_2{=}CHCH{=}CH_2 \xrightarrow[\substack{AcOH,\ H_2SO_4 \\ 8hr,\ 110°}]{SeO_2} AcOCH_2CHOAcCH{=}CH_2 +$$

$$HOCH_2CHOAcCH{=}CH_2 \quad + \quad AcOCH_2CH{=}CHCH_2OAc \quad +$$

$$HOCH_2CH{=}CHCH_2OAc \quad + \quad$$

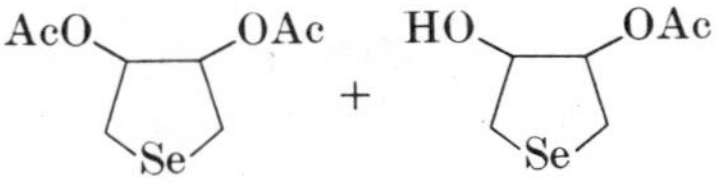

(Ref. 27)

The effect of mineral acids on the selenium dioxide oxidation of acetylenes is demonstrated with phenylacetylene and diphenylacetylene.[56]

[144] F. Dallacker, K. W. Glombitza, and M. Lipp, *Ann. Chem.*, **643**, 67 (1961).

$$C_6H_5C{\equiv}CH \begin{cases} \xrightarrow[\text{AcOH}]{\text{SeO}_2} C_6H_5COCHO \\ \\ \xrightarrow[\text{AcOH, H}_2SO_4]{\text{SeO}_2} C_6H_5COCO_2H \end{cases}$$

$$C_6H_5C{\equiv}CC_6H_5 \xrightarrow[\text{H}_2SO_4]{\text{SeO}_2, \text{AcOH}} C_6H_5COCOC_6H_5$$

The dehydrogenating ability of selenium dioxide has been employed in a number of interesting reactions besides those described elsewhere in this chapter (p. 273) and previously.* Cyclic olefins and alcohols such as **52** and **53** have been aromatized by selenium dioxide in good yields.

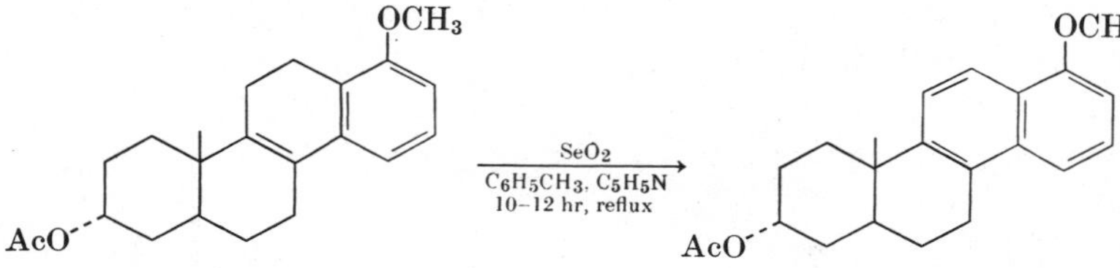

(79%)   (Ref. 145)

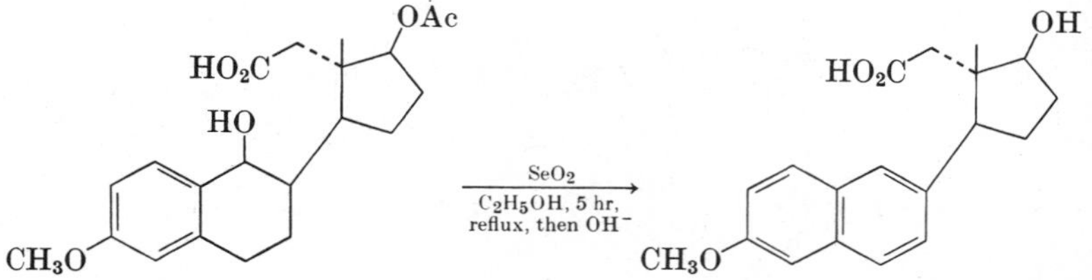

(Ref. 146)

Ethyl 2-oxo-5,5-diethoxycyclohexanecarboxylate (**54**) undergoes a novel aromatization when heated with selenium dioxide. Interesting results also were obtained in attempts to introduce oxygen at C-2 of ethyl α-safranate (**55**).[147]

*Ref. 1, pp. 340—341.

[145] W. S. Johnson, J. Ackerman, J. F. Eastham., and H. A. DeWalt, Jr., *J. Amer. Chem. Soc.*, **78**, 6302 (1956).
[146] J. H. Dygos and L. J. Chinn, *J. Org. Chem.*, **38**, 4319 (1973).
[147] G. Büchi, W. Pickenhagen and H. Wüest, *J. Org. Chem.*, **37**, 4192 (1972).

(Ref. 148)

It has been observed that in the presence of selenium dioxide the steroidal triketone **56** combined with methanol to give the 3,3-dimethoxyketal.[149]

Pyrolysis of the ketal caused the elimination of methanol, and the 3-ene resulted. Investigation of the action of selenium dioxide and methanol or ethylene glycol on a number of other steroid ketones showed that ketal formation is limited to saturated 3-keto derivatives.[150–154] Apparently

[148] K. Lempert, K. Simon-Ormai, and R. Markovits-Kornis, *Acta Chim. Acad. Sci. Hung.* **51**, 305 (1967) [*C.A.*, **66**, 115420s (1967)].

[149] E. P. Oliveto, C. Gerold, and E. B. Hershberg, *J. Amer. Chem. Soc.*, **76**, 6113 (1954).

[150] J. H. Fried, A. N. Nutile, and G. E. Arth, *J. Amer. Chem. Soc.*, **82**, 5704 (1960).

[151] B. J. Magerlein, *J. Org. Chem.*, **24**, 1564 (1959).

[152] A. L. Nussbaum, T. L. Popper, E. P. Oliveto, S. Friedman, and I. Wender, *J. Amer. Chem. Soc.*, **81**, 1228 (1959).

[153] E. P. Oliveto, H. Q. Smith, C. Gerold, R. Rausser, and E. B. Hershberg, *J. Amer. Chem. Soc.*, **78**, 1414 (1956).

[154] E. P. Oliveto, H. Q. Smith, C. Gerold, L. Weber, R. Rausser, and E. B. Hershberg, *J. Amer. Chem. Soc.*, **77**, 2224 (1955).

selenium dioxide functions as an acid catalyst for ketal formation and removes the water formed to drive the reaction to completion.

A highly sensitive test has been developed for the diagnosis of unsaturation types in steroids based upon oxidation by selenium dioxide at 0–25°.[21, 155] The test is specific for 5α- or Δ^5-steroids having a double bond adjacent to C-14;* however, it has been reported that some of the unsaturated bile acids (5β-) apparently give positive tests.[156]

When allyl and propargyl ethers are allowed to react with selenium dioxide, they undergo an oxidative cleavage which is illustrated in the three accompanying equations.[157]

$$CH_2{=}CHCH_2OC_6H_5 \xrightarrow[\substack{AcOH, \text{ dioxane} \\ 1 \text{ hr, reflux}}]{SeO_2} C_6H_5OH + \underset{(59\%)}{CH_2{=}CHCHO}$$

$$HC{\equiv}CCH_2OC_6H_5 \xrightarrow{SeO_2} C_6H_5OH + \underset{(79\%)}{HC{\equiv}CCHO}$$

$$C_6H_5CH{=}CHCH_2OCH_3 \xrightarrow{SeO_2} C_6H_5CH{=}CHCHO + CH_3OH$$

Alkyl or aryl aldehyde and ketone semicarbazones and bis(semicarbazones) have been converted to 1,2,3-selenadiazoles by oxidation with selenium dioxide. Thermolysis of the selenadiazoles leads to alkynes in good yields, as shown for the preparations of cyclooctyne and methylphenylacetylene.[99–102, 158–162]

*For details consult Ref. 5, Vol. I, p. 998.

155 L. F. Fieser, J. Amer. Chem. Soc., 75, 4395 (1953).

156 R. Osawa, Bull. Chem. Soc. Jap. 35, 158 (1962) [C.A., 57, 912f (1962)].

157 K. Kariyone and H. Yazawa, Tetrahedron Lett., 1970, 2885.

158 I. Lalezari, A. Shafiee, and H. Golgolab, J. Heterocycl. Chem., 10, 655 (1973).

159 I. Lalezari, A. Shafiee, and S. Yazdany, J. Pharm. Sci., 63, 628 (1974) [C.A. 80, 146083c (1974)].

160 I. Lalezari, N. Sharghi, A. Shafiee, and M. Yalpani, J. Heterocycl. Chem., 6, 403 (1969).

161 I. Lalezari, A. Shafiee, and M. Yalpani, Angew. Chem., Int. Ed. Engl., 9, 464 (1970).

162 H. Meier and I. Menzel, J. Chem. Soc., D, 1971, 1059.

Selenium dioxide has been used for the demethylation of nicotine (57) and the replacement of mercapto groups by hydroxyl in a number of N-heterocycles as shown for 2-mercapto-4-methylquinoline (58). It also has been found to convert thiourea derivatives (59) to the corresponding substituted ureas.[163]

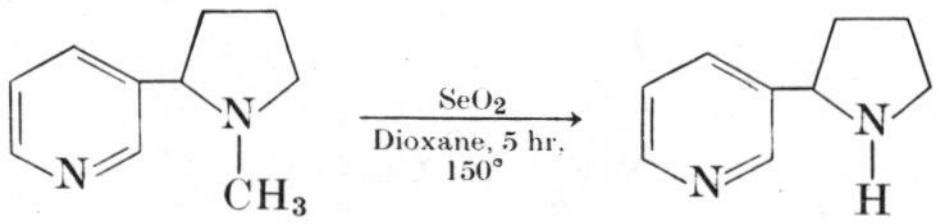

57 → (40%) (Ref. 164)

SeO₂, Dioxane, 5 hr. 150°

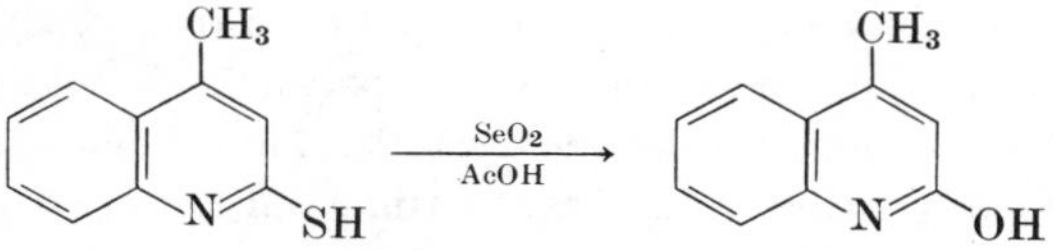

58 → (Ref. 165)

SeO₂, AcOH

$$p\text{-}CH_3C_6H_4SO_2NHCSNHC_4H_9 \xrightarrow{SeO_2} p\text{-}CH_3C_6H_4SO_2NHCONHC_4H_9$$

59

Tri-p-tolylphosphine is oxidized by selenium dioxide to the phosphine oxide and either selenium or the phosphine selenide, depending on the relative proportions of the reagents.[166] Phosphoranes, such as 60, are reported to give unsaturated 1,4-diketones.[167]

$$(p\text{-}CH_3C_6H_4)_3P \xrightarrow[C_6H_6,\,reflux]{SeO_2} (p\text{-}CH_3C_6H_4)_3PO$$

$$(C_6H_5)_3P{=}CHCOR \xrightarrow[Dioxane]{SeO_2} RCOCH{=}CHCOR \quad (70\text{--}87\%)$$

60

$$(R = aryl,\ OC_2H_5)$$

Natural amino acids, with the exception of cysteine and tryptophan, are resistant to oxidation by selenium dioxide in aqueous medium, but

---

[163] T. Kodama, K. Uehara, and S. Shinohara, *Yuki Gosei Kagaku Kyohai Shi*, **25**, 498 (1967) [*C.A.*, **68**, 12621u (1968)].

[164] A. Sadykov, *J. Gen. Chem. USSR*, **17**, 1710 (1947) [*C.A.*, **42**, 2610 (1948)].

[165] L. Monti and G. Franchi, *Gazz. Chim. Ital.*, **81**, 764 (1951).

[166] S. I. A. El Sheikh, B. C. Smith, and M. E. Sobeir, *Angew. Chem., Int. Ed. Engl.*, **9**, 308 (1970).

[167] M. I. Shevchuk, A. F. Tolochko, and A. V. Dombrovskii, *Zh. Org. Kkim.*, **7**, 1692 (1971) [*C.A.*, **75**, 140781d (1971)].

are more easily attacked in acetic acid.[168, 169] The rate of oxidation of hydroxy acids such as citric, malic, tartaric, lactic, and glycolic acids as well as malonic and maleic acids by selenium dioxide is increased after preliminary treatment with potassium permanganate.[170] Also, mono esters of ascorbic acid reduce selenious acid quantitatively, and the reaction has been established as a method for determining such compounds.[171]

EXPERIMENTAL CONSIDERATIONS

Selenium dioxide oxidations are relatively simple to perform, but a number of factors must be considered before undertaking such a procedure. Since the yields of desired oxidation products are frequently low it is conceivable that they could be improved if experimental conditions were optimized. It appears that few detailed studies have been carried out on variables such as proportions of reactants, time and temperature of reaction, solvents, presence of other materials in the system, and workup procedures. A number of these factors are discussed in detail below.

As stated earlier, selenium dioxide and selenious acid are used interchangeably to denote the reactant needed to effect the oxidations under consideration in this chapter. Although there appear to be no clearly established examples of the superiority of one reagent over the other, the dioxide is referred to in most experimental procedures.

Selenium Dioxide

Many of the physical and chemical properties of selenium dioxide have been recorded.[172, 173] It is a white crystalline solid that melts at 340° in a closed tube and sublimes at 1 atm around 317°. It is readily soluble in water (70 % SeO_2 is dissolved at 20° on a weight basis), with which it reacts to form selenious acid, which dissociates to a lesser extent than does sulfurous acid. Selenium dioxide decomposes at about 1000°, whereas sulfur dioxide is stable to approximately 2800°.

[168] E. Neuzil, M. Labadie, and J. C. Breton, *Bull. Soc. Pharm. Bordeaux*, **104**, 200 (1965) [*C.A.*, **65**, 13816p (1966)]; J. C. Breton, M. Labadie, and E. Neuzil, *ibid.*, **104**, 206 (1965) [*C.A.*, **65**, 13817a (1966)].

[169] P. Saumande, M. Labadie, J. C. Breton, and E. Neuzil, *Bull. Soc. Pharm. Bordeaux*, **111**, 69 (1972) [*C.A.*, **78**, 84773a (1973)].

[170] T. N. Srivastava and S. P. Agarwal, *J. Prakt. Chem.*, [4] **4**, 319 (1957).

[171] T. Tukamoto, S. Ozeki, H. Kaga, and M. Taniguchi, *Yakugaku Zasshi*, **90**, 73 (1970) [*C.A.*, **72**, 93310a (1970)].

[172] D. M. Chizhikov and V. P. Shchastlivyi, *Selenium and Selenides*, trans. from the Russian by E. M. Elkin, Collet's Publishers Ltd., London, 1968, p. 353.

[173] K. B. Bagnall, *The Chemistry of Selenium, Tellurium, and Polonium*, Elsevier Publishing Co., Amsterdam, 1966, p. 57.

The dioxide behaves like a weak base in sulfuric acid and in oleum. It is soluble in fused antimony tribromide, selenium oxychloride, and in benzene. It has limited solubility in a number of organic solvents, and its solubility in methanol, ethanol, acetone, acetic acid, and acetic anhydride is about 10% ($11.8°$), 7% ($14°$), 4% ($15.3°$), 1% ($13.9°$), and slight ($12°$), respectively, at the temperatures indicated. It has been shown by Raman spectroscopy that alcohol solutions of both selenium dioxide and selenious acid at room temperature contain water and selenites such as $(RO)_2SeO$ and $ROSeO_2H$.[174] The equilibrium position depends on the concentrations of the alcohol and the water. Also, selenium dioxide has been found to react with benzoic anhydride to give a product, $C_{14}H_{10}O_5Se$, presumed to be dibenzoyloxyselenium oxide, which can be used as an oxidizing agent.[175] It seems reasonable that selenium dioxide should react similarly with other hydroxylic compounds used as solvents to give selenites and mixed anhydrides which then are the reactants involved in the oxidation reactions.

Other recent investigations of the properties of selenium dioxide include those on its gaseous thermal stability,[176, 177] a mass-spectrometric study of its sublimation,[178] its saturated vapor pressure and molecular composition in the gas phase,[179] the structure of selenious acid,[180] and of its aminolysis products,[94] and complexes with nitrogen bases.[103]

Analytical methods employed for the determination of selenium dioxide include its reaction with potassium iodide, followed by titration of the liberated iodine with sodium thiosulfate,[181] and a spectrophotometric method based upon the oxidation of p-sulfophenylhydrazine to a diazonium oxidation product which couples with 1-naphthylamine to form an azo dye with an absorbance peak at 520 nm.[182, 183]

Although both selenium dioxide and selenious acid are commercially available from a number of chemical suppliers, a discussion of the methods of synthesis and purification of the dioxide is appropriate. Many procedures

[174] A. Simon and R. Paetzold, Z. Anorg. Allg. Chem., 303, 53 (1960).

[175] F. Nerdel and J. Kleinwächter, Naturwissenschaften, 42, 577 (1955).

[176] V. I. Sonin, G. I. Novikov, and O. G. Polyachenok, Zh. Fiz. Khim., 43, 2980 (1969) [C.A.. 72, 93861f (1970)].

[177] V. I. Sonin and O. G. Polyachenok, Vestsi Akad. Navuk Belarus. SSR, Ser. Khim. Navuk, 1971, 121 [C.A., 75, 81007f (1971)].

[178] P. J. Ficalora, J. C. Thompson, and J. L. Margrave, J. Inorg. Nucl. Chem., 31, 3771 (1969).

[179] N. N. D'yachkova, E. N. Vigdorovich, G. P. Ustyugov, and A. A. Kudryavtsev, Izv. Akad. Nauk. SSSR, Neorg. Mat., 5, 2219 (1969) [C.A., 72, 93586v (1970)].

[180] A. Simon and R. Paetzold, Z. Anorg. Allg. Chem., 301, 246 (1959).

[181] E. S. Gould, Anal. Chem., 23, 1502 (1951).

[182] G. F. Kirkbright and John H. Yoe, Anal. Chem., 35, 808 (1963).

[183] I. I. Nazarnko and A. N. Ermakov, Analytical Chemistry of Selenium and Tellurium, Halsted Press, New York, 1973.

indicate that freshly prepared, or resublimed, selenium dioxide was employed as the oxidizing agent. The need for this in some cases might be questioned, but as indicated earlier (p. 286) it has been reported that aged dioxide may give results different from those obtained with new material.

The standard methods of preparation of selenium dioxide involve the oxidation of selenium with nitric acid, or its combustion in oxygen and nitrogen dioxide. The selenium dioxide may be purified by a wet process or by sublimation, the latter apparently being preferred.* More concise directions for the preparation and purification of selenium dioxide are given in *Organic Syntheses*.[184]

Other strong oxidizing agents such as dichromate and permanganate oxidize selenium, and recent studies have been made on oxidizing it with manganese dioxide in aqueous solutions containing sulfuric acid.[185] The patent literature refers to the use of nitrogen dioxide and hydrogen peroxide to oxidize selenium in the presence of inert carriers, and the resulting solutions of dioxide are used to oxidize organic substances present in the solutions.[186-188] An apparatus has been described for the continuous production of selenium dioxide.[189]

CAUTION: Selenium dioxide, selenious acid, and selenium-containing products obtained from oxidations must be used with considerable care.[190] Sax states: "The physiological properties of selenium compounds are similar to those of arsenic compounds. Some organoselenium compounds have the high toxicity of other organometals. Inorganic selenium compounds can cause dermatitis. Garlic odor of breath is a common symptom. Pallor, nervousness, depression, and digestive disturbances have been reported from chronic exposure. Any selenium dioxide solid or solutions spilt on the skin should be removed immediately by washing under the tap."[191] In the event that it comes in contact with the tissues around and under the fingernails, considerable pain is experienced and a red discoloration of the affected parts develops from the precipitated selenium.

---

* These procedures are described in detail in Ref. 1 pp. 344–346.

[184] H. A. Riley and A. R. Gray in *Org. Syn. Coll.* Vol. **2**, 509 (1943).

[185] P. P. Tsyb and T. D. Shulgina, *Zh. Prikl. Khim.*, **45**, 1442 (1972) [*C.A.*, **78**, 45749c (1973)].

[186] E. S. Roberts and L. J. Christmann, U.S. Pat. 3,268,294 [*C.A.*, **65**, 16903c (1966)].

[187] E. S. Roberts and L. J. Christmann, U.S. Pat. 3,405,171 [*C.A.*, **70**, 77664x (1969)].

[188] J. P. Zumbrunn, Fr. Pat. 2,038, 575 [*C.A.*, **75**, 87658r (1971)].

[189] V. G. Alekseev, *Tr. Vsesoyuz. Nauch.—Issledovatel. Inst. Khim. Reaktivov*, **1959**, 47 [*C.A.*, **55**, 2332f (1961)].

[190] D. L. Klayman and W. H. H. Günther, Eds., *Organic Selenium Compounds: Their Chemistry and Biology*, Wiley-Interscience, New York, 1973.

[191] N. R. Sax, *Dangerous Properties of Industrial Materials*, 3rd ed., Reinhold Book Corp. New York. 1968, p. 1086.

Selenium dioxide oxidations must be carried out in efficient hoods; proper care in disposing of all solvents and by-products from such reactions is necessary. Since selenium dioxide and selenious acid are reasonably expensive reagents, there may be times when it is appropriate to recover the selenium from oxidation reactions. It should be pulverized if necessary, freed from organic impurities by washing with suitable solvents, and dried for several hours in an oven before it is used for conversion to the dioxide.

Solvents

Table I shows the number and diversity of solvents that have been employed in selenium dioxide oxidations. In addition, combinations of solvents, e.g., acetic acid-acetic anhydride, dioxane-acetic acid, benzene-ethanol, t-butyl alcohol-pyridine, etc., are often more desirable than individual solvents. Frequently, varying amounts of water are added to the reaction mixtures to increase solubility.

TABLE I. SOLVENTS EMPLOYED IN SELENIUM DIOXIDE OXIDATIONS[a]

Acetic acid	Diethyl ether	3-Picoline
Acetic anhydride	Dimethylformamide	Propionic acid
Amyl alcohols; n-, i-, t-	Dioxane	Propyl alcohols, n-, i-
Benzene	Diphenyl ether	Pyridine
Bromobenzene	Ethanol	Sulfuric acid
n-Butyl acetate	Ethyl acetate	Tetrahydrofuran
Butyl alcohols, n-, t-	Ethylene glycol di-	Toluene
Carbon tetrachloride	methyl ether	1,2,4-Trichlorobenzene
Chlorobenzene	Isoquinoline	Water
Deuterium oxide	Methanol	Xylene
Diethylene glycol	Nitrobenzene	
dimethyl ether	Phenetole	

[a] Solvent combinations are not included; consult the Tabular Survey.

The most commonly used solvents are dioxane, acetic acid, acetic anhydride, ethanol, t-butyl alcohol, pyridine and combinations thereof. It should be recalled (p. 293) that selenium dioxide reacts with alcohols, acids, acid anhydrides, and possibly other solvents to give new selenium compounds that are now the oxidizing agents. A patent describes the use of the monomethyl ester of selenious acid for the dehydrogenation of steroids.[192]

[192] G. Langhein, M. Meyer, E. Menzer, and J. Zaumseil, Ger. (East) Pat. 42,959 [C.A., 65, 2329e (1966)].

A number of studies have shown that the nature of the solvent can have considerable effect upon both the products and yields from selenium dioxide oxidations. It must be recognized, though, that some of the products may result from secondary reactions of the solvents with intermediates, or with primary products of the oxidation reactions. Olefin oxidations usually are carried out in solutions of ethanol, ethanol-water, acetic acid, acetic anhydride, or mixtures of the latter two; absence of solvent may lead to explosions. Aldehydes and alcohols are the normal products in the first two solvents, while the others afford acetates. Aliphatic carbonyl compounds generally are oxidized to dicarbonyls in dioxane, ethanol and higher alcohols, and aromatic hydrocarbons. They are dehydrogenated in solvents such as acetic acid, and higher alcohols to which acetic acid or pyridine may be added. Toluene, xylene, dioxane, acetic acid, and higher alcohols have been used in benzylic-type oxidations to produce aldehydes or acids, particularly in heterocyclic systems.

It was observed quite early* that the oxidation of 1-methylcyclohexene in ethanol affords a mixture of 2-methylcyclohexen-3-ol (in 35% yield) and 2-methyl-2-cyclohexen-1-one (in 27% yield), whereas in water the ketone accounts for 90% of the mixture obtained. When the reaction was run in acetic acid, 1-acetoxy-2-methyl-2-cyclohexene was isolated in 40% yield. Dihydro-α-dicyclopentadiene gave the allylic acetate or ethers, depending upon whether the oxidation was performed in acetic acid and acetic anhydride or alcohols. Also, it was reported that the yield of camphorquinone from the oxidation of camphor varied between 27 and 95% depending upon whether the reaction was run in ethanol, toluene, xylene, acetic anhydride, or without a solvent.

The digitoxigenin 3-acetate (50, OAc) was obtained on oxidative cyclization of the steroidal ester (p. 284) in 30% yield when benzene was the solvent and the reaction time was 10 hours;[131, 132] however, the same reaction in boiling dioxane for 16 hours gave 17α-hydroxydigitoxigenin 3-acetate (61).[193] The report that treatment of diketodihydrolanosteryl

61

* See Ref. 1, p. 343.
[193] N. Danieli, Y. Mazur, and F. Sondheimer, *Tetrahedron*, **23**, 715 (1967).

acetate with selenium dioxide in a mixture of acetic acid and acetic anhydride for 2–5 hours gave a product from which a triketone, triketolano-
steryl acetate,[194] could be isolated was not confirmed. Instead, a dike-
tolanostadienyl acetate was obtained. The reaction was reinvestigated and
it was determined that the controlling factor was the quantity of water
used to dissolve the selenium dioxide.[195] Under anhydrous conditions
consistent yields (30%) of triketolanostadienyl acetate were obtained;
however, as the amount of water used to dissolve the selenium dioxide
increased, the yields of the triacetate decreased and the quantity of dike-
tolanostadienyl acetate increased until it became the main product.

The oxidation of 2,4-cycloheptadienone has been studied in acetic
acid, water, pyridine, and ethanol and the highest yield (70%) of tropone
(48) was obtained in the latter solvent.[196] Also, oxidation of oleic acid by
selenium dioxide while varying solvents, time, temperature, and concentrations has been investigated.[197] Spectroscopic analysis of the reaction mixtures showed that their compositions varied with the dielectric
constant of the solvent. Yields of hydroxylated products were approximately 35% in benzene, 19% in ethyl acetate, 34% in t-butyl alcohol,
and 45% in dimethylformamide.

The oxidation of 19-norsteroids, substituted at the 3 position by hydroxyl, acetate, or fluorine, and including a double bond at the 5(10) position, by selenious acid yielded the 5(10), 9(11)-dienic systems when carried
out in acetic acid, ethanol, dioxane, and a 50/50 mixture of dioxane-
acetic acid, in the presence of a little water.[198] It was fastest in acetic
acid. A systematic study of the syntheses of a series of coumarin aldehydes
by oxidation of the corresponding methyl derivatives indicated that solvent, concentration of reagents, position of the substituents in the reactants, temperature, and time of heating play important parts in the
reactions.[199] In these syntheses, xylene was chosen as the solvent,
despite the fact that selenium dioxide is insoluble in it, because the starting materials were recovered in all cases when ethanol, amyl alcohol,
water, benzene, dioxane, or toluene was used.

It has been reported that freshly sublimed selenium dioxide does not
react with anisole at 130–150° (20 hours), but that ordinary dioxide, or
that to which water had been added, afforded bis-(p-methoxyphenyl)

[194] C. Dorée, J. F. McGhie, and F. Kurzer, *J. Chem. Soc.*, **1949**, 570.
[195] J. F. Cavalla and J. F. McGhie, *J. Chem. Soc.*, **1951**, 744.
[196] E. E. Van Tamelen and G. T. Hildahl, *J. Amer. Chem. Soc.*, **78**, 4405 (1956).
[197] A. Tubul-Peretz, M. Naudet, and E. Ucciana, *Rev. Fr. Corps Gras*, **13**, 155 (1966)
[*C.A.*, **65**, 613f (1966)].
[198] A. Guida and M. Mousseroncanet, *Bull. Soc. Chim. Fr.*, **1971**, 1098.
[199] A. Schiavello and E. Cingolani, *Gazz. Chim. Ital.*, **81**, 717 (1951).

selenide in fairly low yield.[200] Selenious acid has been shown to convert (diphenylmethylene)cyclopropane to 2,2-diphenylcyclobutanone when refluxed in dioxane, while the same reaction in acetic acid-acetic anhydride gave 4,4-diphenyl-3-buten-1-ol acetate as well.[201]

The composition of the products obtained from the oxidation of isoprene by selenium dioxide was found to be dependent on the molar ratio of the reactants and solvents.[86] The solvents employed were acetic acid and acetic anhydride and, although ten products were identified from the original reaction mixture, it was possible to control conditions so that a few principal products could be obtained.

An additional example of the effect of solvent on a reaction system is given on p. 289; it describes the different results obtained in efforts to introduce oxygen at C-2 of ethyl α-safranate by selenium dioxide oxidation in acetic acid or dioxane.[147] Obviously the solvent for a selenium dioxide oxidation can be selected so that it undergoes oxidation itself to yield products which may react with other substances in the system. This was demonstrated by treating a solution of guaiazulene (62) in acetone with selenium dioxide. The acetone was oxidized to dihydroxyacetone which *in situ* underwent condensation with 2 mols of starting material to produce the product. [202]

The foregoing discussion indicates that considerable care should be exercised in the selection of a solvent for a selenium dioxide oxidation so that it does not enter into reaction with the intermediates or products of the oxidation. The greatly increased knowledge of the mechanisms of the action of selenium dioxide on organic molecules can be of invaluable aid in this matter. Instead of merely following experimental procedures developed many years ago, when little was known about the nature of selenium dioxide oxidation, one can use mechanistic information now available to design rational procedures for the main types of reactions effected by this agent.

[200] G. V. Boyd, M. Doughty, and J. Kenyon, *J. Chem. Soc.*, **1949**, 2196.
[201] E. V. Dehmlow, *Z. Naturforsch, B.* **24**, 1197 (1969).
[202] W. Treibs and R. Vogt, *Chem. Ber.*, **94**, 1739 (1961).

Temperature, Reaction Time, and Other Variables

It was stated earlier* that temperature, reaction time, and other variables can have important effects on the course of selenium dioxide oxidations. In general, vigorous or extended conditions lead to oxidation beyond the primary stages, e.g., aldehydes are converted to acids, dihydroxy or polyhydroxy derivatives or mixtures of them with monohydroxy compounds can result, and dehydrogenation may occur. The accompanying equations summarize some older results obtained with $\Delta^{9(10)}$-octalin.†

It has been found that the oxidation of 17α, 21-dihydroxy-3,11,20-trioxopregnane 21-acetate (**56**) with about a 10% excess of selenium dioxide in a nitrogen atmosphere, with t-butyl alcohol and acetic acid as the solvent, for 0.75 hour, afforded the Δ^4-derivative.[203] Upon approximately doubling the amount of oxidizing agent and extending the time of

* Ref. 1, p. 344.
† Ref. 1, p. 344.
[203] D. A. van Dorp and S. A. Szpilfogel, Dutch Pat. 86,368 [$C.A.$, **53**, 6295d (1959)].

reaction to 8 hours, the corresponding $\Delta^{1,4}$-derivative was obtained. Similar results were observed with the allo isomer of **56** and with $11\beta,17\alpha,21$-trihydroxy-3,20-dioxopregnane 21-acetate.

The effects of the amounts of water and oxidant and the reaction time on the oxidation of diketodihydrolanosteryl acetate are discussed on p. 297.

The reaction of methyl 3,11-diketo-(Z)-4,17(20)-pregnadien-21-oate (p. 304) with selenium dioxide in t-butyl alcohol and acetic acid gave two products in yields of 10–20% and 20–25%.[204] One was assigned the 16β-hydroxy-(E)-structure (20–25% yield) and the other was formulated as the 16α-hydroxy-(Z)-compound (10–20% yield). By shortening the reaction time and lowering the temperature, selective hydroxylation at C-16 could be achieved without affecting the Δ^4-3-ketone system. When the solvent was changed to tetrahydrofuran and the reaction time reduced to 5 hours, the original ester gave methyl $3,11$-diketo-16β-hydroxy-(E)-4,17(20)-pregnadien-21-oate and methyl $3,11$-diketo-16α-hydroxy-(E)-4,17(20)-pregnadien-21-oate in 31% and 17% yields, respectively.

The oxidations of 2-, 3-, and 4-pyridinemethanol have been studied with 0.5 and 1 mol of selenium dioxide in dioxane at 80°, pyridine at 90° or without solvent in the range 110–200° for variable periods of time.[28] With no solvent and excess oxidant, the 2 derivative gave the corresponding acid (in 80% yield) in 5 minutes at 150°, or in 85% yield in 90 minutes at 110°; with an equivalent amount or excess of selenium dioxide, picolinaldehyde was formed (in 86–90% yield) in the presence of dioxane or pyridine. The aldehyde was obtained in 100% yield when the ratio of selenium dioxide to base was 0.5, without solvent, at 160° for 3 minutes. The 4-pyridinemethanol behaved similarly, while the 3 derivative failed to react in solution and afforded low yields of the aldehyde in the absence of a solvent.

A number of variables have been investigated in the oxidation of ethylene with selenium dioxide at 50 psi in acetic acid at 110–125° (pp. 266, 277).[50–52] When sodium acetate was added to the reaction mixture, ethylene glycol diacetate was not formed; however, addition of a strong mineral acid resulted in a shift of product distribution to favor the latter and to produce ethylene glycol monoacetate also. Not only did the acidity have an effect on the product distribution, but also it resulted in a greatly increased reaction rate. Other workers have reported a similar effect of a strong mineral acid on the selenium dioxide oxidation of other acyclic and cyclic olefins.[26, 27, 48, 49] Likewise, the catalytic effect of hydrochloric

[204] J. E. Pike, F. H. Lincoln, G. B. Spero, R. W. Jackson, and J. L. Thompson, *Steroids*, **11**, 755 (1968).

acid on the oxidation of 1,2-diaroylethanes to the corresponding ethylenes has been noticed,[205] and some use has been made of it for oxidation of steroids.[206]

Mercury or its salts are employed fairly often during selenium dioxide oxidations,[207-210] and iron powder has been added to oxidation mixtures for the dehydrogenation of steroids.[192] Sodium acetate has been shown to change the products of oxidation of ethylene (p. 300), and it has been added to steroidal oxidation mixtures.[211] Oxidation of 1,3,5-cycloheptatriene with selenium dioxide to tropone has been carried out in aqueous dioxane solution buffered with potassium dihydrogen phosphate.[127] It has been reported further that the dehydrogenation of 3-oxosteroids by selenium dioxide can be improved by the presence of an ion exchanger, preferably an alkaline one used in a ratio of 1–5 parts to 1 part of the oxidizing agent.[212]

Explosions which occur during the selenium dioxide oxidation of compounds such as isonitrosocamphor have been controlled by addition of sand.[213] Selenium dioxide also has been modified by dispersal on kieselguhr.[214] The latter assisted the removal of the deposited selenium by simple filtration.

Workup Procedures

Perhaps the most annoying aspect of selenium dioxide oxidations involves the removal of the precipitated selenium, or unreacted dioxide and selenium-containing compounds from reaction mixtures. The selenium usually precipitates in the red form, which then may change to the gray allotrope. The latter can cause operational difficulties with stirring of reaction mixtures and removal from reaction vessels when it forms a large stonelike mass. If unreacted selenium dioxide is not removed from a reaction mixture, selenium may continue to precipitate during purification

[205] N. Campbell and N. M. Khanna, J. Chem. Soc., **1949**, Suppl. Issue, No. 1, S33.

[206] Z. Hodinář and B. Pelc, Chem. Listy, **49**, 1733 (1955) [C.A., **50**, 5704e (1956)]; Collect. Czech. Chem. Commun., **21**, 264 (1956) [C.A., **50**, 10117i (1956)].

[207] R. E. Beyler, A. E. Oberster, F. Hoffman, and L. H. Sarett, J. Amer. Chem. Soc., **82**, 170 (1960).

[208] L. Canonica, G. Jommi, F. Pelizzoni, and C. Scolastico, Gazz. Chim. Ital., **95**, 138 (1965).

[209] J. A. Cella and R. C. Tweit, J. Org. Chem., **24**, 1109 (1959).

[210] G. Jommi, P. Manitto, and C. Scolastico, Gazz. Chim. Ital., **95**, 151 (1965).

[211] Upjohn Co., Neth. Pat. Appl., 6,414,319 [C.A., **64**, 3645b (1966)].

[212] N. V. Organon, Neth, Pat. 98,950 [C.A., **60**, 616c (1964)].

[213] J. Vène, Bull. Soc. Sci. Bretagne, **19**, 14 (1943) [C.A., **41**, 739h (1947)].

[214] C. Shen, Y. Chen, H. Chang, A. Yue, Y. Chang, Y. Tsai, P. Sun. F. Hou, and Y. Liu, Yao Hsueh Hsueh Pao, **11**, 242 (1964) [C.A., **61**, 8363f (1964)].

of the product and become a real nuisance. It is eliminated often by pouring the reaction mixture, after filtration, into water, extracting with an inert solvent, and washing the solution of the product with a base such as sodium bicarbonate. Other reagents have been employed also for ridding oxidation mixtures of excess selenium dioxide or selenious acid. The *Organic Syntheses* procedure for preparing glyoxal bisulfite from paraldehyde states that lead(II) acetate is more satisfactory than sulfur dioxide for the removal of selenious acid, provided that the solution is kept cool and a large excess is avoided.[215] Sulfur dioxide,[73] or sodium bisulfite in acidic medium,[216] thiourea,[217] and a bicarbonate type of anion exchange resin[83] have been used for eliminating excess selenium dioxide or selenious acid.

Elimination of traces of selenium and selenium-containing organic compounds is perhaps more difficult. One of the best methods appears to be to reflux a filtered, washed, and dried solution of product in an inert solvent with precipitated silver (p. 304).[21, 218] Mercury,[219–221] deactivated Raney nickel,[77, 206, 222, 223] zinc dust,[224] and sodium borohydride in aqueous ethanol[225] have been employed similarly, and aqueous solutions of chromic anhydride,[206] ammonium sulfide,[226] and potassium cyanide[227] have been recommended for the same purpose. Dilute aqueous hydrogen peroxide also has been reported to remove alkyl selenium by-products.[228]

Frequently oxidation reaction mixtures have been filtered through filter aids, such as diatomaceous earth,[196, 229] and then chromatographed to effect further purification.[194, 230, 231] Sometimes it is sufficient to reflux the reaction mixture in ethanol for about 10 minutes; then the selenium can be removed by filtration.

[215] A. R. Ronzio and T. D. Waugh, *Org. Syn., Coll. Vol.*, **3**, 438 (1955).
[216] M. Levi and I. Pesheva, *Farmatsiya* (Sofia), **15**, 266 (1965) [*C.A.*, **64**, 17531c (1966)].
[217] Y. Watanabe, Y. Ito, and T. Matsuura, *J. Sci. Hiroshima Univ.*, Ser. A, **20**, 203 (1957) [*C.A.*, **52**, 6816i (1958)].
[218] C. H. Issidorides, M. Fieser, and L. F. Fieser, *J. Amer. Chem. Soc.*, **82**, 2002 (1960).
[219] J. H. Fried, G. E. Arth, and L. H. Sarett, *J. Amer. Chem. Soc.*, **81**, 1235 (1959).
[220] J. Jacques, G. Ourisson, and C. Sandris, *Bull. Soc., Chim., Fr.*, **1955**, 1293.
[221] R. Rambaud and M. Vessiere, *Bull. Soc. Chim. Fr.*, **1961**, 1567.
[222] M. Heller, S. M. Stolar, and S. Bernstein, *J. Org. Chem.*, **26**, 5044 (1961).
[223] A. Zürcher, H. Heusser, O. Jeger, and P. Geistlich, *Helv. Chim. Acta*, **37**, 1562 (1954).
[224] W. M. Hoehn, C. R. Dorn, and B. A. Nelson, *J. Org. Chem.* **30**, 316 (1965).
[225] V. Viswanatha and G. S. K. Rao, *Indian J. Chem.*, **10**, 763 (1972).
[226] M. Kocór and M. Tuszy-Maczka, *Bull. Acad. Polon. Sci. Ser. Sci. Chim.*, **9**, 405 (1961) [*C.A.*, **60**, 6910e (1964)].
[227] H. Watanabe, *Pharm. Bull.* (Tokyo), **5**, 426, 431 (1957) [*C.A.*, **52**, 9059e (1958)].
[228] R. F. Lauer, Hoffmann-LaRoche, personal communication.
[229] C. Djerassi and A. Bowers, U.S. Pat. 3,051,703 [*C.A.*, **58**, 6904f (1963)].
[230] Ciba Ltd., Swiss Pat. 255,306 [*C.A.*, **44**, 3043a (1950)].
[231] Chas. Pfizer and Co., Inc., Brit. Pat. 799,343 [*C.A.*, **53**, 17206f (1959)].

EXPERIMENTAL PROCEDURES

The preceding discussions include many of the general aspects of experimental details employed in selenium dioxide oxidations. The following specific examples have been chosen to illustrate the principal types of reactions effected by selenium dioxide, *i.e.*, oxidations of activated methyl and methylene groups, and aromatic methyl groups; allylic oxidations to alcohols, acetates, and α,β-unsaturated aldehydes; and dehydrogenations of carbonyl compounds. Different solvents, times, and temperatures of reaction and workup procedures are indicated in these examples, but it might be helpful, before undertaking a selenium dioxide oxidation, to consult the tables in the Tabular Survey which refer to the type of compound to be oxidized.

Although most of the yields reported in the following examples are relatively high, this may be atypical. Also, it should be recognized that, with the exception of the *Organic Syntheses* procedures, the others have not been checked, and some were developed before modern instrumentation was used to establish the identity and purity of the products.

1,2-Cyclohexanedione and Phenylglyoxal (Oxidation of —CH$_2$— C=O ⟶ —COCO—). Oxidations of cyclohexanone to 1,2-cyclohexanedione in 60% yield,[232] and of acetophenone to phenylglyoxal in 69–72% yield[184] are described in *Organic Syntheses*.

Orotaldehyde (Oxidation of ArCH$_3$ → ArCHO).[233] A mixture of 63 g (0.5 mol) of 6-methyluracil, 66.6 g (0.6 mol) of selenium dioxide, and 1.5 l of acetic acid was refluxed with mechanical stirring for 6 hours. During this time the white suspension of selenium dioxide was gradually replaced by gray selenium. The hot reaction mixture was fitered and the selenium cake was extracted with 2 × 250 ml of boiling acetic acid. The combined yellow filtrate and extracts were evaporated to dryness under reduced pressure, giving 60 g of a yellow solid, which gave a positive 2,4-dinitrophenylhydrazone test. The crude orotaldehyde, which still contained some selenium and excess selenium dioxide, was purified as follows. The solid was dissolved in 600 ml of warm water, and an aqueous solution of sodium bisulfite (30 g of sodium bisulfite in 60 ml of water) was cautiously added in small portions to the stirred mixture which was boiled with active charcoal and Celite for 10 minutes, then filtered. The filtrate was acidified with concentrated hydrochloric acid to pH 1. On cooling, 25 g of pure orotaldehyde was collected, mp 273–275° dec (slower

[232] C. C. Hach, C. V. Banks, and H. Diehl in *Org. Syn., Coll. Vol.*, **4**, 229 (1963).
[233] K.-Y. Zee-Cheng and C. C. Cheng, *J. Heterocycl. Chem.*, **4**, 163 (1967).

heating caused carbonization at 273–275° without melting). An additional 16 g of product was obtained from the concentrated mother liquor which brought the total yield to 58%.

$\Delta^{8(14)}$-Cholestene-3β,7α-diol Diacetate (Oxidation of —CH$_2$CH=CH— → —CHOAcCH=CH—).[21] A solution of 2 g (4.8 mmol) of Δ^7-cholestenyl acetate in 50 ml of absolute diethyl ether and 40 ml of acetic acid was treated at 25° with a mixture of 40 ml of 0.1 M selenious acid in acetic acid (made from selenious acid) and 8 ml of water. The solution (25°) turned yellow in a minute or two, and when left overnight had deposited a large amount of red selenium. The solution was filtered, diluted with water, extracted with diethyl ether, and the extract was washed with soda solution, dried, and stirred for 2 hours at 25° with precipitated silver. The filtered solution was light yellow but completely free of selenium. Evaporation gave a yellow glass that solidified at once when rubbed with methanol. The material was brought into solution and allowed to crystallize; there resulted 1.01 g (44%) of slightly yellow diacetate, mp 134–137°. Short treatment with Norit in diethyl ether removed the color, and crystallization from methanol gave 0.7 g, mp 138.5–139.5°, [α]$_D^{25}$ — 2.1° (chloroform).

Methyl 3,11-Diketo-16α-hydroxy-(Z)- and 3,11-Diketo-16β-hydroxy-(E)-1,4,17,(20)-pregnatrien-21-oate (Oxidation of —CH$_2\overset{|}{C}$=$\overset{|}{CH}$CO$_2$R → —CHOHC=CHCO$_2$R).[204] A mixture of 50 g (0.14 mol) of methyl 3,11-diketo-(Z)-4,17(20)-pregnadien-21-oate, 50 g (0.45 mol) of selenium dioxide, 5 ml of acetic acid, and 1.5 l of t-butyl alcohol was heated at reflux for 24 hours. After cooling, the reaction mixture was filtered through Celite: Magnesol and the filter cake was washed with ethyl acetate. The filtrate and wash were evaporated to dryness, and the residue was taken up in ethyl acetate which was washed successively with sodium bicarbonate solution, freshly prepared ice-cold ammonium polysulfide solution, aqueous dilute ammonia, dilute hydrochloric acid, sodium bicarbonate solution, and water. The organic phase was dried over sodium sulfate and evaporated to dryness. The residue (51.0 g) was dissolved in methylene chloride and chromatographed over 1.5 kg of Florisil. The column was developed with increasing percentages of acetone in Skellysolve B.

The first fraction from the column, 13.4 g, was crystallized from acetone-Skellysolve B to give 12.0 g of the (E)-ester in two crops, mp 197–207°. The analytical sample was recrystallized from acetone: mp 206–208°, ultraviolet (ethanol), nm max (log ϵ): 231 (4.35).

The second fraction from the column, 18.1 g, was crystallized from acetone-Skellysolve B to give 8.6 g of the (Z)-ester, mp 241–246°. The analytical sample was recrystallized from methanol: mp 255–258°, ultraviolet (ethanol), nm max (log ϵ): 233 (4.32).

(E)-2-Methyl-6-methylen-2,7-octadien-1-ol (Oxidation of —CH =C(CH₃)₂ → (E)—CH=C(CH₃)CH₂OH. [58] A mixture of 448 g (3.3 mols) of myrcene and 200 ml of 95% ethanol was heated at 60–70° while 183 g (1.65 mols) of selenium dioxide was added portionwise during 45 minutes. After 1 hour of stirring under reflux, the dark-red solution was steam-distilled. The distillate (approximately 12 l) was extracted with pentane, and the extract was dried over sodium sulfate and concentrated. Distillation of the residue through a short Vigreux column gave the following fractions: I, 130 g, bp 62–68°/9 mm (unchanged myrcene); II, 101 g, bp 56–66°/0.02 mm, consisting principally of the dienol and dienal, based on glc analysis.

Fraction II, 101 g, was dissolved in 500 ml of methanol and the solution was cooled to 0°. It was stirred while 10 g of sodium borohydride was added in portions. Stirring was continued for another hour at room temperature; the reaction mixture was then diluted with water and extracted with pentane. The extract was washed with water, dried over sodium sulfate, and concentrated in a rotary evaporator. On distillation the residue gave 66.9 g (19% yield, based on myrcene) of the (E)-alcohol, bp 60–62°/0.04 mm.

Phenylmaleic Anhydride (Dehydrogenation of $\overset{|}{O}=CCH_2CH_2\overset{|}{C}=O$

$\rightarrow O=\overset{|}{C}CH=CH\overset{|}{C}=O).$ [234] A mixture of 4.9 g (0.025 mol) of phenylsuccinic acid, 3.3 g (0.03 mol) of selenium dioxide, and 20 ml of acetic anhydride was refluxed for 3 hours, filtered hot through a sintered glass funnel, and the residue on the funnel was washed with a little diethyl ether. Concentration of the filtrate under reduced pressure and trituration of the residue with diethyl ether gave 3.8 g (86% yield) of phenylmaleic anhydride, mp 119–120.5°. Recrystallization from benzene-hexane or sublimation under reduced pressure just below its melting point did not raise the melting point.

6α-Fluoroprednisone Diacetate (Dehydrogenation of $O=\overset{|}{C}CH_2^-$

$CH_2— \rightarrow O=\overset{|}{C}CH=CH—).$ [235] A mixture of 5 g (0.011 mol) of 6α-fluorocortisone diacetate and 2.5 g (0.022 mol) of selenium dioxide in 250 ml

---

[234] R. K. Hill, *J. Org. Chem.*, **26**, 4745 (1961).

[235] A. Bowers, E. Denot, M. B. Sanchez, and H. J. Ringold, *Tetrahedron*, **7**, 153 (1959).

of t-butyl alcohol and 0.8 ml of pyridine was heated under reflux in an atmosphere of nitrogen for 24 hours. The reaction mixture was diluted with 250 ml of ethyl acetate and filtered through Celite. After removal of the solvent, the residue was triturated with 500 ml of water, filtered, dried, and crystallized from ethyl acetate-hexane to afford 2.25 g (45% yield) of product, mp 255–258°, raised by crystallizations from ethyl acetate-hexane to 260–262°, $[\alpha]_D$ + 68°, ultraviolet (ethanol), nm max (log ϵ): 236–238 (4.18).

COMPARISON OF SELENIUM DIOXIDE WITH OTHER OXIDIZING AGENTS

A perusal of the Tabular Survey will show that selenium dioxide has been employed for the oxidation of a widely diversified selection of organic compounds. However, in some cases, although products resulted, other oxidizing agents might have been more advantageous. Often the reactions reported were carried out as part of studies which were not designed specifically for synthetic utility. If the latter is an important consideration, then efforts should be made to compare the data available for the reaction of interest with those for a number of different oxidizing agents. Such information is more limited than might be believed, but within the last few years several excellent treatises on the broad aspects of oxidation of organic compounds have been published.[5, 6, 234–239]

Although selenium dioxide attacks a relatively large number of different types of organic molecules, its principal utility derives from its ability to oxidize carbonyl compounds to 1,2-dicarbonyls, to convert olefins to allylic alcohols and related materials, to effect benzylic oxidations, and to cause dehydrogenation of certain structures. Because it is a less vigorous oxidizing agent than permanganate or dichromate ion, it has a degree of selectivity. It is similar to mercuric acetate, lead tetraacetate, and thallium salts which also are not so specific in their actions as agents such as osmium tetroxide, ozone, periodate, chromyl chloride, ruthenium tetroxide, and several other less well-known oxidants.

It should be recalled that selenium dioxide, like some of the other oxidizing agents mentioned, is toxic, often affords only moderate to poor yields of desired products, and may lead to reaction mixtures that present

[236] R. L. Augustine, Ed., Oxidation, Techniques and Applications in Organic Synthesis, Vol. I, Marcel Dekker, New York, 1969; R. L. Augustine and D. J. Trecker, Eds., Oxidation, Vol. II, Marcel Dekker, New York, 1971.

[237] L. J. Chinn, Selection of Oxidants in Synthesis, Marcel Dekker, New York, 1971.

[238] W. S. Trahanovsky, Ed., Oxidation in Organic Chemistry, Part B, Academic Press, New York, 1973.

[239] K. B. Wiberg, Ed., Oxidation in Organic Chemistry, Part A, Academic Press, New York 1965.

purification problems. However, these negative aspects of this unusual agent are outweighed easily when the need arises for certain types of compounds that cannot survive attack by more potent oxidizing agents. Selenium dioxide appears to be the reagent of choice for the oxidation of methyl or methylene groups activated by carbonyls. The aryl glyoxals are obtained in good to excellent yields from the corresponding acetophenones, and α-diketones may be prepared similarly (pp. 279–280). Allylic or benzylic oxidation also is accomplished very effectively by selenium dioxide.

In contrast to other oxidizing agents such as permanganate, osmium tetroxide, permanganate-periodate, ruthenium tetroxide, silver carboxylate-iodine complexes, and palladium chloride, which either add to a double bond to form diols or cause cleavage of the carbon-carbon double bond, selenium dioxide attacks largely at an α position to afford the allylic or benzylic derivatives described previously (pp. 281–286). The methods for direct introduction of the acyloxy group in place of hydrogen attached to carbon have been reviewed.[240]

Chromium(VI) has been used for oxidation of allylic methylene groups to α,β-unsaturated ketones, and results have varied with the chromium reagent and the experimental conditions.[241] The dry chromium trioxide-pyridine complex affords 48–95 % of various enones, but it does not oxidize allylic methyl groups in contrast to the behavior of selenium dioxide.[242]

Dehydrogenation of organic molecules may be accomplished by a number of reagents under quite widely different conditions. The hydrogen may be removed from carbon, hetero atoms, or a combination of carbon and hetero atoms. The common dehydrogenating agents include palladium, platinum, and nickel catalysts as well as sulfur, selenium, and quinones that contain electron-withdrawing substituents. All but the quinones require relatively high reaction temperatures. Although selenium dioxide, mercuric acetate, lead tetraacetate, and manganese dioxide are perhaps not generally thought of as dehydrogenating agents, they are able to remove hydrogen from a relatively large variety of compounds and may be compared with the quinones since these reactions are usually carried out at lower temperatures.

Selenium dioxide dehydrogenated hydroaromatic molecules,[145, 243, 244] but it failed to attack a series of isoxazolines (63) which was converted to

---

[240] D. J. Rawlinson and G. Sosnovsky, *Synthesis*, 1972, 1; 1973, 567.

[241] K. B. Wiberg in *Oxidation in Organic Chemistry*, Part A, K. B. Wiberg, Ed., Academic Press, New York 1965, p. 105.

[242] W. G. Dauben, M. Lorber, and D. S. Fullerton, *J. Org. Chem.*, **34**, 3587 (1969).

[243] Y. Abe, T. Harukawa, H. Ishikawa, T. Miki, M. Sumi and T. Toga, *J. Amer. Chem. Soc.*, **78**, 1422 (1956).

[244] K. Alder and M. Schumacher, *Ann. Chem.*, **570**, 178 (1950).

the corresponding isoxazoles by chromium trioxide in acetic acid or manganese dioxide in acetone.[245] Although selenium dioxide dehydrogenates 1,4-dicarbonyl and related compounds,[69, 234] its real value lies in its ability to dehydrogenate monoketones, particularly steroids and terpenoids. With the former, selenium dioxide and 2,3-dichloro-5,6-dicyano-benzoquinone (DDQ) have had more extensive application than mercuric acetate, lead tetraacetate, manganese dioxide, and other dehydrogenation agents.

The chemistry of DDQ and a comparison of its use with other dehydrogenating agents, including selenium dioxide, have been discussed.[246] Extensive tables of data on selective oxidations of polyhydroxy steroids, as well as methods of introducing double bonds into steroids, have been published.[10, 12, 246] Other examples of dehydrogenations of steroids by selenium dioxide are presented in Table XI.

The conversion of 3-keto- and Δ^4-3-ketosteroids into $\Delta^{1, 4}$-dien-3-ones can be accomplished satisfactorily by use of selenium dioxide in solvents such as tertiary alcohols, usually t-butyl, alone or with acetic acid or pyridine, acetic acid, benzene and water, dioxane and acetic acid, and tetrahydrofuran. Occasionally, mercury or a mercury compound is added to the reaction mixture. Quinones like chloranil or DDQ also are capable of dehydrogenating 3-keto- and Δ^4-3-ketosteroids quite successfully. Chloranil, in solvents such as t-butyl alcohol or xylene, brings about dehydrogenation almost entirely at the 6 position; hydrocortisone 21-acetate is converted in high yield to the 4,6-diene. With DDQ, dehydrogenation of steroidal 4-en-3-ones takes place in the 1,2 position in solvents like dioxane or benzene, and also in the presence of weak acids; however, strong acids such as p-toluenesulfonic acid catalyze the reaction, and 6,7-dehydro derivatives are formed almost exclusively.[247]

The presence of selenium-containing by-products in the selenium dioxide reaction mixtures is definitely an undesirable feature of this reagent and, unless they can be removed by some of the methods described previously, DDQ is preferable for these dehydrogenation reactions.

Although the previous examples of steroidal dehydrogenations are

[245] G. S. D'Alcontres and G. L. Vecchio, *Gazz. Chim. Ital.*, **90**, 337 (1960).
[246] D. Walker and J. D. Hiebert, *Chem. Rev.*, **67**, 153 (1967).
[247] A. B. Turner and H. J. Ringold, *J. Chem. Soc.*, *C*, **1967**, 1720.

based upon activation by ketone functions, selenium dioxide is capable of dehydrogenating a Δ^8-steroid to a diene, e.g., 5α-cholest-8-en-3β-ol affords cholesta-8,14-dien-3β-ol.[248]

In addition to the dehydrogenation reactions described, selenium dioxide has been applied to a number of other ketosteroids, terpenes, hydroaromatic compounds, and diverse structures of a greater variety than has been tried with most agents. In spite of its annoying property of sometimes contaminating reaction mixtures with selenium-containing compounds, which may be difficult to remove, selenium dioxide ranks high with the best reagents for removal of hydrogen from organic molecules.

ACKNOWLEDGMENTS

The author did considerable preliminary work on this review during a sabbatical leave spent at the Department of Chemistry, Massachusetts Institute of Technology. He is deeply grateful for the hospitality and facilities which were made available.

Miss Betty Alston of E. I. du Pont de Nemours and Company was most helpful in surveying the literature for studies on selenium dioxide oxidations, and her assistance is appreciated greatly.

TABULAR SURVEY

The information in the following tables is an extension of that reviewed previously[1] and covers the literature to early 1975. The arrangement of the tables follows that used in *Organic Reactions*, **5**, 349 (1949), with the addition of a few more categories. Also, owing to the difficulties of attempting to categorize multifunctional compounds, arbitrary decisions have been made for location in tables, based upon what seems to be the dominant characteristics of the compound treated with selenium dioxide. It might be helpful to peruse the Table of Contents at the beginning of this chapter before trying to locate a multifunctional compound in the survey tables.

Compounds are placed in the tables according to total carbon content, with the exception of esters which are related to the acids from which they are derived. Within each category based on carbon content, the compounds are arranged alphabetically.

The yields of products are given in parentheses, but dashes do not necessarily mean that the data have not been published since the information often was taken from abstracts. The same is true for the other variables, i.e., solvent, time, and temperature. An attempt has been made to be consistent in nomenclature, and the *Chemical Abstracts* system has been applied when it seemed appropriate; however, common names and formulas have been retained for many simple compounds, steroids, terpenes, other natural products, and complicated structures.

---

[248] W. J. Adams, V. Petrow, and R. Royer, *J. Chem. Soc.*, **1951**, 678.

TABLE II. ACIDS AND ACID DERIVATIVES

No. of C Atoms	Reactant	Solvent/Time/Temperature (°C)	Product(s) and Yield(s) (%)	Refs.
		A. Acids		
2	Acetic acid	H$_2$O/12 hr/200	Succinic acid (—), carbon dioxide (—)	249
4	Crotonic acid	AcOH/5 hr/reflux	γ-Acetoxycrotonic acid (30)	120
	Dihydroxymaleic acid	H$_2$O/2 hr/25	Dihydroxytartaric acid (—)	250
5	(CH$_3$)$_2$C=CHCO$_2$H	AcOH/4 hr/reflux	β-Methyl-γ-crotonolactone (18), AcOCH$_2$CH(CH$_3$)=CHCO$_2$H (22)	120
10	Phenylsuccinic acid	Ac$_2$O/3 hr/reflux	Phenylmaleic anhydride (86)	234
11	p-Methoxyphenylsuccinic acid	Ac$_2$O/30 min/reflux	p-Methoxyphenylmaleic anhydride (80)	234
	3,4-Methylenedioxyphenyl-succinic acid	Ac$_2$O/19 hr/reflux	3,4-Methylenedioxyphenylmaleic anhydride (66)	234
18	Oleic acid	C$_6$H$_6$/—/70	Hydroxy acids (35) (mixture of mono- and dihydroxyelaidic and stearic acids)	197
		CH$_3$CO$_2$C$_2$H$_5$/—/70	Hydroxylated products (19)	197
		t-C$_4$H$_9$OH/—/70	" " (34)	197
		DMF/—/70	" " (45)	197
		AcOH, Ac$_2$O/—/—	Mixture of allylic hydroxy- and dihydroxyoctadecanoic acids, and *vic*-dihydroxyoctadecanoic acids (—)	251
		CCl$_4$/—/—	Mixture, principally unsaturated ketones (—)	251
19	[structure: cyclopentane ring with OH, O(CH$_2$)$_5$CO$_2$H, HO, and C=C–C$_6$H$_{13-n}$ substituents]	Dioxane/—/—	(±)-7-Oxaprostaglandin F1α (60–70)	252

310

B. Anhydrides

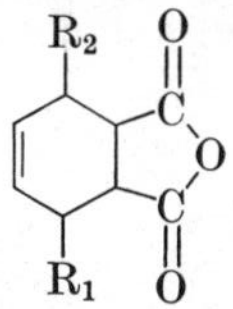

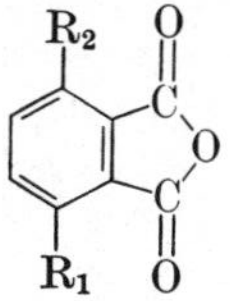

Substituents in reactant:

	Reactant substituents	Conditions	Product	Refs.
28	[cyclohexene anhydride, R2, R1; fluorenylidene structure] $R_1 = $ 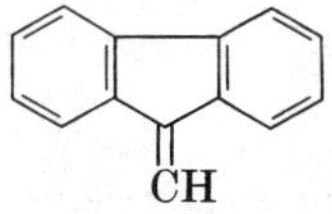 ; $R_2 = C_6H_5$	$Ac_2O/6$ hr/reflux	[phthalic anhydride, R2, R1; fluorenylidene structure] $R_1 = $ 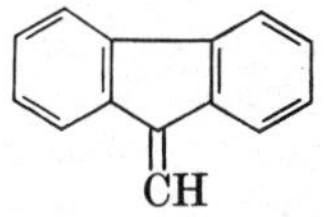 ; $R_2 = C_6H_5$ (—)	244
36	$R_1 = R_2 = -CH=C(C_6H_5)_2$	$Ac_2O/3.5$ hr/reflux	$R_1 = R_2 = -CH=C(C_6H_5)_2$ (—)	
38	$R_1 = -CH=C(C_6H_5)_2$, $R_2 = -CH=CH-CH=C(C_6H_5)_2$	$Ac_2O/4$ hr/reflux	$R_1 = -CH=C(C_6H_5)_2$, $R_2 = $ CH=CH—CH= C(C_6H_5)$_2$ (—)	
40	$R_1 = R_2 = -CH=CH-CH=C-(C_6H_5)_2$	$Ac_2O/3$ hr/reflux	$R_1 = R_2 = -CH=CH-CH=C(C_6H_5)_2$ (—)	

C. Amide

6	2-Ethyl-(E)-crotonamide	—/—/—	2-(1-Hydroxyethyl)-(E)-crotonamide (—)	253

D. Esters

(In acid)

2	cis-8α-Acetoxy-1-methoxy-10a-methyl-5,6,6a,7,8,9,10,10a,11,12-decahydrochrysene	$C_6H_5CH_3$, C_5H_5N/ 10–12 hr/reflux	cis-8α-Acetoxy-1-methoxy-10a-methyl-5,6,6a,7,8,9,10,10a-octahydrochrysene (79)	145
11	Methyl 1-naphthylacetate	None/1.5 hr/190	Methyl 1-naphthylglyoxalate (68)	31, 32

Note: References 249–634 are on pp. 407–415.

311

TABLE II. ACIDS AND ACID DERIVATIVES—(Continued)

No. of C Atoms	Reactant	Solvent/Time/Temperature (°C)	Product(s) and Yield(s) (%)	Refs.
		E. Keto Esters		
(In acid)				
2	5-Acetoxy-4,7,7-trimethyl-2-norbornanone	—/—/—	5-Acetoxy-4,7,7-trimethyl-2,3-norbornanedione (—)	254
	5-Acetoxy-2,3-dimethoxy-5,8,9,10-tetrahydro-1,4-naphthoquinone	C_2H_5OH, H_2O/5 hr/reflux	2,3-Dimethoxy-1,4-naphthoquinone (83)	113
5	$CH_3COCH_2CH_2CO_2R$	None/18 hr/90–95	$CH_3COCH=CHCO_2R$	255
	R = CH_3		R = CH_3 (0–13)	
	R = C_2H_5	..	R = C_2H_5 (0–13)	
	R = $n\text{-}C_4H_9$	..	R = $n\text{-}C_4H_9$ (0–13)	
	R = $C_6H_5CH_2$		R = $C_6H_5CH_2$ (0–13)	
6	2-Carbethoxycyclopentanone	Dioxane/—/—	2-Carbethoxy-2-cyclopenten-1-one (50)	256
	2-Carbomethoxycyclopentanone	Dioxane/—/—	2-Carbomethoxy-2-cyclopenten-1-one (45)	256
7	2-Carbethoxy-2-methylcyclopentanone	Dioxane-H_2O/20 hr/reflux	5-Carbethoxy-5-methylcyclopent-2-en-2-ol-1-one (44)	257
	2-Carbomethoxy-3-methylcyclopentanone	Dioxane/—/—	2-Carbomethoxy-3-methyl-2-cyclopenten-1-one (45)	256
	(structure) R, $CO_2C_2H_5$ cyclohexenone with O; R = CH_3, C_2H_5, C_6H_5, $i\text{-}C_3H_7$, $C_6H_5CH_2$	$t\text{-}C_4H_9OH$/24 hr/reflux	(structure) R, $CO_2C_2H_5$; R = CH_3 (72), C_2H_5 (75), C_6H_5 (91), $i\text{-}C_3H_7$ (83), $CH_2C_6H_5$ (40),	258

312

	2-norbornanone		2,3-norbornanedione	(—)
11	2-Carbethoxymethoxyaceto-phenone	C_5H_5N/—/—	2-Carbethoxymethoxyphenyl-glyoxylic acid (58)	114
12	2-Carbethoxy-4,4-diethoxy-cyclohexanone	Dioxane/2 hr/reflux	Ethyl 5-ethoxysalicylate (27)	148
	1-Carbomethoxy-4,4a,5,6,7,8-hexahydro-4a-methyl-2(3H)-naphthalenone	t-C_4H_9OH,AcOH/72 hr/reflux	1-Carbomethoxy-5,6,7,8-tetrahydro-4a-methyl-2(4aH)-naphthalenone (55)	260
	Ethyl 3-mesityl-3-oxopropionate[3-14C]	Dioxane/18 hr/reflux	Ethyl 3-mesityl-2,3-dioxopropionate-[3-14C] (59)	261
16	Diethyl 3-oxo-4,9-dimethyl-1,2,3,5,6,7,8,9-octahydro-6-naphthylmethylmalonate	$AcOH,H_2O$/30 min/reflux	Diethyl 3-oxo-4,9-dimethyl-3,5,6,7,8,9-hexahydro-6-naphthylmethylmalonate (60)	243
17	5-Benzyloxy-2-carbethoxymeth-oxyacetophenone	C_5H_5N/3 hr/100	5-Benzyloxy-2-carbethoxymethoxy-phenylglyoxylic acid (82)	114

F. Unsaturated Esters

No.	Starting material	Conditions	Product	Ref.
2	3-Acetoxy-1-cyclohexene	Ac_2O/8 hr/reflux	1,4-Diacetoxy-2-cyclohexene (34)	119
	1-Acetoxy-5-hexene	Ac_2O/—/reflux	1,4-Diacetoxy-5-hexene (17), 1,6-diacetoxy-2-hexene (22)	118
	2-Acetoxy-5-hexene	Ac_2O/—/reflux	2,4-Diacetoxy-5-hexene (8), 1,5-diacetoxy-2-hexene (16)	116, 118, 119
	3-Acetoxy-4-hexene	Ac_2O/—/reflux	1,4-Diacetoxy-2-hexene (20–26)	118, 121
	3-Acetoxy-5-hexene	Ac_2O/—/reflux	3,4-Diacetoxy-5-hexene (11), 1,4-diacetoxy-2-hexene (16)	118, 121
	4-Acetoxy-2-methyl-1-pentene	Ac_2O/4 hr/reflux	3,4-Diacetoxy-2-methyl-1-pentene (17), 4-acetoxy-2-acetoxymethyl-1-pentene (16)	119
	4-Acetoxy-2-methyl-1-hexene	Ac_2O/5 hr/reflux	3,4-Diacetoxy-2-methyl-1-hexene (16), 4-acetoxy-2-acetoxymethyl-1-hexene (14)	119

* *Note*: References 249–634 are on pp. 407–415.

TABLE II. ACIDS AND ACID DERIVATIVES—(Continued)

No. of C Atoms	Reactant	Solvent/Time/Temperature (°C)	Product(s) and Yield(s) (%)	Refs.
		F. Unsaturated Esters—(Continued)		
(In acid) 2 (contd.)	4-Acetoxy-2-methyl-2-hexene	AcOH,Ac$_2$O/8 hr/ 90–95	1,4-Diacetoxy-2-methyl-2-hexene (30)	119
	3-Acetoxy-1-cyclooctene	AcOH,Ac$_2$O/13 hr/ 110–120	1,4-Diacetoxy-2-cyclooctene (27)	262
	6-Acetoxy-2-methyl-2-heptene	Ac$_2$O/8 hr/reflux	1,6-Diacetoxy-2-methyl-2-heptene (45)	119
	6-Acetoxy-2-methyl-2-octene	AcOH,Ac$_2$O/7 hr/ 90–95	1,4-Diacetoxy-2-methyl-2-octene (30)	119
	8-Acetoxy-2,6-dimethyl-2-octene	Ac$_2$O/3 hr/reflux	1,8-Diacetoxy-2,6-dimethyl-2-octene (50)	119
	10-Undecenyl acetate	Ac$_2$O/—/reflux	3,11-Diacetoxy-1-undecene (22), 1,11-diacetoxy-2-undecene (—)	118, 121
	4-Acetoxy-2,2,6-trimethyl-(carboxyethylmethylene)-cyclohexane	AcOH/2 hr/reflux	(24)	134
		AcOH,Ac$_2$O/15 min/ reflux	(—)	125
		AcOH/3 hr/reflux	(76)	129, 133

314

No.	Substrate	Conditions	Product	Ref.
	Octahydro-1,2,3-trimethoxybenzo[a]heptalen-9-ol acetate		octahydro-1,2,3-trimethoxybenzo[a]heptalen-7α-ol (—), 9β-acetoxy-8,9,10,11,12,12aβ-hexahydro-1,2,3-trimethoxybenzo[a]heptalene (—)	221
4	$CH_3CH{=}CHCO_2R$ R = CH_3 R = C_2H_5	Dioxane/2 hr/reflux Dioxane/2 hr/reflux	$OHCCH{=}CHCO_2R$ (I) I, R = CH_3 (15) I, R = C_2H_5 (10–15)	221
	R = C_2H_5	AcOH/5 hr/reflux	I, R = C_2H_5 (13), $CH_3CO_2CH_2CH_2CH{=}CHCO_2$-$C_2H_5$ (25)	264, 265
	R = n-C_3H_7 (E)-$CH_3CH{=}CHCO_2CH_3$ $HOCH_2CH{=}CHCO_2C_2H_5$ $CH_2{=}CHCH_2CO_2C_2H_5$	Dioxane/2 hr/reflux Dioxane/4 hr/reflux —/—/— Ac_2O/7 hr/reflux	I, R = n-C_3H_7 (10) (E)-$OHCCH{=}CHCO_2CH_3$ (11) $OHCCH{=}CHCO_2C_2H_5$ (30) $OHCCH{=}CHCO_2C_2H_5$ (—), $CH_3CO_2CH_2CH_2CH{=}CHCO_2C_2H_5$ (25)	221 221
7	Methyl 5-methyl-4-hexenoate	$CH_3OCH_2CH_2OCH_3$/8 hr/reflux	2-Methyl-(E)-5-carbomethoxy-2-pentenal (41)	122
8	Dimethyl 2,5-dimethyl-2-hexene-1,6-dioate	Xylene/9 hr/reflux	Dimethyl 2,5-dimethyl-2,4-hexadiene-1,6-dioate (—)	266
10	$(CH_3)_2C{=}CH$ —[cyclopropane structure]— CO_2R, H, H R = CH_3, t-C_4H_9	—/—/—	OHC $\overset{CH_3}{\underset{}{C}}{=}CH$ —[cyclopropane structure]— CO_2R, H, H R = CH_3, t-C_4H_9	267
	$(CH_3)_2C{=}CH$ —[cyclopropane structure]— $CO_2C(CH_3)_3$, H, H	—/—/—	OHC, $HOCH_2C{=}CH$ —[cyclopropane structure]— $CO_2C(CH_3)_3$, H, H (—)	268
11	Ethyl 10-undecylenate	Ac_2O/20 hr/140	Ethyl 11-acetoxy-9-undecylenate (30)	269

315

* *Note*: References 249–634 are on pp. 407–415.

TABLE II. ACIDS AND ACID DERIVATIVES—(Continued)

No. of C Atoms	Reactant	Solvent/Time/Temperature (°C)	Product(s) and Yield(s) (%)	Refs.
		F. Unsaturated Esters—(Continued)		
(In acid)				
		AcOH,Ac$_2$O/8 hr/reflux	Ethyl 9-acetoxy-10-undecylenate (19), ethyl 11-acetoxy-9-undecylenate (12)	121
13	3-Methoxy-5-hydroxy-7-phenyl-2,6-heptadienoic acid δ-lactone	Dioxane/1.25 hr/100	(two δ-lactone structures joined by +)	270
18	Methyl oleate	C$_2$H$_5$OH/12 hr/reflux	Mixture of methyl (E)-2-octadecenoate and methyl 8-oxooleate (22), methyl stearate (15)	214, 271
	Ethyl 2,2-dimethyl-3-(6-methoxy-2-naphthyl)-3-pentenoate	AcOH/—/reflux	[6,2-CH$_3$OC$_{10}$H$_8$C[C(CH$_3$)$_2$CO$_2$C$_2$H$_5$] = CHCH$_2$]$_2$Se (90)	82

7	C$_6$H$_5$CH$_2$OH	C$_6$H$_6$/—/100, then 195	C$_6$H$_5$CHO (83)	75
8	o-Phthalyl alcohol	—/—/130–140	o-C$_6$H$_4$(CH$_2$O)$_2$SeO (70) (yielded 68% of o-phthaldehyde upon heating)	75
9	1,2-Bishydroxymethyl-3,4-methylenedioxybenzene	Dioxane/45 min/reflux	1,2-Bishydroxymethyl-3,4-methylenedioxybenzeneselenious acid ester (87)	272
	4,5-Methylenedioxy-o-phthalyl alcohol	Dioxane/45 min/reflux	4,5-Methylenedioxy-o-phthalyl alcohol selenious acid ester (97) (converted to phthaldehyde derivative in 89% yield by heating in decalin at 210–220°)	144
12	1,2-Bishydroxymethylnaphthalene	None/20 min/130	1,2-Naphthalic anhydride (—)	75
	2,3-Bishydroxymethylnaphthalene	None/0.5 hr/160	(—) (yielded naphthalene-2,3-dicarboxaldehyde upon heating)	75
20	(C$_6$H$_5$)$_2$C(OH)CHOHC$_6$H$_5$	None/5 min/200	(C$_6$H$_5$)$_2$C(OH)COC$_6$H$_5$ (—)	273
21	(C$_6$H$_5$)$_2$C(OH)CHOHC$_6$H$_4$CH$_3$-o	None/5 min/200	(C$_6$H$_5$)$_2$C(OH)COC$_6$H$_4$CH$_3$-o (65)	273
		C$_2$H$_5$OH/5 hr/reflux	(—)	146

* *Note* References 249–634 are on pp. 407–415

TABLE IV. ALDEHYDE

No. of C Atoms	Reactant	Solvent/Time/ Temperature (°C)	Product(s) and Yield(s) (%)	Refs.
6	Paraldehyde	Dioxane, AcOH/ 6 hr/reflux	Glyoxal (as bisulfite addition compound) (72–74)	215

TABLE V. ETHERS

A. Ether

No. of C Atoms	Reactant	Solvent/Time/ Temperature (°C)	Product(s) and Yield(s) (%)	Refs.
8	1,2-Dimethoxybenzene	H_2O/—/—	$[3,4\text{-}(CH_3O)_2C_6H_3]_2Se$ (—)	274

B. Allyl and Propargyl Ethers[a,157,275]

$$CH_2=CHCH_2OR \xrightarrow{SeO_2} [CH_2=CHCHO] + ROH + Se$$
$$CH\equiv CCH_2OR \xrightarrow{SeO_2} [HC\equiv CCHO] + ROH + Se$$

No. of C Atoms	Reactant	Product(s) and Yield(s) (%)
9	$HC\equiv CCH_2OC_6H_5$	C_6H_5OH (79)
	$CH_2=CHCH_2OC_6H_5$	C_6H_5OH (57)
	$CH_2=CHCH_2OC_6H_4NO_2\text{-}2\text{-}Cl\text{-}4$	$HOC_6H_3NO_2\text{-}2\text{-}Cl\text{-}4$ (52)
10	$CH\equiv CCH_2OC_6H_4CF_3\text{-}3$	$HOC_6H_4CF_3\text{-}3$ (62)
	$CH\equiv CCH_2OC_6H_3CH_3\text{-}3\text{-}Cl\text{-}4$	$HOC_6H_3CH_3\text{-}3\text{-}Cl\text{-}4$ (66)
	$CH_2=CHCH_2OC_6H_5$	$C_6H_5CH_2OH$ (50)
	$CH_2=CHCH_2OC_6H_4CH_3\text{-}2$	$HOC_6H_4CH_3\text{-}2$ (47)
	$CH_2=CHCH_2OC_6H_4CF_3\text{-}3$	$HOC_6H_4CF_3\text{-}3$ (44)
	$CH_2=CHCH_2OC_6H_3CH_3\text{-}3\text{-}Cl\text{-}4$	$HOC_6H_3CH_3\text{-}3\text{-}Cl\text{-}4$ (54)
	$C_6H_5CH=CHCH_2OCH_3$	$C_6H_5CH=CHCHO$ (66) + CH_3OH (—)
11	$CH\equiv CCH_2OC_6H_3OCH_3\text{-}2\text{-}CHO\text{-}4$	$HOC_6H_3OCH_3\text{-}2\text{-}CHO\text{-}4$ (72)
	$CH_2=CHCH_2OC_6H_3OCH_3\text{-}2\text{-}CHO\text{-}5$	$HOC_6H_3OCH_3\text{-}2\text{-}CHO\text{-}5$ (50)

Note: References 249–634 are on pp. 407–415.

[a] All reactions were run in dioxane at the boiling point for 1 hour. The allyl and propargyl ethers were expected to form th...

318

B. Allyl and Propargyl Ethers[a157,275]

No of C Atoms	Reactant	Solvent/Time/Temperature (°C)	Product(s) and Yield(s) (%)	Ref.
12	CH_2=$CHCH_2OC_6H_4OCH_3$-2-CO_2CH_3-4-NO_2-5		$HOC_6H_2OCH_3$-2-CO_2CH_3-4-NO_2-5 (51)	
13			(38)	
16	$C_6H_5C{\equiv}CCH_2OC_6H_3CH_3$-3-Cl-4		$C_6H_5C{\equiv}CCHO$ (35) + $HOC_6H_3CH_3$-3-Cl-4 (—)	

TABLE VI. HYDROCARBONS

No of C Atoms	Reactant	Solvent/Time/Temperature (°C)	Product(s) and Yield(s) (%)	Refs.
			A. Alkane	
4	Butane	None/—/500	Selenophene (3)	84
			B. Alkenes	
2	Ethylene	AcOH, H_2O, HCl/—/125, 50 psig	$CH_3CO_2C_2H_5$ (16.7), C_2H_5OH (2.7), $HOCH_2CH_2OAc$(I) (27.3), $AcOCH_2CH_2OAc$(II) (38.8), $HOCH_2CH_2OH$ (1.8), organo-Se compounds (III) (12.6)	51, 83
		AcOH/—/100–200, 50 psig	II (2.4), $(AcOCH_2CH_2)_2Se$ (35), $(AcOCH_2CH_2)_2Se_2$ (5.9)	50, 52
		AcOH, HCl/—/100–200, 50 psig	I (50), II (45), III (4)	50
		1,2,4-$Cl_3C_6H_3$,H_2O/—/110–120	Glyoxal (—)	276

TABLE VI. HYDROCARBONS—(Continued)

No. of C Atoms	Reactant	Solvent/Time/ Temperature (°C)	Product(s) and Yield(s) (%)	Refs.
		B. Alkenes—(Continued)		
3	Propylene	AcOH/—/—	Bisacetoxypropyl selenides (—)	52
4	(Z)-2-Butene	AcOH,HCl/—/—	meso-AcOCH(CH$_3$)CH(CH$_3$)OAc (—)	54
	(E)-2-Butene	AcOH,HCl/—/—	dl-AcOCH(CH$_3$)CH(CH$_3$)OAc (—)	54
		AcOH,H$_2$SO$_4$/—/—	2,3-Dimethyl-5-vinyl-1,4-oxaselenane (—)	54
	CH$_3$CH=CHCH$_3$	AcOH/—/56–80	AcOCH$_2$CH=CHCH$_3$ (—), CH$_2$=CHCH(OAc)CH$_3$ (—), [CH$_3$CH(OAc)CHCH$_3$]$_2$-Se (—)	23
5	Butenes (mixture)	None/—/500	Selenophene (13)	84
	(CH$_3$)$_2$C=CHCH$_3$	None/—/450–500	3-Methylselenophene (17–19)	84
	i-C$_3$H$_7$CH=CH$_2$	None/—/500	3-Methylselenophene (17)	84
6	1-Hexene	AcOH,H$_2$SO$_4$/10 hr/ 105	n-C$_4$H$_9$CH(OAc)CH$_2$OAc (35), n-C$_3$H$_7$CH(OAc)CH=CH$_2$ (12), CH$_3$(CH$_2$)$_2$CH=CHCH$_2$OAc (5)	24, 49
	(CH$_3$)$_2$C=CHCH$_2$CH$_3$	C$_2$H$_5$OH/2 hr/reflux	CH$_3$CH$_2$CH=C(CH$_3$)CHO (E-, 45)	60
	(CH$_3$)$_2$C=C(CH$_3$)$_2$	None/—/500	3,4-Dimethylselenophene (31)	84
7	Methylenecyclohexane	Ac$_2$O/3 hr/reflux	2-Methylenecyclohexanol (after saponification) (—)	277, 278
8	Allylcyclopentane	AcOH,Ac$_2$O/10 hr/ 115–120	γ-Cyclopentylallyl acetate (29)	279
	n-C$_4$H$_9$CH=C(CH$_3$)$_2$	C$_2$H$_5$OH/1.5 hr/ reflux	n-C$_4$H$_9$CH=C(CH$_3$)CH$_2$OH (70, 98% E-, 2% Z-)	60
	2-Methyl-2-heptene	C$_2$H$_5$OH,H$_2$O/—/—	[ketone structure] (—), [enal CHO structure] (—), [enal CHO structure] (—)	40

1-Octene	AcOH,H₂SO₄/10 hr/115	n-C₆H₁₃CH(OAc)CH₂OAc (35), n-C₄H₉CH=CHCH₂CH₂OAc (Trace)	26, 49, 271
Styrene	AcOH/—/—	C₆H₅COCHO (—), C₆H₅CHOHCH₂OAc (—), C₆H₅CH(OAc)CH₂OAc (—)	48
9 2,4,4-Trimethyl-1-pentene	Ac₂O/10 hr/reflux	t-C₄H₉CH₂C(CH₂OAc)=CH₂ (80)	280, 281
2,4,4-Trimethyl-2-pentene	Ac₂O/10 hr/reflux	t-C₄H₉CH=C(CH₃)CH₂OAc (80)	281
Isopropenylcyclohexane	C₂H₅OH/1.25 hr/reflux	CH₂=C(C₆H₁₁)CHO (4), CH₂=C(C₆H₁₁)CH₂OH (9), HOC₆H₁₀-1-C(CH₃)=CH₂ (37)	60
10 (Z)-3-Methyl-3-octene	C₂H₅OH/10 hr/reflux	n-C₄H₉CH=C(CH₃)COCH₃ (E-, 51), n-C₄H₉CH=C(C₂H₅)CH₂OH (E-, 14)	60
C₆H₅CH=CHCH₃	Ac₂O/5 hr/reflux	C₆H₅CH=CHCH₂OAc (27)	120
p-CH₃OC₆H₄CH₂CH=CH₂	Ac₂O/5 hr/reflux	p-CH₃OC₆H₄CH=CHCH₂OAc (30)	120
p-CH₃OC₆H₄CH=CHCH₃	Ac₂O/8 hr/reflux	p-CH₃OC₆H₄CH(OAc)CH₃ (20)	120
7,7-Dimethyl-2-methylidene-norbornane	Dioxane/20 hr/—	7,7-Dimethyl-3-hydroxy-2-methylidenenorbornane, 7,7-dimethyl-2-methylidene-3-norbornanone (60, mixture)	282
(Z)-2,3-Dimethyl-3-octene	C₂H₅OH/4 hr/reflux	n-C₄H₉CH=C(i-C₃H₇)CHO (Z- and E-, 10), n-C₄H₉CH=C(i-C₃H₇)CH₂OH (Z-, 34; E-, 6)	60
	C₂H₅OH/10 hr/reflux	n-C₄H₉CH=C(C₃H₇-i)CHO (Z-, 12; E-, 46)	60
(E)-2,3-Dimethyl-3-octene	C₂H₅OH/3 hr/reflux	Products same as for Z-isomer refluxed 4 hr (52)	60
	C₂H₅OH/10 hr/reflux	Product identical with that from Z-isomer refluxed 10 hr (—)	60

Note: References 249–634 are on pp. 407–415.

TABLE VI. Hydrocarbons—(Continued)

No. of C Atoms	Reactant	Solvent/Time/Temperature (°C)	Product(s) and Yield(s) (%)	Refs.
		B. Alkenes—(Continued)		
10 (contd.)	6-Methyl-5-hepten-2-one ethylene ketal	C_2H_5OH/10 hr/reflux	$CH_3COCH_2CH_2CH=C(CH_3)CHO$ (E-, 33)	60
11	$p\text{-}CH_3C_6H_4CH=CHCH_3$	Ac_2O/4 hr/reflux	$p\text{-}CH_3C_6H_4CH=CHCH_2OAc$ (29)	120
	(E)-2,2,3-Trimethyl-3-octene	C_2H_5OH/2 hr/reflux	$n\text{-}C_4H_9CH=C(C_4H_9\text{-}t)CH_2OH$ (Z-, 19, E-, 44), 60 $n\text{-}C_4H_9CH=C(C_4H_9\text{-}t)CHO$ (3)	60
13	2-Cyclohexyl-2-heptene		$n\text{-}C_4H_9CH=C(C_6H_{11})CHO$	60
	Z-	C_2H_5OH/10 hr/reflux	Z-, E-, 20–80 (70)	
	E-	C_2H_5OH/5 hr/reflux	Unresolved mixture (41)	
15	$C_6H_5CH_2CH=CHC_6H_4Cl\text{-}p$	AcOH/—/115	$C_6H_5CH(OAc)CH=CHC_6H_4Cl\text{-}p$ (65–73)	39, 283
	$p\text{-}ClC_6H_4CH_2CH=CHC_6H_5$	AcOH/—/115	$p\text{-}ClC_6H_4CH(OAc)CH=CHC_6H_5$ (64–73)	39, 283
	$C_6H_5CH_2CH=CHC_6H_5$	AcOH/—/115	$C_6H_5CH(OAc)CH=CHC_6H_5$ (—)	39, 283
	$C_6H_5CHDCH=CHC_6H_5$ (0.87D)	AcOH/—/115	$C_6H_5CD(OAc)CH=CHC_6H_5$ (0.66D, 63)	39, 283
16	(Diphenylmethylene)cyclopropane	Dioxane/—/reflux	2,2-Diphenylcyclobutanone (—)	201
	$p\text{-}CH_3OC_6H_4CH=CHCH_2C_6H_5$	$AcOH,H_2O$/—/reflux	$(C_6H_5)_2C=CHCH_2OAc$ (—)	201
	$C_6H_5CH=CHCH_2C_6H_4OCH_3\text{-}p$	AcOH/—/115	$p\text{-}CH_3OC_6H_4CH=CHCH(OAc)C_6H_5$ (42–47)	39, 283
		AcOH/—/115	$C_6H_5CH=CHCH(OAc)C_6H_4OCH_3p$ (59–68)	39, 283
18	$2,4,6\text{-}(CH_3)_3C_6H_2CH=CHCH_2C_6H_5$	AcOH/—/115	$2,4,6\text{-}(CH_3)_3C_6H_2CH=CHCH(OAc)C_6H_5$ (82)	39, 283

C. Alkynes

6	C$_4$H$_9$C≡CH	C$_2$H$_5$OH/—/— C$_2$H$_5$OH,H$_2$SO$_4$/—/—	C$_3$H$_7$CHOHC≡CH (27) C$_4$H$_9$COCH(OC$_2$H$_5$)$_2$ (16), C$_4$H$_9$COCO$_2$- C$_2$H$_5$ (8.7), C$_3$H$_7$CHOHC≡CH (6.3)	56 56
8	C$_6$H$_5$C≡CH	AcOH,H$_2$O/—/— AcOH,H$_2$SO$_4$/—/— AcOH,H$_2$O,H$_2$SO$_4$/ 3 hr/reflux	C$_6$H$_5$COCHO (—) C$_6$H$_5$COCO$_2$H (—) ,, (—)	56 284 284
9	C$_6$H$_5$CH$_2$C≡CH	C$_2$H$_5$OH/1 hr/reflux	C$_6$H$_5$COC≡CH (10)	285
12	C$_6$H$_5$CH$_2$C≡CC$_3$H$_7$	C$_2$H$_5$OH/1 hr/reflux	C$_6$H$_5$COC≡CC$_3$H$_7$ (—)	285
14	C$_6$H$_5$C≡CC$_6$H$_5$	AcOH,H$_2$O/—/110	C$_6$H$_5$COCOC$_6$H$_5$ (after H$_2$SO$_4$ addition to reaction mixture) (—)	56
15	C$_6$H$_5$CH$_2$C≡CC$_6$H$_5$	C$_2$H$_5$OH/1 hr/reflux	C$_6$H$_5$COC≡CC$_6$H$_5$ (—)	285

D. Alkadienes

4	1,3-Butadiene	AcOH,H$_2$O,H$_2$SO$_4$/ 8 hr/110	CH$_2$=CHCH(OAc)CH$_2$OAc (I) (—)	27

CH$_2$=CHCH(OAc)CH$_2$OH (II) (—),
AcOCH$_2$CH=CHCH$_2$OAc (III) (—),
AcOCH$_2$CH=CHCH$_2$OH (IV) (—),

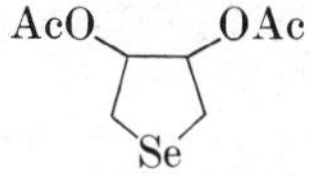

(V) (—), (VI) (—)

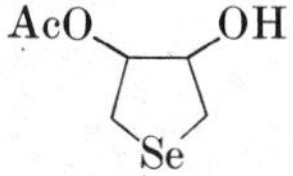

323

Note: References 249–634 are on pp. 407–415.

TABLE VI. Hydrocarbons—*(Continued)*

No. of C Atoms	Reactant	Solvent/Time/ Temperature (°C)	Product(s) and Yield(s) (%)	Refs.
		D. Alkadienes—(Continued)		
4 *(contd.)*		AcOH,H$_2$O,HCl/8 hr/ 110	I, II, III, IV, and AcOCH$_2$CH= CHCH$_2$Cl	27
5	Isoprene	Ac$_2$O or AcOH/—/—	2-Hydroxymethyl-1,3-butadiene, 2-acetoxymethyl-1,3-butadiene, 3-acetoxymethylfuran,1-formyl-3- or 4-methyl-1-vinyl-3-hexene, 2-methyl-1,4-diacetoxy-2-butene, 2-formyl-4-acetoxy-2-butene, 3-hydroxy-4-methylenetetrahydroselenophene, 3-acetoxy-4-methylene-tetrahydroselenophene, 4-acetoxy-3-hydroxy-3-methyltetrahydroselenophene, 3,4-diacetoxy-3-methyltetrahydroselenophene (—)	86
6	2,3-Dimethyl-1,3-butadiene	None/—/400–500	3,4-Dimethylselenophene (27)	84
		—/—/—	4,5-Dimethyl-1-oxa-2-selena-4-cyclohexene-2-oxide (—)	87
10	2,7-Dimethyl-2,6-octadiene	—/—/—	2,7-Dimethyl-(E), (E)-2,6-octadiene-1,8-dial (48)	61

324

TABLE VII. HYDROCARBONS, CYCLIC

No. of C Atoms	Reactant	Solvent/Time Temperature (°C)	Product(s) and Yield(s) (%)	Refs.
		A. Cycloalkane		
10	1,1-Dibromo-2-methyl-3-phenylcyclopropane	AcOH, Ac$_2$O/—/—	(Z)- and (E)-3-Bromo-4-phenyl-3-buten-2-ol acetate (—), (Z)- and (E)-3-bromo-4-phenyl-1,3-butadiene (—)	201
		B. Cycloalkenes		
5	1-Chlorocyclopentene	Ac$_2$O/—/—	3-Chloro-2-cyclopenten-1-yl acetate (—)	278
6	1-Chlorocyclohexene	Ac$_2$O/10 hr/reflux	2-Chloro-2-cyclohexen-1-ol (after saponification) (80)	277, 278
	3-Chlorocyclohexene	Ac$_2$O/—/—	4-Chloro-2-cyclohexen-1-yl acetate (—)	278
	Cyclohexene	AcOH,H$_2$SO$_4$/—/110	1,2-Cyclohexanediol diacetate (32; *cis*:*trans*, 55–45), cyclohexyl acetate (trace), cyclohexenone (trace)	26, 49

Note: References 249–634 are on pp. 407–415.

TABLE VII. HYDROCARBONS, CYCLIC—(Continued)

No. of C Atoms	Reactant	Solvent/Time/Temperature (°C)	Product(s) and Yield(s) (%)	Refs.
		B. Cycloalkenes—(Continued)		
6 (contd.)	Cyclohexene-1-^{13}C	AcOH,Ac$_2$O/12–15 hr/90–95	2-Cyclohexen-1-yl acetate (24)	44
7	2-Chloro-1-methylcyclohexene	Ac$_2$O/10 hr/reflux	2-Methyl-2-cyclohexen-1-one (—), a diene (—)	277
	2-Chloro-4-methylcyclohexene	Ac$_2$O/—/—	2-Chloro-6-methyl-2-cyclohexen-1-yl acetate (—)	278
	1-Methylcyclohexene	Ac$_2$O/3 hr/reflux	2-Methyl-2-cyclohexen-1-ol (—), 3-methyl-2-cyclohexen-1-ol (after saponification) (—)	277
	3-Methylcyclohexene	Dioxane, H$_2$O/12–20 hr/reflux	1-Methyl-2-cyclohexen-1-ol (20), trans-4-methyl-2-cyclohexen-1-ol (5.2), 4:1 mixture of cis-4-methyl-2-cyclohexen-1-ol and trans-6-methyl-2-cyclohexen-1-ol (10)	44
	(+)-3- and 4-Methylcyclohexene (mixture)	Ac$_2$O/3 hr/105	4-Methyl-, 5-methyl-, and 6-methyl-2-cyclohexen-1-yl acetate (—)	8, 39
		n-C$_4$H$_9$OH/30 hr reflux	Mixture of allylic butyl ethers, allylic alcohols, and α,β-unsaturated ketones (15–20)	39
	4-Methylcyclohexene	Ac$_2$O/—/—	6-Methyl-2-cyclohexen-1-yl acetate (—)	278
	(+)-4-Methylcyclohexene	Ac$_2$O/5 hr/reflux	5-Methyl-2-cyclohexen-1-yl acetate (—)	277
	Norbornene	AcOH/—/—	2,3-exo-cis- and 2,7-exo-syn-Norbornanediol diacetate (—)	48
8	Cyclooctene	AcOH,Ac$_2$O/25 hr/100–110	2-Cycloocten-1-yl acetate (38)	262

	Reactant	Reagent/Conditions	Products (%)	Ref.
		AcOH, Ac₂O/11 hr/115–117	2-Cycloocten-1-ol (31), 1-cyclooctene-3,8-diol diacetate (19)	262
	1,4-Dimethylcyclohexene	Dioxane, H₂O/12–20 hr/reflux	2,5-Dimethyl-2-cyclohexen-1-one (13), 2,5-dimethyl-2-cyclohexen-1-ol (cis-, 18; trans-, 41)	44
	2,4-Dimethylcyclohexene	Ac₂O/—/—	2,6-Dimethyl-2-cyclohexen-1-yl acetate (—)	278
	cis-3,5-Dimethylcyclohexene	Dioxane, H₂O/10–20 hr/reflux	1,5-Dimethyl-2-cyclohexen-1-ol (cis, 19; trans, 7), 4,6-dimethyl-2-cyclohexen-1-ol (cis, cis-, 7; trans-, trans-, 5)	44
	1-Ethylcyclohexene	Ac₂O/3 hr/reflux	2-Ethyl-2-cyclohexen-1-ol (after saponification) (—)	277
	1-Nitromethylcycloheptene	Dioxane/2–3 hr/95	1-Formylcycloheptene (I) (35), 1-cycloheptenecarboxylic acid (II) (7)	286
		C₂H₅OH/7 hr/reflux	I (32), II (14)	286
		Ac₂O/11 hr/90	I (16)	286
9	1,5,5-Trimethylcyclohexene	Ac₂O/12 hr/25	2,4,4-Trimethyl-2-cyclohexen-1-yl acetate (33)	16
	3,3,5-Trimethylcyclohexene	Dioxane, H₂O/12–20 hr/reflux	4,4,6-Trimethyl-2-cyclohexen-1-ol (cis-, 35; trans-, 3), 4,6,6-trimethyl-2-cyclohexen-1-ol (cis-, 4.5; trans-, 0.6), 4,4,6-trimethyl-2-cyclohexen-1-one (3.4)	44
10	trans-Δ²-Octalin	Dioxane, H₂O/12–20 hr/reflux	trans-syn-Δ²,1-Octalol (42), trans-anti-Δ²,1-octalol (24)	44
	endo-Dihydrodicyclopentadiene	AcOH, Ac₂O/4–5 hr/40–45	endo-3a,4,5,6,7,7a-Hexahydro-4,7-endomethylene-1-hydroxyindene acetate (I) (50)	287
11	6,7-Dihydro-5H-benzocyclo-heptene	AcOH, Ac₂O/12 hr/25	I (60), isolated as hydroxy compound	288
		C₅H₅N, H₂O/1.75 hr/reflux	Di-(7-oxobenzocycloheptatrien-5-yl) diselenide (23)	85

Note: References 249–634 are on pp. 407–415.

TABLE VII. HYDROCARBONS, CYCLIC—(Continued)

No. of C Atoms	Reactant	Solvent/Time/Temperature (°C)	Product(s) and Yield(s) (%)	Refs.
			B. Cycloalkenes—(Continued)	
12	6,7-Dihydro-6-methyl-5H-benzocycloheptene	C_5H_5N,H_2O/1.75 hr/reflux	6-Methylbenzocyclohepten-7-one (—)	85
	6,7-Dihydro-8-methyl-5H-benzocycloheptene	C_5H_5N,H_2O/1.75 hr/reflux	Di-(8-methylbenzocyclohepten-6-yl) diselenide (—)	85
	1-Phenylcyclohexene	Ac_2O/3 hr/reflux	2-Phenyl-2-cyclohexen-1-ol (after saponification) (—)	277
14	6,7-Dihydro-1,2,3-trimethoxy-5H-benzocycloheptene	C_5H_5N,H_2O/3 hr/reflux	Di-(2,3,4-trimethoxy-7-oxobenzocyclohepten-6-yl) tetraselenide (—)	85
	6,7-Dihydro-2,3,4-trimethoxy-5H-benzocycloheptene	C_5H_5N,H_2O/3 hr/reflux	Di-(1,2,3-trimethoxybenzocyclohepten-6-yl) diselenide (25)	85
15		Xylene/6 hr/reflux	(—)	289
18	1-Ferrocenylcyclopentene	C_2H_5OH,H_2O/16–18 hr/reflux	2-Ferrocenyl-2-cyclopenten-1-one (20), 2-ferrocenyl-3-ethoxy-1-cyclopentene (5)	289a
		C_5H_5N/4 hr/100	(53), (36)	

(5),

291

OCH$_3$

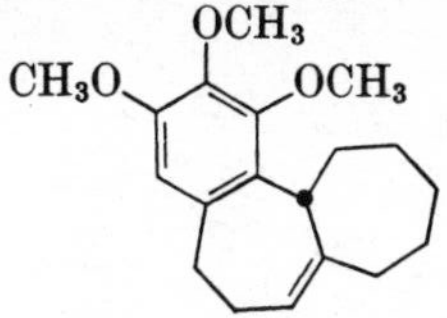

OCH$_3$

OCH$_3$

CH$_3$O

OH

291

(2),

OCH$_3$

OCH$_3$

OCH$_3$

CH$_3$O

OH

291

(17),

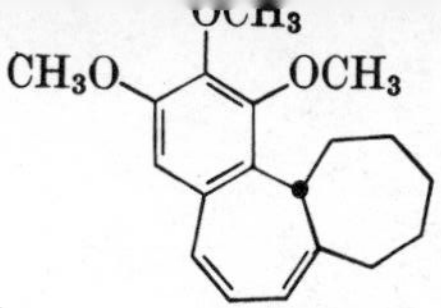

OCH$_3$

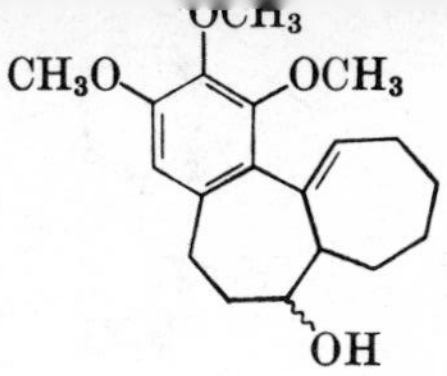

OCH$_3$

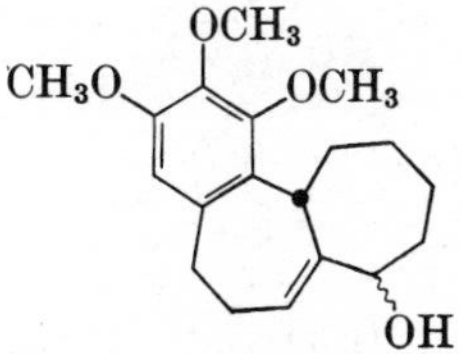

CH$_3$O

(20)

OCH$_3$

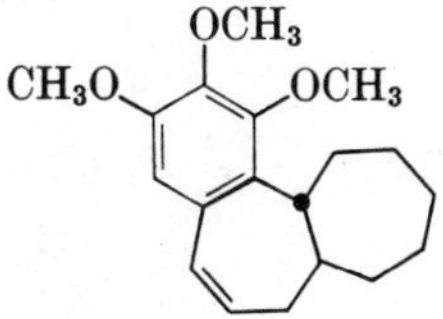

OCH$_3$

CH$_3$O

OH

(54)

OCH$_3$

OCH$_3$

CH$_3$O

OH

(50),

OCH$_3$

OCH$_3$

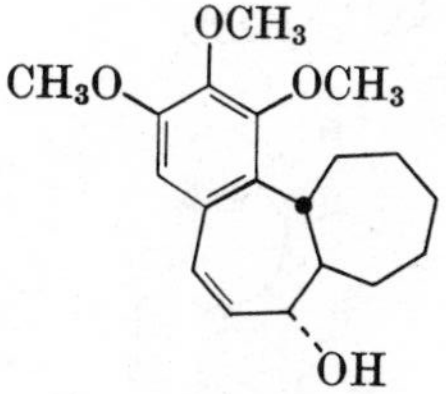

CH$_3$O

(C$_2$H$_5$)$_2$O,AcOH,H$_2$O/ 48 hr/25

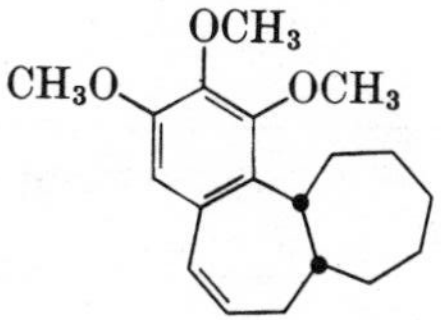

C$_5$H$_5$N/4.5 hr/reflux

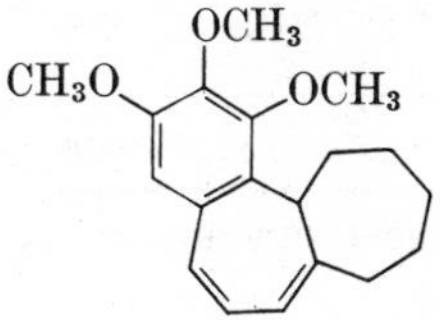

C$_5$H$_5$N/12 hr/85–90

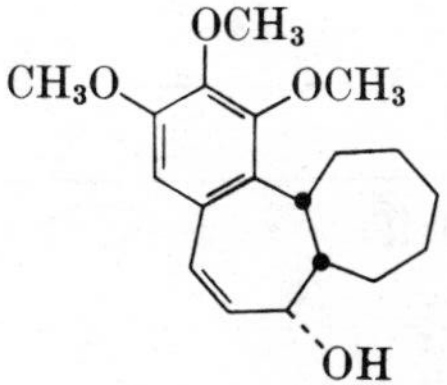

OCH$_3$

OCH$_3$

CH$_3$O

19

OCH$_3$

OCH$_3$

CH$_3$O

OCH$_3$

OCH$_3$

CH$_3$O

Note: References 249–634 are on pp. 407–415.

329

TABLE VII. HYDROCARBONS, CYCLIC—(Continued)

No. of C Atoms	Reactant	Solvent/Time/Temperature (°C)	Product(s) and Yield(s) (%)	Refs.

B. Cycloalkenes—(Continued)

19
(contd.)

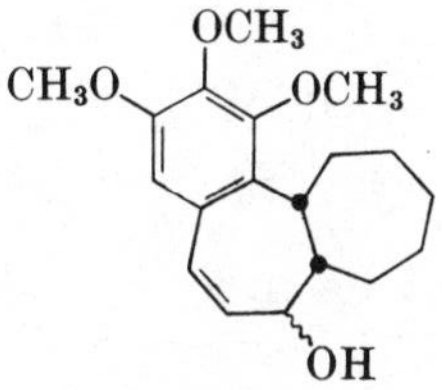

C_5H_5N/10 hr/reflux

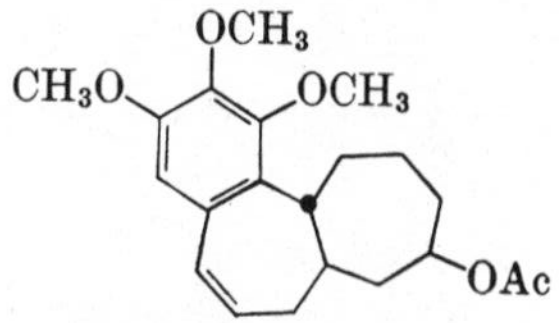

(6), (35), (—)

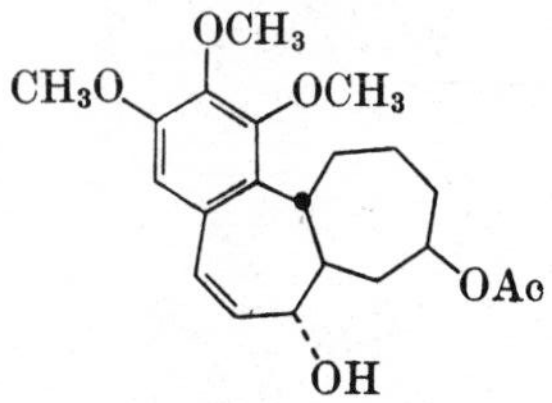

263

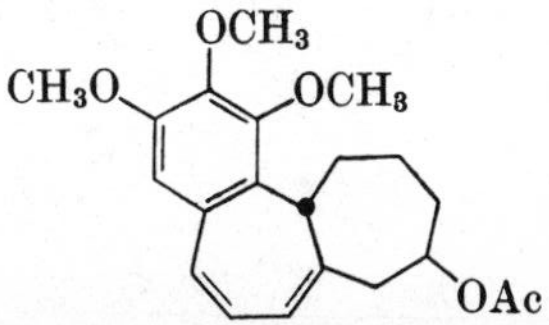

CO$_2$CH$_3$
CHO

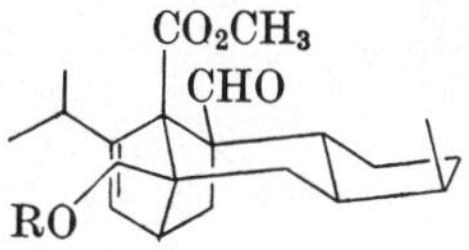

RO

R = H, OAc

21

C$_6$H$_6$/52 hr/reflux

CO$_2$CH$_3$
CHO

O$^-$

$^+$Se

O$^-$

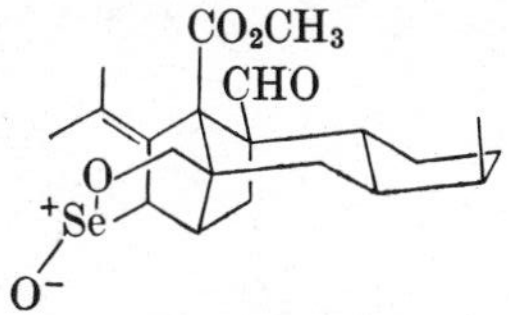

(34, when R = H; no reaction when
R = OAc)

17

C. *Cycloalkadi-, Cycloalkatri-, and Cycloalkatetra-enes*

7	Bromocycloheptatrienes	Dioxane/2 hr/reflux	Tropone (33)	292
	1,3,5-Cycloheptatriene	Dioxane, H$_2$O/15 hr/90	Tropone (25)	127, 128, 293
8	Cyclooctatetraene	—/—/—	o-C$_6$H$_4$(CHO)$_2$ (—), C$_6$H$_5$COCHO (—), cyclooctatrienone (—)	294
10	Dicyclopentadiene	Dioxane, H$_2$O/3 hr/95	endo-3a,4,7,7a-Tetrahydro-4,7-exomethylene-1-hydroxyindene (I) (63)	295, 296, 297
		AcOH, Ac$_2$O/—/40–45 I acetate (50)		298
exo-		AcOH, Ac$_2$O/—/—	exo-3a,4,7,7a-Tetrahydro-4,7-exomethylene-1-hydroxyindene acetate (50)	287
	1,3-Dimethyl-1-ethyl-3,5-cyclohexadiene	C$_5$H$_5$N/7 hr/85–90	1,3-Dimethyl-2-ethylbenzene (I) (—), 1,3-dimethyl-4-ethylbenzene (II) (—), 3-ethyl-3-methyl-1-hydroxymethyl-1,5-cyclohexadiene (—)	299
		AcOH/—/—	I (—), II (—), III (—), and organo-selenium compounds (—)	299

Note: References 249–634 are on pp. 407–415.

TABLE VII. HYDROCARBONS, CYCLIC—(*Continued*)

No. of C Atoms	Reactant	Solvent/Time/ Temperature (°C)	Product(s) and Yield(s) (%)	Refs.
			C. Cycloalkadi-, Cycloalkatri-, and Cycloalkatetra-enes—(Continued)	
10 (*contd.*)	1,2,6,6-Tetramethyl-1,3-cyclohexadiene	C₂H₅OH/6 hr/80 C₂H₅OH/8 hr/82	II (—), organoselenium compounds (—) Prehnitene (—), 3,3-dimethyl-2-methylene-6-cyclohexene-1-carboxalde-hyde (—) and Se compounds (—)	300 301, 302, 303
	1,5,5,6-Tetramethyl-1,3-cyclohexadiene	C₅H₅OH/—/—	Prehnitene (—)	303, 304
17	2-Benzyloxymethyl-1,3-dimethyl-1,3-cyclohexadiene	C₂H₅OH/—/—	2-Benzyloxymethyl-3-hydroxymethyl-1-methyl-1,3-cyclohexadiene (62)	305
18		C₅H₅N/5 hr/reflux	(25)	291
			D. Aromatic	
7	Toluene	None/1.5 hr/340	C₆H₅CHO (35), C₆H₅CO₂H (10)	306
8	Xylene			
	ortho-	None/1.5 hr/250	*o*-CH₃C₆H₄CHO (6), *o*-CH₃C₆H₄CO₂H (—)	306
	meta-	None/1.5 hr/250	*m*-CH₃C₆H₄CHO (44), *m*-CH₃C₆H₄CO₂H (—)	306
	para-	None/1.5 hr/250	*p*-CH₃C₆H₄CHO (10), *p*-CH₃C₆H₄CO₂H (—)	306

9	Mesitylene	None/4 hr/reflux	3,5-$(CH_3)_2C_6H_3CHO$ (21)	306
10	1,4-Dihydronaphthalene	—/—/—	Naphthalene (—)	307
11	2-Methylnaphthalene	None/15–20 min/reflux	2-Naphthaldehyde (34), 2-naphthoic acid (—)	306
14	Anthracene, deuterated	D_2O/—/220–230	Anthraquinone-d_8 (—)	308
20		—/—/—	(—)	309
32		$C_6H_5NO_2$/—/—	(—)	310
40		$C_6H_5NO_2$/1 hr/reflux	(—)	311

Note: References 249–634 are on pp. 407–415.

333

TABLE VIII. KETONES

No. of C Atoms	Reactant	Solvent/Time/Temperature (°C)	Product(s) and Yield(s) (%)	Refs.
			A. Monoketones	
3	Acetone	Xylene, H_2O/—/reflux	Glyoxal (—)	188
6	Cyclohexanone	Dioxane, H_2O/6–8 hr/reflux	1,2-Cyclohexanedione (60)	175, 231, 312
7	*syn*-7-Chloro-2-norbornanone	C_6H_5Br/12 hr/150–155	*syn*-7-Chloro-2,3-norbornanedione (58)	313
	2,4-Cycloheptadienone	AcOH, H_2O, C_5H_5N or C_2H_5OH/2.5 hr/reflux	Tropone (70)	196
	Cycloheptanone	C_2H_5OH/6 hr/reflux	1,2-Cycloheptanedione (90)	107
	6,7-Dichlorobicyclo[3.2.0]-2-heptanone	t-C_4H_9OH/—/—	6,7-Dichlorobicyclo-[3.2.0]-3-hepten-2-one (—)	314
	trans-5,6-Dichloro-3-norbornanone	—/—/—	*trans*-5,6-Dichloro-2,3-norbornanedione (50–60)	315
	1,3-Dimethyl-1-cyclopenten-5-one	AcOH/0.5 hr/100–110	1,3-Dimethyl-1-cyclopentene-4,5-dione (35)	316
	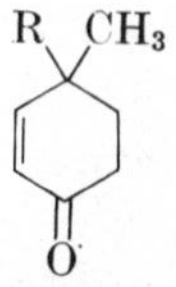 [structure with R, CH_3] R = CH_3, C_2H_5	t-C_4H_9OH/24 hr/reflux	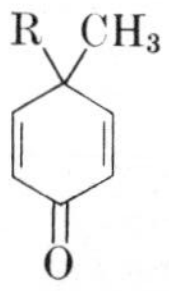 (I), 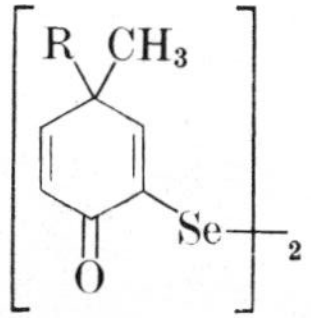 [structure with R, CH_3] (II), [structure with R, CH_3, Se] R = CH_3, I (—), II (14); C_2H_5, I (70), II (14)	258

334

8

Reactant	Conditions	Product (% Yield)	Refs.
$C_6H_5COCH_3$	Xylene,H_2O/—/reflux	C_6H_5COCHO (—)	175, 188
p-RC$_6$H$_4$COCH$_3$ R = C$_2$H$_5$ to n-C$_{10}$H$_{21}$	Dioxane,H_2O/4 hr/reflux	p-RC$_6$H$_4$COCHO (20–82) R = C$_2$H$_5$ to n-C$_{10}$H$_{21}$	108
p-ROC$_6$H$_4$COCH$_3$ R = C$_2$H$_5$ to n-C$_5$H$_{11}$	Dioxane,H_2O/4 hr/reflux	p-ROC$_6$H$_4$COCHO (20–80) R = C$_2$H$_5$ to n-C$_5$H$_{11}$	108
$C_6H_5COCH_2Br$	C_2H_5OH/12 hr/reflux	$C_6H_5COCO_2C_2H_5$ (70)	38
p-BrC$_6$H$_4$COCH$_3$	C_2H_5OH,H_2O/6 hr/reflux	p-BrC$_6$H$_4$COCHO (52)	205
1-Cyclohexenyl methyl ketone	Dioxane/3.5 hr/	1-Cyclohexenylglyoxal (—)	317
3,5-Diiodo-4-hydroxy-acetophenone	—/—/—	3,5-Diiodo-4-hydroxyphenylglyoxal (—)	318
4-Dimethylcyclohexanone	Dioxane,H_2O/6–8 hr/reflux	4,4-Dimethyl-1,2-cyclohexanedione (—)	319
4,4-Dimethyl-2-cyclohexen-1-one	t-C$_4$H$_9$OH/—/reflux	4,4-Dimethyl-2,5-cyclohexadien-1-one (I) (70), bis (I)-2,2-diselenide (14)	70
o-HOC$_6$H$_4$COCH$_3$	Dioxane,H_2O/4 hr/reflux	o-Hydroxyphenylglyoxylic acid γ-lactone (30)	320
p-HOC$_6$H$_4$COCH$_3$	C_2H_5OH/10 hr/reflux	p-HOC$_6$H$_4$COCHO (I) (95)	321
	H_2O/8 hr/reflux	I (44)	321
o-O$_2$NC$_6$H$_4$COCH$_3$	Dioxane,H_2O/7 hr/reflux	o-O$_2$NC$_6$H$_4$COCHO (—)	322
p-O$_2$NC$_6$H$_4$COCH$_3$	C_2H_5OH/9 hr/reflux	p-O$_2$NC$_6$H$_4$COCHO (—)	323
X,Y-C$_6$H$_3$COCH$_3$ X = H, Y = 2,3,4-OH	Dioxane/4–20 hr/reflux	YC$_6$H$_4$COCHO (72–87)	324
X = H; Y = 4-OCH$_3$	Dioxane/48 hr/reflux	4-CH$_3$OC$_6$H$_4$COCHO (82)	
X = 3-OH; Y = 4-OH	Dioxane/30 hr/reflux	3,4-(HO)$_2$C$_6$H$_3$COCHO (—)	

Note: References 249–634 are on pp. 407–415.

TABLE VIII. Ketones—(Continued)

No. of C Atoms	Reactant	Solvent/Time/Temperature (°C)	Product(s) and Yield(s) (%)	Refs.
			A. Monoketones—(Continued)	
8 (contd.)	X = 3-OCH₂C₆H₅, Y = 4-OCH₂C₆H₅	C_2H_5OH/—/—	$3,4\text{-}(C_6H_5CH_2O)_2C_6H_3COCHO$ (82)	
9	(1S)-1-Bromo-α-fenchocamphorone	Ac_2O/—/—	(1S)-1-Bromo-α-fenchocamphorone-quinone (31.2)	325
	3,5-Diiodo-4-hydroxypropiophenone	—/—/—	3,5-Diiodo-4-hydroxyphenyl methyl diketone (—)	318
	3,5-Diiodo-2-methoxyacetophenone	—/—/—	3,5-Diiodo-2-methoxyphenylglyoxal (—)	318
	(1R)-[2-¹⁸O]-7,7-Dimethyl-2-norbornanone	Ac_2O/—/—	(1R)-[2-¹⁸O]-7,7-Dimethyl-2,3-norbornanedione (—)	326
	4-Ethyl-4-methyl-2-cyclohexen-1-one	$t\text{-}C_4H_9OH$/—/reflux	4-Ethyl-4-methyl-2,5-cyclohexadien-1-one (I) (70), bis (I)-2,2-diselenide (14)	70
	(1R)-α-Fenchocamphorone	Ac_2O/—/—	α-Fenchocamphoronequinone (68)	325
	α-Fenchocamphorone-¹⁸O	Ac_2O/4 hr/150	α-Fenchocamphoronequinone-¹⁸O (32)	327
	5-Iodo-2-methoxyacetophenone	—/—/—	5-Iodo-2-methoxyphenylglyoxal (—)	318
	$p\text{-}IC_6H_4COC_2H_5$	—/—/—	$p\text{-}IC_6H_4COCOCH_3$ (—)	318
	$m\text{-}CH_3C_6H_4COCH_3$	Dioxane,H_2O/4 hr/reflux	$m\text{-}CH_3C_6H_4COCHO$ (62)	273
	β-Methyltropolone methyl ether-A	Dioxane/18 hr/reflux	β-Formyltropolone (25)	328, 329
	β-Methyltropolone methyl ether-B	Dioxane/8 hr/reflux	β-Formyltropolone methyl ether-B (37), β-carboxytropolone methyl ether-B (15)	328, 329

336

	Reactant	Conditions	Product	Ref.
10	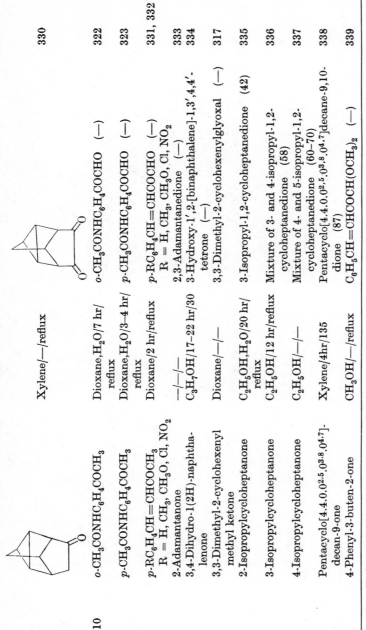	Xylene/—/reflux	(two-ketone structure)	330
	o-CH₃CONHC₆H₄COCH₃	Dioxane,H₂O/7 hr/reflux	o-CH₃CONHC₆H₄COCHO (—)	322
	p-CH₃CONHC₆H₄COCH₃	Dioxane,H₂O/3—4 hr/reflux	p-CH₃CONHC₆H₄COCHO (—)	323
	p-RC₆H₄CH=CHCOCH₃ R = H, CH₃, CH₃O, Cl, NO₂	Dioxane/2 hr/reflux	p-RC₆H₄CH=CHCOCHO (—) R = H, CH₃, CH₃O, Cl, NO₂	331, 332
	2-Adamantanone	—/—/—	2,3-Adamantanedione (—)	333
	3,4-Dihydro-1(2H)-naphthalenone	C₃H₇OH/17–22 hr/30	3-Hydroxy-1',2'-[binaphthalene]-1,3',4,4'-tetrone (—)	334
	3,3-Dimethyl-2-cyclohexenyl methyl ketone	Dioxane/—/—	3,3-Dimethyl-2-cyclohexenylglyoxal (—)	317
	2-Isopropylcycloheptanone	C₂H₅OH,H₂O/20 hr/reflux	3-Isopropyl-1,2-cycloheptanedione (42)	335
	3-Isopropylcycloheptanone	C₂H₅OH/12 hr/reflux	Mixture of 3- and 4-isopropyl-1,2-cycloheptanedione (58)	336
	4-Isopropylcycloheptanone	C₂H₅OH/—/—	Mixture of 4- and 5-isopropyl-1,2-cycloheptanedione (60–70)	337
	Pentacyclo[4.4.0.0²·⁵.0³·⁸.0⁴·⁷]decan-9-one	Xylene/4hr/135	Pentacyclo[4.4.0.0²·⁵.0³·⁸.0⁴·⁷]decane-9,10-dione (87)	338
	4-Phenyl-3-buten-2-one	CH₃OH/—/reflux	C₆H₅CH=CHCOCH(OCH₃)₂ (—)	339

Note: References 249–634 are on pp. 407–415.

337

TABLE VIII. KETONES—(Continued)

No. of C Atoms	Reactant	Solvent/Time/Temperature (°C)	Product(s) and Yield(s) (%)	Refs.
		A. Monoketones—(Continued)		
11	4-t-Butylcycloheptanone	C_2H_5OH/—/—	Mixture of 4- and 5-t-butylcyclo-heptanedione (—)	340
12	Cyclopent[cd]azulenone	Dioxane/3 hr/60	Cyclopent[cd]azulene-1,2-dione (10)	341
	3,4-Dihydro-7-methoxy-3-methyl-1-(2H)-naphthalenone	C_2H_5OH/6 hr/reflux	2-Hydroxy-7-methoxy-3-methyl-1,4-naphthoquinone (—)	342
13	4'-Chloro-3-(2-furyl)-2'-hydroxyacrylophenone	i-$C_5H_{11}OH$/3 hr/reflux	2-(2-Furyl)-7-chlorochromone (55)	112
	β-Damascenone	Dioxane/45 min/60	2-Oxo-β-damascenone (70), 2,3,6-trimethyl-1-crotonylbenzene (18)	343
	Dihydroxyperinaphthindenone	C_2H_5OH/—/—	Perinaphthindanetrione (—)	250
	2'-Hydroxy-3-(4-nitro-2-pyrryl)acrylophenone	$C_5H_{11}OH$/3 hr/reflux	2-(4-Nitro-2-pyrryl)chromone (20)	110
	Methyl 2-(4,4,6-trimethyl-5,6-epoxy-5-cyclohexenyl)vinyl ketone	THF/12 hr/reflux	Methyl 2-(2,4,4-trimethyl-3,6-dihydroxy-3-cyclohexenyl)vinyl ketone (—)	344
14	2,4-BrX$C_6H_3COCH_2C_6H_4$-4-Y X, Y = H or F	Dioxane/—/—	2,4-BrX$C_6H_3COCOC_6H_4$-4-Y X, Y = H or F (—)	345
	p-HO$C_6H_4COCH_2C_6H_5$	—/—/—	p-HO$C_6H_4COCOC_6H_5$ (—)	346
	p-RO$C_6H_4COCH_2C_6H_5$ R = C_2H_5, i-C_3H_7, C_4H_9, C_5H_{11}, i-C_5H_{11}, C_6H_{13}, C_7H_{15}, C_8H_{17}, and $C_{10}H_{21}$	Dioxane,H_2O/12 hr/reflux	p-RO$C_6H_4COCOC_6H_5$ (97, when R = C_6H_{13})	347
	p-(4-ClC_6H_4)$C_6H_4COCH_3$	Dioxane,H_2O/5 hr/90	p-(4-ClC_6H_4)C_6H_4COCHO (—)	348
	7,7a-Dihydro-4-methoxy-7a-methyl-3-isopropyl-5-(6H)-indenone	t-C_4H_9OH,AcOH/96 hr/reflux	(—), (—)	349

338

No.	Substrate	Conditions	Product (yield)	Ref.
	methoxyacrylophenone p-$C_6H_5C_6H_4COCH_3$	reflux	chromone (55)	
		Dioxane, H_2O/5 hr/90	p-$C_6H_5C_6H_4COCHO$ (—)	348
			p-$C_6H_5C_6H_4COCHO\cdot C_2H_5OH$ (—)	348
15	4'-Acetamido-3(2-furyl)-2'-hydroxyacrylophenone	C_2H_5OH/—/—	7-Acetamido-2-(2-furyl)chromone (55)	112
	5'-Acetamido-3(2-furyl)-2'-hydroxyacrylophenone	i-$C_5H_{11}OH$/3 hr/reflux	6-Acetamido-2-(2-furyl)chromone (55)	112
	Benzyl o-selenomethylphenyl ketone	—/—/—	o-Selenomethylbenzil (—)	350
	4-Bromo-3-nitro-2'-hydroxy-chalcone	i-$C_5H_{11}OH$/48 hr/reflux	3'-Nitro-4'-bromoflavone (81)	111
16	Dibenzo[a,e]cycloocten-5(6H)-one	Dioxane, H_2O/127 hr/reflux	Dibenzo[a,e]cyclooctene-5,6-dione (24)	351
	2'-Hydroxy-5'-carboxychalcone	$C_5H_{11}OH$/12hr/reflux	Flavone-6-carboxylic acid (40)	352
	2'-Hydroxy-5'-carboxy-3-hydroxychalcone	$C_5H_{11}OH$/12 hr/	3'-Hydroxyflavone-6-carboxylic acid (20)	352
	3-(p-Methoxyphenyl)propiophenone	$C_5H_{11}OH$/10–15 hr/reflux	4-Methoxychalcone (—)	353
	3-Phenyl-2'-hydroxy-4'-methoxypropiophenone	$C_5H_{11}OH$/15 hr/reflux	7-Methoxyflavone (—)	353
17	3'-Acetamido-5'-chloro-2'-hydroxychalcone	i-$C_5H_{11}OH$/18 hr/reflux	8-Acetamido-6-chloroflavone (40)	354
	Benzyl 5-(2-oxobenzimidazolyl) ketone	—/—/—	Phenyl 5-(2-oxobenzimidazolyl) diketone (—)	355
	(structure: OH C_6H_5 ...)	$C_5H_{11}OH$/—/—	(structure: C_6H_5 ...) (—)	356

Note: References 249-634 are on pp. 407-415.

TABLE VIII. KETONES—(*Continued*)

A. Monoketones—(*Continued*)

No. of C Atoms	Reactant	Solvent/Time/Temperature (°C)	Product(s) and Yield(s) (%)	Refs.
17 (*contd.*)	4β,8α-Dimethyl-4α-phenyl-hexahydro-3(9)-inden-2-one	AcOH/4 hr/reflux	4β,8α-Dimethyl-4α-phenylhexahydro-3(9)-indene-1,2-dione (97)	357
	2′-Hydroxy-5′-carboxy-4-methoxychalcone	$C_5H_{11}OH$/12 hr/reflux	4′-Methoxyflavone-6-carboxylic acid (30)	352
	3-(4-Methoxyphenyl)-2′-hydroxy-4-methoxypropiophenone	$C_5H_{11}OH$/15 hr/reflux	4′,7-Dimethoxyflavone (—)	353
18	3′-Acetamido-5′-chloro-2′-hydroxy-4-methoxychalcone	i-$C_5H_{11}OH$/18 hr/reflux	8-Acetamido-6-chloro-4-methoxyflavone (—)	354
	1-Benzoyl-2-methylnaphthalene	H_2O/10 hr/250	1-Benzoyl-2-naphthoic acid (76)	139
	3-(3,4-Dimethoxyphenyl)-2′-hydroxy-4′-methoxypropiophenone	$C_5H_{11}OH$/15 hr/reflux	3′,4′,7-Trimethoxyflavone (—)	353
	3-(3,4-Dimethoxyphenyl)-3′,4′-methylenedioxypropiophenone	$C_5H_{11}OH$/10–15 hr/reflux	3,4-Dimethoxy-3′,4′-methylenedioxychalcone (—)	353
	1-Mesityl-1-phenylpropanone	Dioxane,H_2O/6 hr/reflux	Mesitylphenylpyruvaldehyde (—), mesityl phenyl diketone (—)	358
	3-(4-Methoxyphenyl)-2′-hydroxy-3′,4′-dimethoxy-propiophenone	$C_5H_{11}OH$/15 hr/reflux	4′,7,8-Trimethoxyflavone (—)	353
19	Benzyl 2-methoxy-1-naphthyl ketone	—/—/—	2-Methoxy-1-naphthyl phenyl diketone (—)	359
	Benzyl 4-methoxy-1-naphthyl ketone	Ac_2O/2 hr/150	4-Methoxy-1-naphthyl phenyl diketone (80)	359
	2-Methylnaphthyl p-tolyl ketone	H_2O/4 hr/230–240	1-p-Toluyl-2-naphthoic acid (52)	140

	Reactant	Conditions	Product (%)	Ref.
20	3-(3,4,5-Trimethoxyphenyl)-3',4'-methylenedioxypropiophenone	C$_5$H$_{11}$OH/10–15 hr/ reflux	3,4,5-Trimethoxy-3',4'-methylene-dioxychalcone	353
		Xylene/—/reflux	(88)	360
24	4'-Benzyloxy-3-(2-furyl)-2'-hydroxyacrylophenone	i-C$_5$H$_{11}$OH/3 hr/ reflux	7-Benzyloxy-2-(2-furyl)chromone (45)	112
	3'-i-Amyl-2-hydroxy-2,4,4',6'-tetramethoxychalcone	C$_5$H$_{11}$OH/24 hr/140	8-i-Amyl-2',4',5,7-tetramethoxyflavone (50)	361
	5'-i-Amyl-2'-hydroxy-2,4,4',6'-tetramethoxychalcone	C$_5$H$_{11}$OH/24 hr/140	6-i-Amyl-2',4',5,7-tetramethoxyflavone (40)	361
25	4-Benzyloxy-2'-hydroxy-4',5',6'-trimethoxychalcone	—/—/—	4'-Benzyloxy-5,6,7-trimethoxyflavone (—)	362

Note: References 249–634 are on pp. 407–415.

341

TABLE VIII. KETONES—(Continued)

No. of C Atoms	Reactant		Solvent/Time/Temperature (°C)	Product(s) and Yield(s) (%)		Refs.

B. Diketones

	RCOCH₂COR'		Dioxane/—/—	RCOCOCOR'		363
	R = C₆H₅	R' = CH₃	`: : /12 hr/reflux`	R = C₆H₅	R' = CH₃ (46)	
	CH₃	CH₃	`: : /20 hr/90°`	CH₃	CH₃ (39)	
	C₂H₅	C₂H₅	`: : /100°`	C₂H₅	C₂H₅ (44)	
	(CH₃)₂CH	(CH₃)₂CH	`: : / : :`	(CH₃)₂CH	(CH₃)₂CH (60)	
	(CH₃)₃C	(CH₃)₃C	`: : /90°`	(CH₃)₃C	(CH₃)₃C (57)	
	(CH₃)₃C	CH₃	`: : / : :`	(CH₃)₃C	CH₃ (41)	
	C₆H₅	OC₂H₅	`: : /12 hr/reflux`	C₆H₅	OC₂H₅ (66)	
	p-O₂NC₆H₄	OC₂H₅	`: : / : :`	p-O₂NC₆H₄	OC₂H₅ (88)	
	p-CH₃C₆H₄	OC₂H₅	`: : / : :`	p-CH₃C₆H₄	OC₂H₅ (69)	
	p-CH₃OC₆H₄	OC₂H₅	`: : / : :`	p-CH₃OC₆H₄	OC₂H₅ (65)	
	Mesityl	OC₂H₅	`: : /18 hr/`	Mesityl	OC₂H₅ (64)	
	C₆H₅	NH₂	`: : / : :`	C₆H₅	NH₂ (10)	
6	1,3-Cyclohexanedione		CH₃OH/—/reflux	1,2,3,4,6,7,8,9-Octahydrophenoxaselenin-1,9-dione 10-oxide (78)		96
	[1,4-¹³C₂]-1,4-Cyclohexanedione		H₂O/50 min/60-70	[1,4-¹³C₂]-p-Benzoquinone (72)		364
	1,4-Cyclohexanedione		Dioxane/2 hr/reflux	Hydroquinone (14)		148
7	5-Methyl-1,3-cyclohexanedione		CH₃OH/—/reflux	3,7-Dimethyl-1,2,3,4,6,7,8,9-octahydro-phenoxaselenin-1,9-dione-10-oxide (—)		96
9	Polyfluoro-1,3-indandione		C₆H₆/—/—	1,2-Diperfluorophthaloylethane (—)		365

10	1,3-Bis(bromoacetyl)benzene	C_2H_5OH/17 hr reflux	Diethyl m-phenylenediglyoxylate (83)	366
	1,4-Bis(bromoacetyl)benzene	C_2H_5OH/17 hr/reflux	Diethyl p-phenylenediglyoxylate (100)	366
11	o-Diacetylbenzene	$i\text{-}C_3H_7OH$/3–4 hr/reflux	2-Hydroxy-1,4-naphthoquinone (—)	113
	o-Acetylpropiophenone	$i\text{-}C_3H_7OH$/3 hr/reflux	2-Hydroxy-3-methyl-1,4-naphthoquinone, 2-methyl-1,4-naphthoquinone (—)	113
	2-Methyl-1-phenyl-1,3-butanedione	—/—/—	$C_6H_5COCOCH_3$ (—), AcOH (—), $C_6H_5COCO_2H$ (—)	117
12	o-Acetyl-n-butyrophenone	$i\text{-}C_3H_7OH$/4 hr reflux	2-Hydroxy-3-ethyl-1,4,-naphthoquinone (40), 2-ethyl-1,4-naphthoquinone (—)	113
	5-Phenyl-1,3-cyclohexanedione	CH_3OH/—/reflux	3,7-Diphenyl-1,2,3,4,6,7,8,9-octahydrophenoxaselenin-1,9-dione 10-oxide (—)	96
14	5,5-Dimethyl-4-phenyl-1,3-cyclohexanedione	CH_3OH/—/reflux	2,8-(or 4,6-)Diphenyl-3,3,7,7-tetramethyl-1,2,3,4,6,7,8,9-octahydrophenoxaselenin-1,9-dione 10-oxide (—)	96
		AcOH/1 hr/reflux	(—)	367, 368

Note: References 249–634 are on pp. 407–415.

343

TABLE VIII. KETONES—(Continued)

No. of C Atoms	Reactant	Solvent/Time/Temperature (°C)	Product(s) and Yield(s) (%)	Refs.
			B. Diketones—(Continued)	
16	$(p\text{-BrCH}_2\text{COC}_6\text{H}_4)_2$	C_2H_5OH/17 hr/reflux	$(p\text{-C}_2\text{H}_5\text{O}_2\text{CCOC}_6\text{H}_4)_2$ (35)	366
	$p\text{-BrCH}_2\text{COC}_6\text{H}_4\text{OC}_6\text{H}_4\text{COCH}_2\text{-Br-}p$	C_2H_5OH/17 hr/reflux	$p\text{-C}_2\text{H}_5\text{O}_2\text{CCOC}_6\text{H}_4\text{OC}_6\text{H}_4\text{COCO}_2\text{C}_2\text{H}_5\text{-}p$ (100)	366
	Dibenzo[a,e]cyclooctene-5,11,(6H,12H)-dione	Dioxane/84 hr/ reflux	$C_{16}H_{12}O_4$ (70), $C_{16}H_{10}O_2Se$ (20), $C_{31}H_{16}O_3$ (4)	351
	$(C_6H_5COCH_2)_2$	$AcOH,H_2O$/21 hr/90	(E)-$C_6H_5COCH{=}CHCOC_6H_5$ (I) (75–85)	69
		CH_3OH, H_2O/24 hr/ reflux	I (56)	205
		C_2H_5OH, HCl/16 hr/ reflux	I (73)	205
	$(p\text{-BrC}_6\text{H}_4\text{COCH}_2)_2$	C_2H_5OH, HCl/20 hr/ reflux	(E)-$p\text{-BrC}_6\text{H}_4\text{COCH}{=}\text{CHCOC}_6\text{H}_4\text{Br-}p$ (64)	205
	$(p\text{-ClC}_6\text{H}_4\text{COCH}_2)_2$	C_2H_5OH, HCl/36 hr/ reflux	(E)-$p\text{-ClC}_6\text{H}_4\text{COCH}{=}\text{CHCOC}_6\text{H}_4\text{Cl-}p$ (58)	205

18	4,5-Diphenyl-1,3-cyclohexane-dione	CH_3OH/—/reflux	2,3,7,8-(or 3,4,6,7-)Tetraphenyl-1,2,3,4,6,7,8,9-octahydrophenoxaselenin-1,9-dione 10-oxide (—)	96
	$(p\text{-}CH_3C_6H_4COCH_2)_2$	C_2H_5OH/72 hr/reflux	(E)-$p\text{-}CH_3C_6H_4COCH{=}CHCOC_6H_4CH_3\text{-}p$ (62)	205
19	2-Aceto-3,4,5,6-tetramethoxy-propiophenone	$i\text{-}C_3H_7OH$/6 hr/reflux	2-Hydroxy-3-methyl-5,6,7,8-tetramethoxy-1,4-naphthoquinone (—), 2-methyl-5,6,7,8-tetramethoxy-1,4-naphthoquinone (—)	113
20	2-Aceto-3,4,5,6-tetramethoxy-butyrophenone	$i\text{-}C_3H_7OH$/5 hr/reflux	3-Ethyl-2-hydroxy-5,6,7,8-tetramethoxy-1,4-naphthoquinone (—), 2-ethyl-5,6,7,8-tetramethoxy-1,4-naphthoquinone (—)	113
21	1,2,3-Triphenyl-1,3-propane-dione	—/—/—	Benzil (—), $C_6H_5CO_2H$ (—)	117
23	α-Benzoyl-2'-hydroxy-2-methoxychalcone	$C_5H_{11}OH$/12 hr/reflux	3-Benzoyl-2'-methoxyflavone (60)	109
25	α-(3,4,5-Trimethoxybenzoyl)-2'-hydroxychalcone	$C_5H_{11}OH$/17 hr/reflux	3-(3,4,5-Trimethoxybenzoyl)flavone (60)	109

C. Tetraketone

18	1,6-Diphenyl-1,3,4,6-hexane-tetrone	Dioxane/24 hr/reflux	2,5-Dibenzoyl-3,4-dihydroxyselenophene (29)	369

Note: References 249–634 are on pp. 407–415.

345

TABLE IX. NITROGEN COMPOUNDS

A. Amines

No. of C Atoms	Reactant	Solvent/Time/Temperature (°C)	Product(s) and Yield(s) (%)	Refs.
2	(CH₃)₂NH	(C₂H₅)₂O/—/−20	[(CH₃)₂N]₂SeO (—), [(CH₃)₂CH₂]₂Se₂O₅ (—)	94
3	H₂NCH₂CH₂NH₂	DMF/40 min/reflux	1,2,5-Selenadiazole (43)	89
	H₂N(CH₂)₃NH₂	DMF/40 min/reflux	3-Methyl-1,2,5-selenadiazole (35)	89
	(structure with NHCH₃ and NH₂, R and R')	AcOH/—/—	(selenadiazole structure) R = H R' = Cl; Cl H; H Br; Br H; H CH₃; CH₃ H; H CH₃O; CH₃O H	370
8	2,3-Diamino-1,4-dimethoxybenzene	H₂O/0.5 hr/reflux	4,7-Dimethoxybenzoselenadiazole (79)	92
	4,5-Diaminobenzo[b]selenophene	—/—/—	Seleno[3,2-e]benzo-2,1,3-selenadiazole (—)	371
10	2,3-Diaminonaphthalene	H₂O/—/—	2,1,3-Naphtho(2,3-c)selenadiazole (—)	91

11	1,2-Diamino-8-methoxynaphthalene	H_2O/—/—	9-Methoxynaphtho[1,2-c][1,2,5]-selenadiazole (24)	372
14	3-Chloro-1,2-diaminoanthraquinone	DMF/2–3 min/reflux	4-Chloroanthra[1,2-c][1,2,5]selenadiazole-6,11-dione (100)	93
	4-Chloro-1,2-diaminoanthraquinone	DMF/2–3 min/reflux	5-Chloroanthra[1,2,c]selenadiazole-6,11-dione (100)	93
	2,3-Diaminoanthraquinone	—/—/—	Anthra[2,3-c][1.2.4]selenadiazole-5,10-dione (93)	373
	9,10-Diaminophenanthrene	C_2H_5OH/—/reflux	Phenanthro[9,10-c][1,2,5]selenadiazole (—)	90

B. Amino Acids

8	5-Bromo-2,3-diamino-4-methyl-benzoic acid	—/—/—	6-Bromo-4-carboxy-7-methyl-2,1,3-benzoselenadiazole (—)	95
	2,3-Diamino-4-methylbenzoic acid	—/—/—	4-Carboxy-7-methyl-2,1,3-benzo-selenadiazole (—)	95

C. Imides

15	N,N-Phthaloyl-m-toluidine	None/50 min/250	N,N-Phthaloyl-m-aminobenzoic acid (70), N,N-phthaloyl-m-aminobenzaldehyde (8)	374
	N,N-Phthaloyl-p-toluidine	None/50 min/250	N,N-Phthaloyl-p-aminobenzoic acid (37), N,N-phthaloyl-p-aminobenzaldehyde (35)	374

D. Nitriles

4	(Z)-$CH_3CH=CHCN$	Dioxane/20 hr/reflux	$OHCCH=CHCN$ (—)	221
	$CH_2=CHCH_2CN$	AcOH/6 hr/reflux	$CH_3CO_2CH_2CH=CHCN$ (—)	221
		Ac_2O/6 hr/reflux	$CH_2=CHCH(OCOCH_3)CN$ (—),(Z)—CH_3-$CO_2CH_2CH=CHCN$ (—)	221
8	o-$ClC_6H_4CH_2CN$	—/—/180–190	o-$ClC_6H_4CO_2H$(12), o-ClC_6H_4CN (66)	33

Note: References 249–634 are on pp. 407–415.

TABLE IX. Nitrogen Compounds—(*Continued*)

No. of C Atoms	Reactant	Solvent/Time/ Temperature (°C)	Product(s) and Yield(s) (%)	Refs.
		D. Nitriles—(*Continued*)		
8(*contd.*)	m-ClC$_6$H$_4$CH$_2$CN	—/—/180–190	m-ClC$_6$H$_4$CO$_2$H(11), m-ClC$_6$H$_4$CN (62)	33
	p-ClC$_6$H$_4$CH$_2$CN	—/—/180–190	p-ClC$_6$H$_4$CO$_2$H (—), p-ClC$_6$H$_4$CN (—)	33
10	Camphoric mononitrile	C$_6$H$_5$CH$_3$/4 hr/reflux	Camphoric anhydride (11)	213
12	1-C$_{10}$H$_7$CH$_2$CN	—/—/180–200	1-C$_{10}$H$_7$COCN (47)	31, 32
		E. Heterocyclic Compounds		
5	6-Methylpyridazine 1-oxide	C$_5$H$_5$N/—/—	Pyridazine-6-carboxaldehyde 1-oxide (—)	375
6	6-Methyluracil	AcOH/6 hr/reflux	Orotaldehyde (58)	233
	5,5-Dimethyl-1-pyroline 1-oxide	CH$_3$OH/1.5 hr/ reflux	5,5-Dimethyl-1-pyrrolin-3-one 1-oxide (as 2,4-dinitrophenylhydrazone) (—)	376
	6-Methylthymine	AcOH/5 hr/reflux	Thymine-6-carboxaldehyde (60)	233
	3-Nitro-2-picoline	Dioxane/—/—	3-Nitropicolinaldehyde (—)	377
	4-Nitro-3-picoline 1-oxide	Dioxane/6.5 hr/101	No reaction	73
		C$_5$H$_5$N/16 hr/117	No reaction	73
	3-Nitro-2-pyridinemethanol	Dioxane/—/—	3-Nitropicolinaldehyde (—)	377
	2-Picoline	None/30 min/reflux	Picolinic acid, I (64)	73, 74
		None/2 hr/110–120	I (74)	73
		Dioxane/2 hr/100	I (32)	73
		C$_5$H$_5$N/70 min/115	I (66)	73
		3-Picoline/80 min/110	I (59)	73
		Isoquinoline/80 min/ 110	I (47)	73
		(C$_6$H$_5$)$_2$O/3 hr/130	I (26)	378
		Xylene/—/reflux	I (—)	188

348

Starting Material	Reaction Conditions (Solvent/Time/Temp)	Products (%)	Refs.
3-Picoline	—/—/—	Picolinaldehyde, II (50); I (Trace)	379
	C_5H_5N/2 hr/115	No reaction	73, 379
4-Picoline	None/30 min/reflux	Isonicotinic acid (I) (62)	74
	None/2 hr/110–120	I (74)	74
	Dioxane/2 hr/100	I (53)	73
	C_5H_5N/70 min/115	I (83)	73
	3-Picoline/80 min/110	I (82)	73
	Isoquinoline/20 min/110	I (71)	73
	$(C_6H_5)_2O$/30 min/185	I (57)	378
	H_2O/—/140	I (80–88)	216
	—/—/—	I (40), isonicotinaldehyde (25)	379
2-Picoline 1-oxide	Dioxane/8 hr/101	Picolinic acid (I) (16), picolinaldehyde 1-oxide (II) (19)	73
3-Picoline 1-oxide	C_5H_5N/4 hr/117	II (59)	73
	C_5H_5N/10 hr/116	No reaction	73
4-Picoline 1-oxide	Dioxane/7 hr/101	Isonicotinic acid (40)	73
	C_5H_5N/2.5 hr/118	Isonicotinic acid 1-oxide (75)	73
2-Pyridinemethanol	Dioxane/2.5 hr/80; 0.5 mol SeO_2: base	Picolinaldehyde (I) (90)	28
	Dioxane/2.5 hr/80; 1.0 mol SeO_2: base	I (48), aldehyde-SeO_2 compound (25)	28
	None/3 min/160; 0.5 mol SeO_2: base	I (100)	28
	None/90 min/110; 1.0 mol SeO_2: base	Picolinic acid (II) (85)	28
	None/3 min/200; 0.5 mol SeO_2: base	I (75), II (Trace)	28
	None/5 min/150; 1.0 mol SeO_2: base	II (80)	28

Note: References 249–634 are on pp. 407–415.

349

TABLE IX. Nitrogen Compounds—(Continued)

No. of C Atoms	Reactant	Solvent/Time/ Temperature (°C)	Product(s) and Yield(s) (%)	Refs.
		E. Heterocyclic Compounds—(Continued)		
6 (*contd.*)	2-Pyridinemethanol (*contd.*)	C_5H_5N/1 hr/90 0.5 mol SeO_2: base	I (86)	28
		C_5H_5N/2.5 hr/90; 1.0 mol SeO_2: base	I (94)	28
		C_2H_5OH/2 hr/reflux	I (80)	380
		Dioxane/5 hr/reflux	I (95)	380
	3-Pyridinemethanol	None/10 min/160; 0.5 mol SeO_2: base	Nicotinaldehyde (I) (60), nicotinic acid (II) (Trace)	28
		None/3 min/200 0.5 mol SeO_2: base	I (27), II (Trace)	28
		None/90 min/110; 1 mol SeO_2: base	I (5), I (Trace)	28
		None/20 min/150; 1 mol SeO_2: base	I (43), II (Trace)	28
		Dioxane/2.5 hr/80; 0.5 mol SeO_2: base	No reaction	28
		Dioxane/2.5 hr/80; 1 mol SeO_2: base	No reaction	28
		C_5H_5N/1 hr/90; 0.5 mol SeO_2: base	No reaction	28
		C_5H_5N/2.5 hr/90; 1 mol SeO_2: base	No reaction	28
	4-Pyridinemethanol	None/3 min/160; 0.5 mol SeO_2: base	Isonicotinaldehyde (I) (100)	28
		None/5 min/150; 1 mol SeO_2: base	I (37), isonicotinic acid (II) (41)	28

350

	Substrate	Conditions	Product (%)	Ref.
6		Dioxane/2.5 hr/80; 0.5 mol SeO₂: base	I (89)	28
		Dioxane/2.5 hr/80; 1 mol SeO₂: base	I (76)	28
		C₅H₅N/2.5 hr/90; 0.5 mol SeO₂: base	I (100)	28
		C₅H₅N/2.5 hr/90 1 mol SeO₂: base	I (72)	28
7	2-Acetylpyridine	Dioxane/5 hr/reflux	I (80)	380
		Dioxane/105 min/80	Picolinic acid (I) (63)	73, 381
		Dioxane/130 min/116	I (73)	73
		C₅H₅N/105 min/80	I (67)	73
		C₅H₅N/130 min/116	I (73)	73
	3-Acetylpyridine	Dioxane/120 min/75	3-Pyridineglyoxylic acid (I) (32)	73
		Dioxane/120 min/95	I (54)	73
		C₅H₅N/120 min/80	I (70)	73
	4-Acetylpyridine	Dioxane/170 min/70	Isonicotinic acid (I) (40)	73
		C₅H₅N/165 min/70	I (43)	73
	2,5-Dimethyl-4-acetyloxazole	Dioxane/20 hr/reflux	2,5-Dimethyl-4-oxazolylglyoxal (95)	382
	2-Ethylpyridine	Dioxane/2 hr/90	Picolinic acid (I) (1)	73
		Dioxane/4 hr/101	I (48)	73
		C₅H₅N/2 hr/90	I (6)	73
		C₅H₅N/70 min/117	I (31)	73
	3-Ethylpyridine	—/—/—	Nicotinic acid (Trace)	74
	4-Ethylpyridine	Dioxane/2 hr/85	Isonicotinic acid (I) (—)	73
		Dioxane/2 hr/101	I (—)	73
		C₅H₅N/2 hr/85	I (—)	73
		C₅H₅N/70 min/117	I (—)	73
	Imidazo [1,2-a]pyridine	AcOH/23 hr/reflux	3,3'-Di(imidazo[1,2-a]pyridyl) selenide (55), 3,3'-di(imidazo[1,2-a]pyridyl) diselenide (21)	383

Note: References 249–634 are on pp. 407–415.

351

NITROGEN COMPOUNDS—(*Continued*)

No. of C Atoms	Reactant	Solvent/Time/Temperature (°C)	Product(s) and (Yield(s)) (%)	Refs.
		E. Heterocyclic Compounds—(Continued)		
7 (*contd.*)	2,4-Lutidine	None/30 min/reflux	2,4-Pyridinedicarboxylic acid (I) (39)	74
		None/2 hr/reflux	I (60)	74
		C_5H_5N/70 min/90	2-Methylisonicotinic acid (44)	73
		3-Picoline/30 min/110	I (67)	73
		—/—/—	I (30), 2,4-Pyridinedicarboxaldehyde (1)	379
	2,5-Lutidine	C_5H_5N/2 hr/115	5-Methylpicolinic acid (79)	73
	2,6-Lutidine	C_5H_5N/70 min/115	2,6-Pyridinedicarboxylic acid (I) (69)	73, 384
		3-Picoline/30 min/110	I (61)	73
		—/—/—	I (63), 2,6-pyridinedicarboxaldehyde (0.5)	379
	3,5-Lutidine	C_5H_5N/2.5 hr/115	No reaction	73
	2,4-Lutidine 1-oxide	C_5H_5N/6 hr/119	2-Formylisonicotinic acid (I) (31)	73
	2,5-Lutidine 1-oxide	C_5H_5N/7 hr/117	I (43)	73
	2,5-Lutidine 1-oxide	C_5H_5N/4 hr/116	5-Methyl-2-pyridinecarboxaldehyde 1-oxide (66)	73
	2,6-Lutidine 1-oxide	C_5H_5N/4.5 hr/117	2,6-Pyridinedicarboxaldehyde 1-oxide (I) (42)	73
	3,5-Lutidine 1-oxide	C_5H_5N/7 hr/117	I (58)	73
	5-Methyl-4-nitro-2-pyridine-methanol	C_5H_5N/10 hr/117	No reaction	73
		Dioxane/4 hr/80	5-Methyl-4-nitro-2-pyridinecarboxaldehyde (85)	385
	6-Methyl-2-pyridinemethanol	Dioxane/5 hr/reflux	6-Methyl-2-pyridinecarboxaldehyde (90)	380
	3-Nitro-2,5-lutidine	—/—/—	5-Methyl-3-nitro-2-pyridinecarboxalde-hyde (—)	386
	4-Nitro-2,5-lutidine	—/—/—	5-Methyl-4-nitro-2-pyridinecarboxalde-hyde ( )	385

352

2,6-Pyridinedimethanol	AcOEt/5 hr/reflux Dioxane,C₅H₅N, or AcOH	2,6-Diformylpyridine (I) (72) I (76, 67, or 76)	387 387
2,4,4-Trimethyl-1-pyrroline 1-oxide	CH₃OH/2 hr/reflux	3,3-Dimethyl-5-oxo-1,2,3,4-tetrahydropyridine 1-oxide (28) (after treatment with hydrochloric acid)	388
8			
4,5-Diamino-2-phenyl-v-triazole	—/—/—	5-Phenyl-5H[1,2,3]triazolo[4,5-c][1,2,5]selenadiazole (—)	389
5-Ethyl-2-methylpyridine	None/30 min/reflux None/2 hr/110–120 3-Picoline/2 hr/115 H₂SO₄/40–90 min/250–275	5-Ethyl-2-pyridinecarboxylic acid (I) (58) I (76) I (64) 2,5-Pyridinedicarboxylic acid (75)	74 74 73 390
Indole	C₆H₆/2 hr/reflux	3,3'-Diindolyl selenide (23)	391, 392
2-Methylbenzothiazole	Dioxane/1 hr/reflux	2-Benzothiazolecarboxaldehyde (—)	393
2-Methylbenzoxazole	Dioxane/1 hr/reflux	o-Acetamidophenol (—)	393
7-Methylimidazo [1,2-a]-pyridine	AcOH/23 hr/reflux	3,3'-Di(7-methylimidazo [1,2-a-pyridyl) selenide (—), 3,3'-di(7-methylimidazo [1,2-a-pyridyl) diselenide (—)	383
5-Nitro-2,3,6-trimethylpyridine	—/—/—	3,6-Dimethyl-5-nitro-2-pyridinecarboxaldehyde (—)	386
2,4,6-Trimethylpyridine	None/30 min/reflux None/2 hr/110–120 C₅H₅N/70 min/90 C₅H₅N/70 min/117	2,4,6-Pyridinetricarboxylic acid (I) (31) I (59) 2,4-Dimethylisonicotinic acid (31) I (71)	74 74 73 73
(pyridine with CH₃, C=NNHCONH₂ substituent)	—/—/—	(selenadiazole-pyridine structure) (—)	159

Note: References 249–634 are on pp. 407–415.

No. of C Atoms	Reactant	Solvent/Time/Temperature (°C)	Product(s) and Yield(s) (%)	Refs.
		E. Heterocyclic Compounds—(Continued)		
9	3-Acetamido-6-methyl-2-pyridinemethanol	—/—/—	3-Acetamido-6-methyl-2-pyridine-carboxaldehyde (—), 3-acetamido-6-methylpicolinic acid (—)	394
	6-Carboxyoxindole	AcOH/45 min/reflux	6-Carboxyisatin (40)	395
	2-Methylindole	C_6H_6/2 hr/reflux	3,3'-Di(2-methylindolyl) triselenide (16–22)	391, 392
10	1-Phenyl-3-pyrazolidone	HCl/—/—	1-Phenyl-3-hydroxypyrazole (62)	396
	5-Propyl-2-picoline	C_5H_5N/3.5 hr/reflux	5-Propylpicolinic acid (—)	397
	1-p-Bromophenyl-2-(hydroxymethyl)imidazole	Dioxane/4–8 hr/reflux	1-p-Bromophenyl-2-imidazolecarboxaldehyde (50)	398
	5-Butyl-2-picoline	C_5H_5N/3.5 hr/reflux	5-Butylpicolinic acid (43)	397
	1-Butyl-2-picolinium chloride	Dioxane/90 min/100	1-Butyl-2-carboxypyridinium chloride (31)	73
	1-Butyl-3-picolinium chloride	C_5H_5N/18 hr/117	No reaction	73
	1-Butyl-4-picolinium chloride	C_5H_5N/20 min/116	1-Butyl-4-carboxypyridinium chloride (72)	73
	4-Carbethoxy-6-trifluoromethyl-2-pyridinemethanol	Dioxane/17 hr/reflux	4-Carbethoxy-6-trifluoromethyl-2-pyridinecarboxaldehyde (—)	399
	5-Chloro-1-methylisoquinoline	—/—/—	5-Chloro-1-isoquinolinecarboxaldehyde (—)	400
	2,3-Dihydro-2-oxo-1-benzazepine	—/—/—	2,3-Dihydro-2,3-dioxo-1-benzazepine (—)	116
	1,2-Dimethylindole	C_6H_6/2 hr/reflux	3,3'-Di-(1,2-dimethylindolyl)triselenide (23)	391
	2,3-Dimethylindole	AcOEt/—/—	2-Formyl-3-methylindole (22),3-methyl-indole-2-carboxylic acid (7), (2)	401

(2)

2,3-Dimethylquinoxaline 1,4-dioxide	AcOEt/1.5 hr/reflux	3-methyl-2-quinoxalinecarboxylic acid (—), 2,3-quinoxalinedicarboxaldehyde (—), 2-Methyl-3-formylquinoxaline 1,4-dioxide (70)	403, 375
1,2-Di-(6-pyridazinyl)ethane	—/—/—	1,2-Di-(6-pyridazinyl)ethylene (—)	141
Lepidine	Ac_2O,AcOH/2.75 hr/85–90	4-Quinolinecarboxaldehyde (50–60)	
3-Methylisoquinoline	None/30 min/170	3-Isoquinolinecarboxaldehyde (I) (25–37), 3-isoquinolinecarboxylic acid (II) (trace), di(3-isoquinolinyl)glyoxal (3)	404
3-Methylquinoline	None/10 min/220	I (48)	405
	C_5H_5N/20 hr/115	II (52)	73
	C_5H_5N/5 hr/115	No reaction	73
Nicotine	Dioxane/5 hr/150	Nornicotine (40)	164
1-m-Nitrophenyl-2-(hydroxymethyl)imidazole	Dioxane/4–8 hr/reflux	1-m-Nitrophenyl-2-imidazolecarboxaldehyde (—)	398
1-Phenyl-2-(hydroxymethyl)-imidazole	Dioxane/4–8 hr/reflux	1-Phenyl-2-imidazolecarboxaldehyde (36)	398
Quinaldine	C_5H_5N/70 min/115	Quinaldic acid (75)	73
	Dioxane/—/—	Quinaldaldehyde (I) (—)	406
	C_6H_6/—/—	I (—)	175
	—/—/—	I (68)	407
Quinaldine 1-oxide	C_5H_5N/4 hr/reflux	Quinaldaldehyde-N-oxide (54)	408
	C_2H_5OH/—/—	1,2-Di-(2-quinolyl)ethylene 1,1'-dioxide (20)	408
2,4,6-Trimethylpyridine-3,5-dicarboxylic acid	C_5H_5N/2 hr/100	Pyridinium pyridine-2,3,4,5,6-pentacarboxylate (67)	73
	3-Picoline/2 hr/100	3-Picolinium pyridine-2,3,4,5,6-pentacarboxylate (39)	73
	Isoquinoline/2 hr/100	Isoquinolinium pyridine-2,3,4,5,6-pentacarboxylate (31)	73

Note: References 249–634 are on pp. 407–415.

No. of C Atoms	Reactant	Solvent/Time/ Temperature (°C)	Product(s) and Yield(s) (%)	Refs.
		E. Heterocyclic Compounds—(Continued)		
11	3-Acetyl-4-hydroxy-1,2-dihydro-1-isoquinolone	—/—/100	1,2,3,4-Tetrahydro-1,3,4-isoquino-linetrione (89)	117
	3-Acetyl-5-phenylisoxazole	Dioxane/—/reflux	5-Phenyl-3-isoxazolylglyoxal (—)	382
	4-Benzylpyridazine	AcOH/1 hr/100	4-Benzoylpyridazine (87)	409
	1,4-Dimethylcarbostyril	Xylene/45 min/150	1-Methyl-4-carbostyrilcarboxaldehyde (I) (98)	410
		None/1.25 hr/150–175	I (70)	143
	4,8-Dimethyl-2-hydroxy-quinoline	m-Xylene/8 hr/reflux	2-Hydroxy-8-methyl-4-quinolinecarboxalde-hyde (37)	411
	2,3-Dimethylquinoline	m-Xylene/1.5 hr/reflux	3-Methylquinaldaldehyde (50)	412
	2,4-Dimethylquinoline	m-Xylene/1.5 hr/reflux	4-Methylquinaldaldehyde (0.5), 2,4-quinolinedicarboxaldehyde (4), 4-methylquinaldic acid (10)	413
	2,5-Dimethylquinoline	Dioxane/1.5 hr/reflux	5-Methylquinaldaldehyde (I) (—)	414
		m-Xylene/1 hr/reflux	I (—), 5-methylquinaldic acid (—)	414
	2,6-Dimethylquinoline	Dioxane/1 hr/reflux	6-Methylquinaldaldehyde (40)	415
	2,7-Dimethylquinoline	Dioxane/2.5 hr/reflux	7-Methylquinaldaldehyde (32), 7-methylquinaldic acid (14)	416
	2,8-Dimethylquinoline	Dioxane/1.5 hr/reflux	8-Methylquinaldaldehyde (47)	414
	6,7-Dimethylquinoline	—/—/—	Quinoline-6,7-dicarboxaldehyde (—), 6-methylquinoline-7-carboxaldehyde (—), 7-methylquinoline-6-carboxaldehyde (—)	417

	Reactant	Conditions	Product (Yield %)	Ref.
	N-Ethyl-1,3-dioxotetrahydro-isoquinoline	$C_6H_5CH_3$/8–12 hr/reflux	N-Ethyl-1,3,4-trioxotetrahydro-isoquinoline (70)	115
	1-p-Methoxyphenyl-2-(hydroxy-methyl)imidazole	Dioxane/4–8 hr/reflux	1-p-Methoxyphenyl-2-imidazolecar-boxaldehyde (34)	398
12	N-Methylanabasine	Dioxane/4 hr/150	Anabasine (48)	164
	5-(4-Acetoxybutyl)-2-picoline	C_5H_5N/8 hr/reflux	5-(4-Acetoxybutyl)picolinic acid (47)	418
	4-Acetyl-5-methyl-3-(p-nitrophenyl)isoxazole	Dioxane/24 hr/reflux	5-Methyl-3-(p-nitrophenyl)-4-isoxazolyl-glyoxal (95)	382
	4-Acetyl-5-methyl-3-phenyl-isoxazole	Dioxane/—/reflux	5-Methyl-3-phenyl-4-isoxazolylglyoxal (95)	382
	4-Benzylpyridine	AcOH/0.5 hr/reflux	4-Benzoylpyridine (81)	419
	4,4'-Dimethyl-2,2'-bipyridyl	Dioxane/24 hr/reflux	4-Carboxy-4'-formyl-2,2'-bipyridyl (9)	420
	6-Ethoxyquinaldine	Dioxane/1.5 hr/reflux	5-Ethoxyquinaldaldehyde (43), 6-ethoxyquinaldic acid (8)	411
	1-Ethyl-4-methylcarbostyril	Xylene/45 min/150	1-Ethyl-4-formylcarbostyril (97)	410
	Methyl 4-methylquinoline-7-carboxylate	—/—/—	7-Carbomethoxyquinoline-4-carboxaldehyde (—)	417
	N-i-Propyl-1,3-dioxotetra-hydroisoquinoline	$C_6H_5CH_3$/8–12 hr/reflux	N-i-Propyl-1,3,4-trioxotetrahydro-isoquinoline (70)	115
	1,2,3,4-Tetrahydrocarbazole	AcOEt/—/—	1-Oxo-1,2,3,4-tetrahydrocarbazole (44)	401
13	N-Butyl-1,3-dioxotetrahydro-isoquinoline	$C_6H_5CH_3$/8–12 hr/reflux	N-Butyl-1,3,4-trioxotetrahydroiso-quinoline (70)	115
	N-i-Butyl-1,3-dioxotetrahydro-isoquinoline	$C_6H_5CH_3$/8–12 hr/reflux	N-i-Butyl-1,3,4-trioxotetrahydro-isoquinoline (70)	115
	5-Keto-7-methyljuloline	Xylene/45 min/150	7-Formyl-5-ketojuloline (84)	410
	N-Methyl-1,2,3,4-tetrahydro-carbazole	AcOEt/—/—	N-Methyl-1-oxo-1,2,3,4-tetrahydro-carbazole (24), N-methylcarbazole (16), Se compound (4)	401

Note: References 249–634 are on pp. 407–415.

357

NITROGEN COMPOUNDS—(Continued)

No. of C Atoms	Reactant	Solvent/Time/Temperature (°C)	Product(s) and Yield(s) (%)	Refs.
		E. Heterocyclic Compounds—(Continued)		
12 (contd.)	2-Styrylpyridine	None/—/200–210	Phenyl-2-pyridylglyoxal (31), 2-α-pyridylselenonaphthene (20)	88
	4-Styrylpyridine	None/—/200	2-α-Pyridylselenonaphthene (—)	88
14	3,6-Dimethyl-4,5-phenanthroline hemihydrate	Dioxane/2 hr/reflux	6-Carboxy-3-formyl-4,5-phenanthroline (45)	420
	9-Methyl-3-nitrophenanthridine	Dioxane/6 hr/reflux	3-Nitrophenanthridine-9-carboxaldehyde (63), 1,2-di-(3-nitro-9-phenanthridyl)-ethylene (—)	142
	9-Methylphenanthridine	Dioxane/6.5 hr/reflux	Phenanthridine-9-carboxaldehyde (70), 1,2-di-(9-phenanthridinyl)ethylene (7)	142
15	7-Chloro-5-phenyl-1,3,4,5-tetrahydro-2H-1,4-benzodiazepin-2-one	t-C_4H_9OH, pyridine/30 min/60	7-Chloro-5-phenyl-1,3-dihydro-2H-1,4-benzodiazepin-2-one (70)	421
	N-Cyclohexyl-1,3-dioxotetrahydroisoquinoline	$C_6H_5CH_3$/8–12 hr/reflux	N-Cyclohexyl-1,3,4-trioxotetrahydroisoquinoline (70)	115
	2-Methyl-3-phenylquinoxaline	$AcOCH_3$/4 hr/reflux	3-Phenyl-2-quinoxalinecarboxaldehyde (66), 2-cyano-3-phenylquinoxaline (38)	422
	2-Methyl-3-phenylquinoxaline 1-oxide	$AcOC_2H_5$/3 hr/reflux	3-Phenyl-2-quinoxalinecarboxaldehyde 1-oxide (70)	423

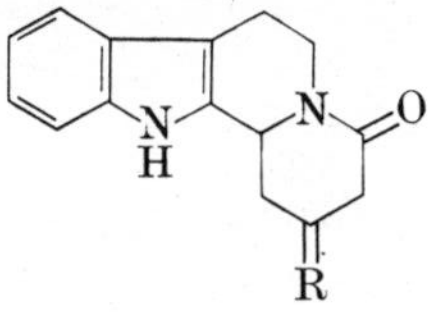

401

$R = H_2$, $(CO_2C_2H_5)_2$ (62)

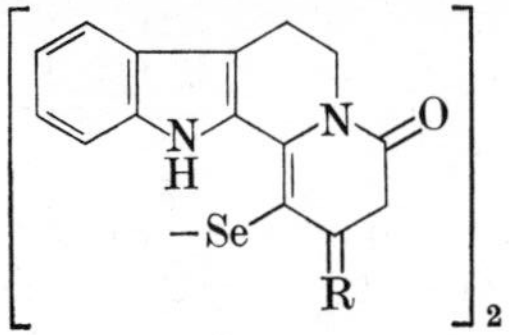

$R = H_2$ (63), $(CO_2C_2H_5)_2$ (62)

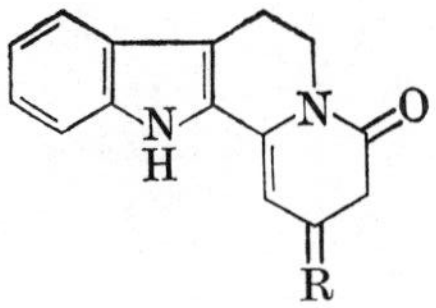

—/—/—

$R = H_2$, $(CO_2C_2H_5)_2$

$R = (CO_2C_2H_5)_2$ (25)

16	6-Methyl-1-(tri-O-acetyl-β-D-ribofuranosyl)uracil 2-Phenylacetylquinoxaline	Dioxane, AcOH/14 hr/reflux	6-Formyl-1-(tri-O-acetyl-β-D-ribofuranosyl)uracil (—)	424

Ac$_2$O/4 hr/reflux   Phenyl-2-quinoxalinylglyoxal   (86)   425

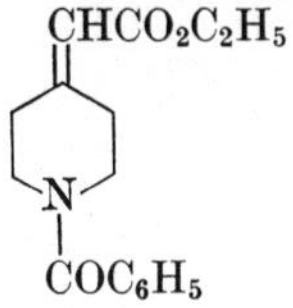

C_6H_6/54 hr/reflux

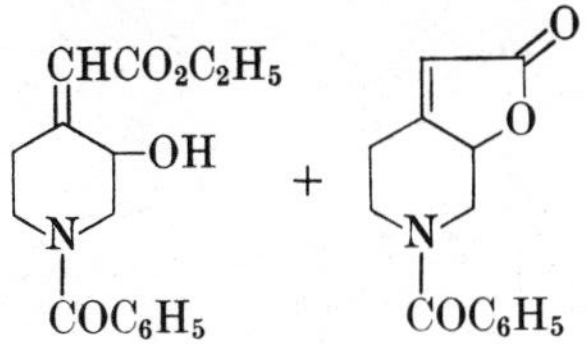

(—, mixture)

426

359

Note: References 249–634 are on pp. 407–415.

No. of C Atoms	Reactant	Solvent/Time/Temperature (°C)	Product(s) and Yield(s) (%)	Refs.
		E. Heterocyclic Compounds—(Continued)		
16 (contd.)	I, R = H₂	AcOEt/—/—	(I) R = O (26)	401
	I, R = H₂	AcOH/—/—	(I) R = O (—)	401
		—/—/—	(—)	427
17	Benzyl-4-methylcarbostyril	Xylene/45 min/150	1-Benzyl-4-formylcarbostyril (56)	410
	X-9-Methylphenanthridine	Dioxane/6 hr/reflux	X-Phenanthridine-9-carboxaldehyde (I) (—), 1,2-di-(X-9-phenanthridyl)ethylene (II) (—) (I,II) (—)	142
	X = NHCO₂C₂H₅-2		(I,II) (3)	
	X = NHCO₂C₂H₅-3		I (64), II (9)	
	X = NHCO₂C₂H₅-4		I (70), II (7)	

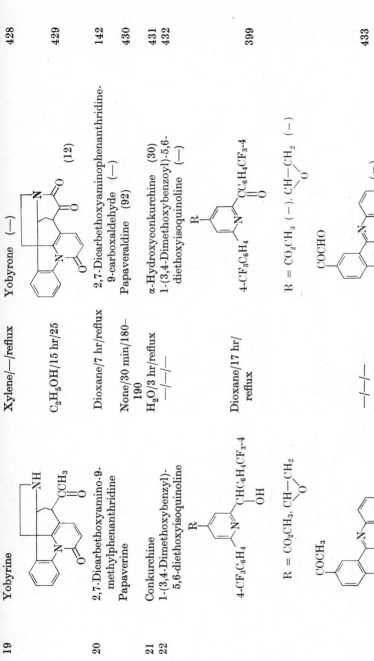

19	Yobyrine	Xylene/—/reflux	Yobyrone (—)	428
		C$_2$H$_5$OH/15 hr/25	(12)	429
20	2,7-Dicarbethoxyamino-9-methylphenanthridine	Dioxane/7 hr/reflux	2,7-Dicarbethoxyaminophenanthridine-9-carboxaldehyde (—)	142
	Papaverine	None/30 min/180–190	Papaveraldine (92)	430
21	Conkurchine	H$_2$O/3 hr/reflux	α-Hydroxyconkurchine (30)	431
22	1-(3,4-Dimethoxybenzyl)-5,6-diethoxyisoquinoline	—/—/—	1-(3,4-Dimethoxybenzoyl)-5,6-diethoxyisoquinoline (—)	432
		Dioxane/17 hr/reflux		399
	R = CO$_2$CH$_3$, CH—CH$_2$	—/—/—	R = CO$_2$C'H$_3$ (—), CH—CH$_2$ (—)	433

Note: References 249–634 are on pp. 407–415.

361

No. of C Atoms	Reactant	Solvent/Time/ Temperature (°C)	Product(s) and Yield(s) (%)	Refs.
		E. Heterocyclic Compounds—(Continued)		
24	Conessine	H_2O/3 hr/reflux	α-Hydroxyconessine (60)	434
	(E)-1,2-Dianilino-1,2-di-(2-pyridyl)ethylene	AcOH/—/—	2,2′-Pyridil (60)	435
25	2,5-Dimethyl-4-triphenylsilylpyridine	—/—/—	5-Methyl-4-triphenylsilylpicolinic acid (—)	436
27	1-(4-Chlorophenyl)-2,3-diphenyl-1,4-dihydroquinoline	—/—/—	1-(4-Chlorophenyl)-2,3-diphenyl-4-oxo-1,4-dihydroquinoline (—)	437
		F. Miscellaneous Nitrogen Compounds		
3	$RC(=NOH)CH_2CR=NOH$	Ac_2O or CH_3OH/—/reflux	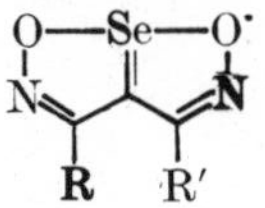	98
	R = H; R′ = C_6H_5		R = H, R′ = C_6H_5 (6)	
	R = H; R′ = p-$CH_3C_6H_4$		R = H; R′ = p-$CH_3C_6H_4$ (5)	
	R = H; R′ = p-ClC_6H_4		R = H; R′ = p-ClC_6H_4 (6)	
	R, R′ = —$(CH_2)_4$—		R, R′ = —$(CH_2)_4$— (60)	
	R, R′ = 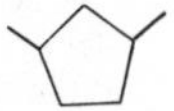		R, R′ = 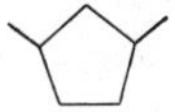 (75)	
	R, R′ = 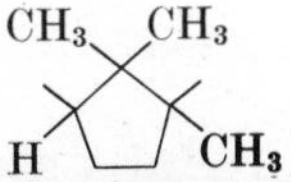		R, R′ = 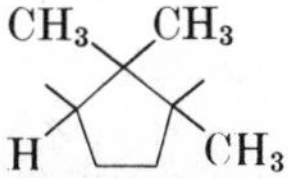 (65)	

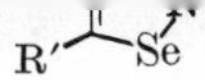

$$R' \overset{\parallel}{\underset{}{Se}}$$

R = H, CH₃, C₂H₅, C₃H₇, C₄H₉, C₆H₁₃,
C₆H₁₃, C₆H₅CH₂, CN, CO₂H, CO₂C₂H₅

R = H, CH₃, C₂H₅, C₃H₇, C₄H₉, C₆H₁₃,
C₆H₅CH₂ . CN, CO₂H, CO₂C₂H₅

R' = H, CH₃, C₆H₅

R' = H, CH₃, C₆H₅

RR' = (CH₂)₃₋₆, (CH₂)₁₀

RR' = (CH₂)₃₋₆, (CH₂)₁₀

4	Biacetyl dioxime	DMF/40 min/reflux	3,4-Diphenyl-1,2,5-selenadiazole (49)	89
5	Ethyl pyruvate semicarbazone	Dioxane/—/—	4-Carbethoxy-1,2,3-selenadiazole (33)	439
	CH₃C(=NOC₂H₅)CH₃	C₂H₅OH/—/reflux	CH₃C(=NOC₂H₅)CO₂C₂H₅ (—)	440
6	C₃H₇CH=NOC₂H₅	Dioxane/4.5 hr/reflux	C₂H₅COCH=NOC₂H₅ (33)	440
7	(CH₃)₂CHC(=NOC₂H₅)CH₃	C₂H₅OH/—/reflux	(CH₃)₂CHC(=NOC₂H₅)CO₂C₂H₅ (—)	440
8	R'CH₂CR=NNHC₆H₃X-2-Y-4	C₂H₅OH, AcOH/2–40 hr/—	R'COCR=NNHC₆H₃X-2-Y-4 (—)	441
	R = H, C₆H₅, substituted C₆H₅, alkyl		RCCR=NNHC₆H₃-2-Y-4 (—)	
	R' = H, substituted C₆H₅ X, Y = H, NO₂		∥ NNHC₆H₃X-2-Y-4 R = H, substituted C₆H₅, alkyl R' = H, substituted C₆H₅ X, Y = H, NO₂	
9	Acetophenone guanylhydrazone	AcOH/2 hr/reflux	3-Amino-6-phenyl-as-triazine (62)	160
	Cyclooctanone semicarbazone	Dioxane/—/—	Octahydrocycloocta-[1,2,3]-selenadiazole (—)	162
10	C₆H₅C(=NOCH₃)CH₃	Dioxane/4.5 hr/reflux	C₆H₅C(=NOCH₃)CHO (49)	440
	C₆H₅C(=NOC₂H₅)CH₃	Dioxane/4.5 hr/reflux	C₆H₅C(=NOC₂H₅)CHO (60)	440

Structure (row 10): R with NNHC₆H₃X-2-Y-4 and =O cyclohexanone (—)

Structure (left, row 12): R with =NNHC₆H₃X-2-Y-4 cyclohexane

R = H, CH₃; X = Y = NO₂

| 12 | Diglyme/10 hr/— | | | 441 |

Structure (row 12 right): R with =NNHC₆H₃X-2-Y-4 cyclohexene (—)

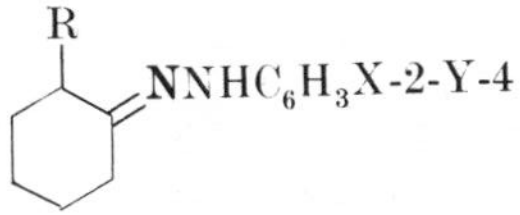

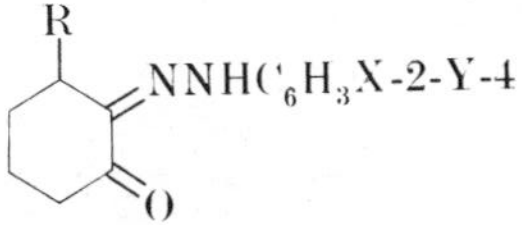

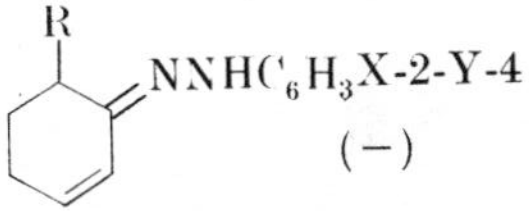

R = H, CH₃; X = Y = NO₂

Note: References 249–634 are on pp. 407–415.

NITROGEN COMPOUND—(Continued)

No. of C Atoms	Reactant	Solvent/time/Temperature (°C)	Product(s) and Yield(s) (%)	Refs.
			F. Miscellaneous Nitrogen Compounds—(Continued)	
13	NNHCONH$_2$	—/—/—	(—)	442
14	Benzil dioxime	DMF/40 min/reflux	3,4-Diphenyl-1,2,5-selenadiazole (49)	89
15	Acetophenone 4-phenylsemicarbazone	Dioxane/—/—	sym-Diphenylurea (—)	99
16	NNHC$_6$H$_3$X-2-Y-4 X, Y = H, NO$_2$	AcOH/6–12 hr/—	NNHC$_6$H$_3$X-2-Y-4 (10–95) X, Y = H, NO$_2$	441
17	2,4-Pentanedione bis-(2,4-dinitrophenylhydrazone)	—/—/—	(—)	98
23	NNHC$_6$H$_3$X-2-Y-4 C$_6$H$_5$ X = Y = NO$_2$	Glyme, AcOH/6–12 hr/—	NNHC$_6$H$_3$X-2-Y-4, C$_6$H$_5$ C$_6$H$_5$ (—) NNHC$_6$H$_3$X-2-Y-4 C$_6$H$_5$ C$_6$H$_5$ (—) X = Y = NO$_2$	441

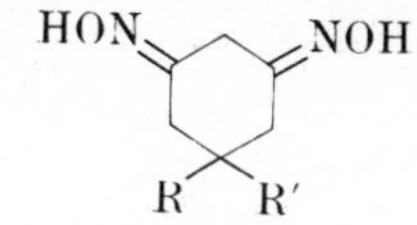

HON=⟨ ⟩=NOH

Ac_2O or CH_3OH/—/ reflux

96,97

$R = R' = H$ (—); $R = R' = CH_3$ (—)

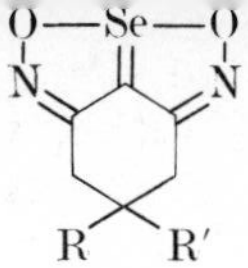

$R = R' = H$; $R = R' = CH_3$

NNHC$_6$H$_3$X-2-Y-4

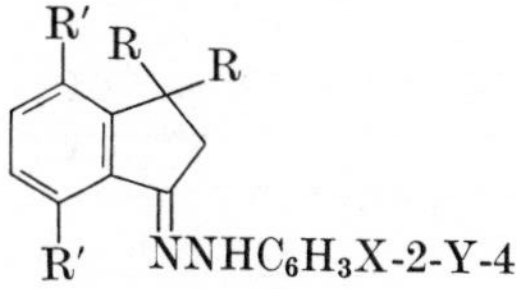

$R = H$, CH_3, C_6H_5; $R' = H$, CH$_3$
$X = Y = NO_2$

AcOH/60–170 hr/—

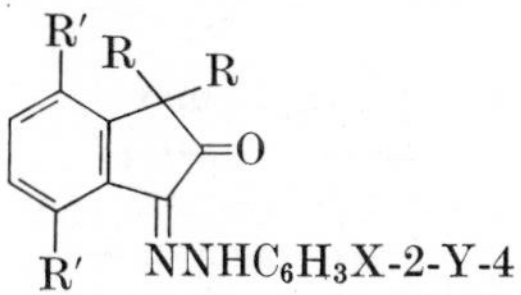

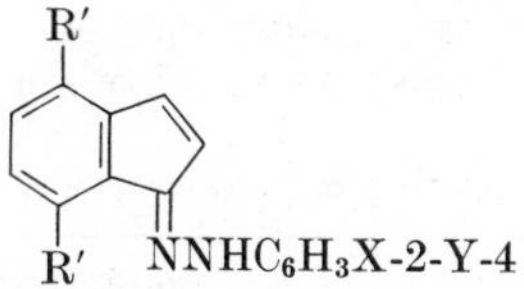

441

NNHC$_6$H$_3$X-2-Y-4 (—),

NNHC$_6$H$_3$X-2-Y-4 (—)

$R = H$, CH_3 C_6H_5; $R' = H$, CH$_3$
$X = Y = NO_2$

H_2NCNHN=C—$(CH_2)_n$CO$_2$R
‖ |
O   C$_6$H$_5$

$R = H$, C_2H_5; $n = 3, 5, 6, 7$

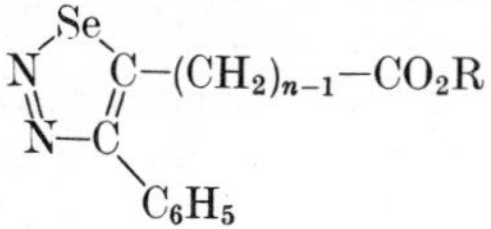

AcOH/—/—

—(CH$_2$)$_{n-1}$—CO$_2$R (—)

158

H_2NCNHN=C—$(CH_2)nC$=NNHCNH$_2$
‖ | | ‖
O   C$_6$H$_5$ C$_6$H$_5$ O

$n = 2, 4, 5, 6, 7, 8, 10$

AcOH/—/—

—(CH$_2$)$_{n-2}$— (—)

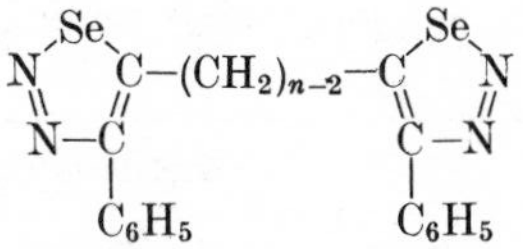

158

Note: References 249–634 are on pp. 407–415.

TABLE X. OXYGEN COMPOUNDS

No. of C Atoms	Reactant	Solvent/Time/ Temperature (°C)	Product(s) and Yield(s) (%)	Refs.

A. Phenols

No. of C Atoms	Reactant	Solvent/Time/ Temperature (°C)	Product(s) and Yield(s) (%)	Refs.
6	Hydroquinone	Xylene, H$_2$O/—/ reflux	p-Benzoquinone (—)	188
12	4-Chloro-2-hexylphenol	Dioxane, H$_2$O/6 hr/ reflux	4-Chloro-2-hexanoylphenol (—)	443

B. Heterocyclic Compounds

No. of C Atoms	Reactant	Solvent/Time/ Temperature (°C)	Product(s) and Yield(s) (%)	Refs.
5	6-Hydroxy-1,3-dioxepan-5-one	C$_6$H$_5$CH$_3$, AcOH/4 hr/reflux	1,3-Dioxepane-5,6-dione (41)	444
7	4-Methoxy-6-methyl-2-pyrone	Dioxane/1 hr/180 (sealed tube)	6-Formyl-4-methoxy-2-pyrone (65), 6-hydroxymethyl-4-methoxy-2-pyrone	445 (25)
8	2,2,5,5-Tetramethyltetrahydro-3-furanone	Dioxane/10 hr/reflux	2,2,5,5-Tetramethyltetrahydro-3,4-furandione (83)	446
9	2-Acetyl-5-trimethylsilylfuran	Dioxane/2 hr/reflux	5-Trimethylsilyl-2-furanylglyoxal (40)	447
	4-Methoxy-3,5,6-trimethyl-2-pyrone	Dioxane/1 hr/165 (sealed tube)	3,5-Dimethyl-6-formyl-4-methoxy-2-pyrone (52)	445
	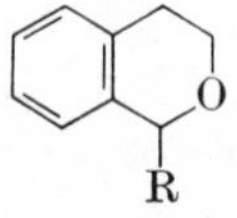	Xylene/12 hr/reflux	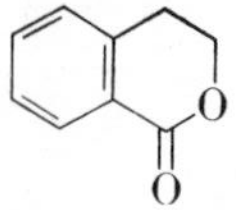 I (85)	138
	R = H	None/2 hr/140-160	I (65)	137, 448
	R = CH$_3$, C$_2$H$_5$, C$_3$H$_7$, CH$_2$C$_6$H$_5$	Xylene/6 hr/reflux	I (—)	449

R = CH₃, CH₃; C₂H₅, C₂H₅ → R = CH₃, CH₃; C₂H₅, C₂H₅

	Reactant	Conditions	Product	Ref.
10	R = CH₃, CH₃; C₂H₅, C₂H₅		R = CH₃, CH₃; C₂H₅, C₂H₅	
	2-Benzofuryl methyl ketone	Xylene/6 hr/reflux	2-Benzofurylglyoxal (—)	449
	2,5-Diethyl-2,5-dimethyl-tetrahydro-3-furanone	—/—/—	2,5-Diethyl-2,5-dimethyltetrahydro-3,4-furandione (86)	433 446
		Dioxane/10 hr/reflux		
	Ethyl 3-methyl-2-benzo-furancarboxylate	AcOH/72 hr/reflux	Ethyl 3-formyl-2-benzofurancarboxylate (—), ethyl 3-acetoxymethyl-2-benzo-furancarboxylate (—)	450
11	3-Methylisochroman	None/2 hr/140–160	3-Methyl-3,4-dihydroisocoumarin (69)	137
	7-Methylisochroman	None/2 hr/140–160	7-Methyl-3,4-dihydroisocoumarin (68)	137
	5,8-Dimethylisochroman	None/2 hr/180	5,8-Dimethyl-3,4-dihydroisocoumarin (59)	137
	4-Methyl-7-methoxycoumarin	Xylene/8 hr/reflux	4-Formyl-7-methoxycoumarin (84)	199
12	2,2,5,5-Bis(tetramethylene)-tetrahydrofuran-3-one	Dioxane, H₂O/12 hr/reflux	2,2,5,5-Bis(tetramethylene)tetrahydro-furan-3,4-dione (83)	451
	4-Methyl-5,7-dimethoxycou-marin	Xylene/6 hr/reflux	4-Formyl-5,7-dimethoxycoumarin (66)	199
	4-Methyl-6,7-dimethoxycou-marin	Xylene/8 hr/reflux	4-Formyl-6,7-dimethoxycoumarin (34)	199
	4-Methyl-7,8-dimethoxycou-marin	Xylene/8 hr/reflux	4-Formyl-7,8-dimethoxycoumarin (71)	199

Note: References 249–634 are on pp. 407–415.

B. Heterocyclic Compounds—(Continued)

No. of C Atoms	Reactant	Solvent/Time/Temperature (°C)	Product(s) and Yield(s) (%)	Refs.
14	2,2,5-Bis(pentamethylene)-tetrahydrofuran-3-one	Dioxane, H_2O/12 hr/reflux	2,2,5,5-Bis(pentamethylene)tetrahydrofuran-3,4-dione (85)	451
	Khellin	AcOEt/3 hr/reflux	4,9-Dimethoxy-5-oxo-5H-furo-[3,2-g][1]benzopyran-7-carboxylic acid (16), corresponding 7-carboxaldehyde (—)	452
15	5,6-Benzoisochroman	None/2 hr/140–160	5,6-Benzo-3,4-dihydroisocoumarin (43)	137
	Flavanone *(structure)*	Ac_2O/—/reflux	Flavone (—), flavanol (—)	453
17	*(structure)*	Xylene/1 hr/180	*(structure)* (100)	454
18	5-Benzyloxy-7-methoxy-4-methylcoumarin	Xylene/5 hr/reflux	5-Benzyloxy-7-methoxy-4-formylcoumarin (93)	455, 456
19	4,7-Dimethoxy-5-cinnamoyl-6-hydroxycoumarone	i-$C_5H_{11}OH$/16 hr/reflux	5,8-Dimethoxy-2-phenylfuro-[2',3':6,7]chromone (—)	457
20	4,7-Dimethoxy-5-(p-methoxy-cinnamoyl)-6-hydroxycoumarone	i-$C_5H_{11}OH$/16 hr/reflux	5,8-Dimethoxy-2-(p-methoxyphenyl)furo-[2',3':6,7]chromone (—)	457

4,7-Dimethoxy-5-(3,4-dimethoxycinnamoyl)-6-hydroxycoumarone | i-C$_5$H$_{11}$OH/16 hr/ reflux | 5,8-Dimethoxy-2-(3,4-dimethoxyphenyl)furo[2',3':6,7]chromone (—) | 457

CH$_2$OH

OH

(CH$_2$)$_4$CH$_3$

O

CH$_3$ CH$_3$

Δ^1-Tetrahydrocannabinol | —/—/— | | 458

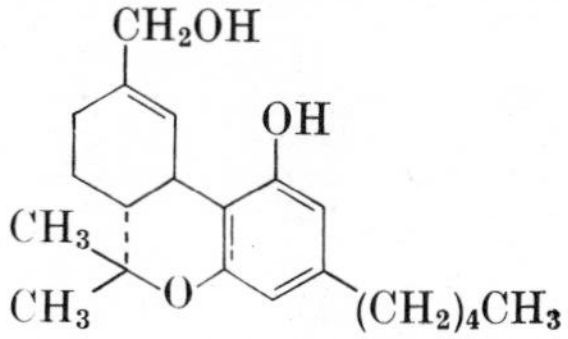

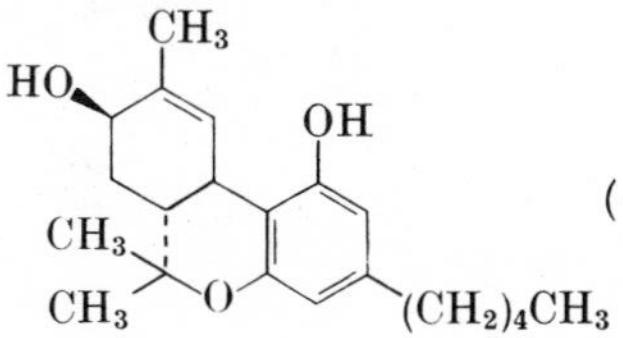

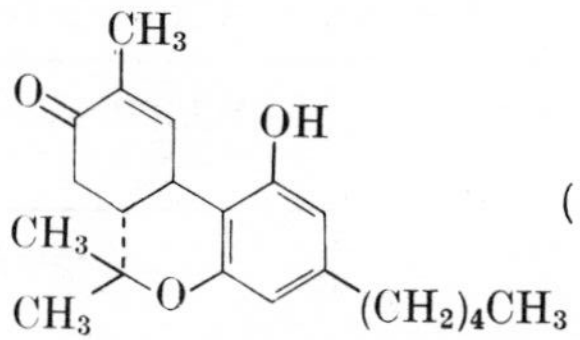

(—),

CH$_3$

HO

OH

(CH$_2$)$_4$CH$_3$

O

CH$_3$ CH$_3$

(—),

CH$_3$

O

OH

(CH$_2$)$_4$CH$_3$

O

CH$_3$ CH$_3$

(—),

Note: References 249–634 are on pp. 407–415.

OXYGEN COMPOUNDS—(Continued)

No. of C Atoms	Reactant	Solvent/Time/ Temperature (°C)	Product(s) and Yield(s) (%)	Refs.

B. Heterocyclic Compounds—(Continued)

21
(contd.)

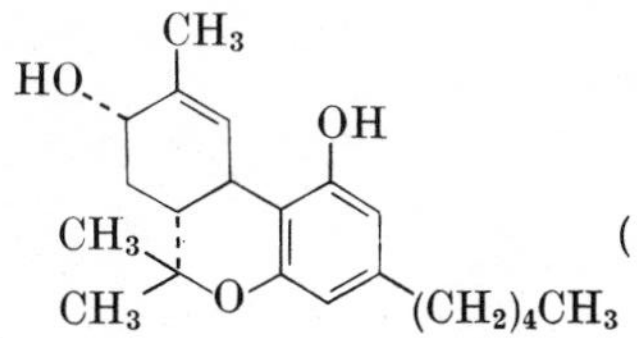

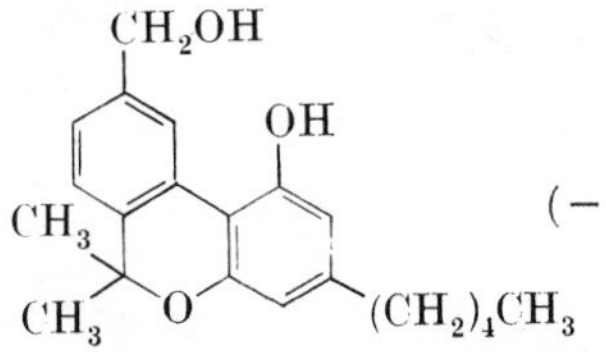

(−),

(−)

Δ⁶-Tetrahydrocannabinol —/—/—

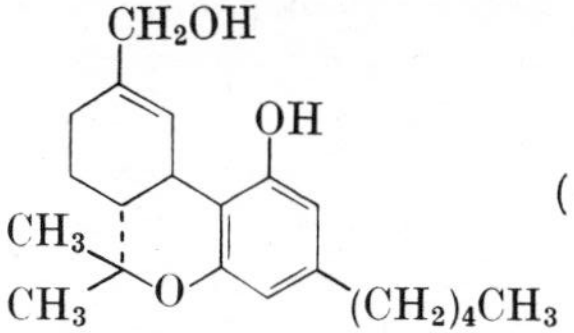

(–), 458

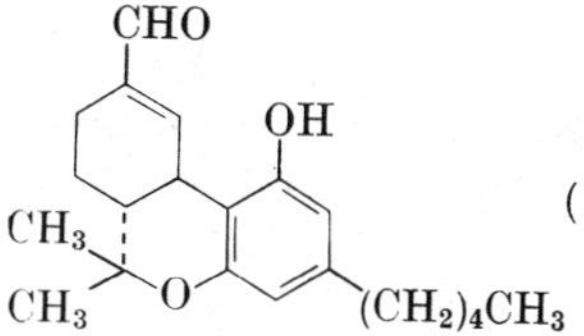

(–) 459

| 22 | CONHC$_6$H$_4$Cl-p | Dioxane/—/— | 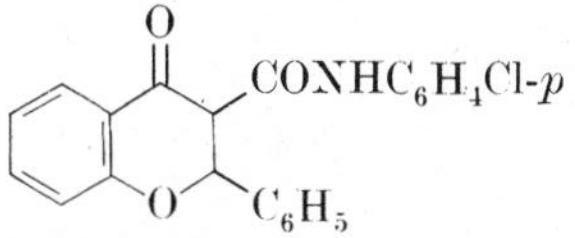 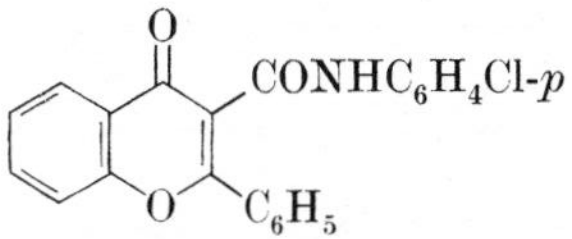 CONHC$_6$H$_4$Cl-p (64.5) | 459 |

5,7-Dibenzyloxy-4-methyl-coumarin   24   Xylene/5 hr/reflux   5,7-Dibenzyloxy-4-formylcoumarin   (—)   455, 456

4'-Benzyloxy-5,6,7-trimethoxyflavanone   25   —/—/—   4'-Benzyloxy-5,6,7-trimethoxyflavone   (—)   362

Note: References 249–634 are on pp. 407–415.

371

TABLE XI. Steroids

No. of C Atoms	Reactant	Solvent/Time/ Temperature (°C)	Product(s) and Yield(s) (%)	Refs.
19	5α-Androstane-3,17-dione	$t\text{-}C_5H_{11}OH$, AcOH/ 17.5 hr/110–120	5α-Androst-1-ene-3,17-dione (13), androst-4-ene-3,17-dione (8), androsta-1,4-diene-3,17-dione (5)	29
	1-Androstene-3,17-dione	$C_6H_5OC_2H_5$, C_5H_5N/5 hr/reflux	1,4-Androstadiene-3,17-dione (—)	231
	1α-Deuterio-5α-androstane-3,17-dione	$t\text{-}C_5H_{11}OH$, AcOH/19 hr/110–120	5α-Androst-1-ene-3,17-dione (—)	29
	17β-Hydroxy-5-androsten-3-one	$t\text{-}C_4H_9OH$, AcOH/—/ reflux	1-Dehydro-17β-hydroxy-5-androsten-3-one (80)	67
	5-Methyl-10-norandrost-8(9)-ene-3,6-diol-17-one	C_2H_5OH/7 d/25	5-Methyl-10-norandrost-8(9)-ene-3,6,11-triol-17-one (—)	460
	Retrotestosterone	C_6H_6, H_2O/48 hr/ reflux	1-Dehydroretrotestosterone (—)	461
	Testosterone	$t\text{-}C_4H_9OH$, AcOH/5 hr/reflux	1-Dehydrotestosterone (I) (53), seleno-1-dehydrotestosterone (II) (16), 2,17β-dihydroxyandrosta-1,4-diene-3-one (2)	67, 79, 226, 462
		C_6H_6, H_2O/55 hr/ reflux	I (12), II (6)	78
20	17β-Acetoxy-3β-fluoro-5(10)-estrene	AcOH/3 hr/60	17β-Acetoxy-3β-fluoro-5(10), 9(11)-estradiene (33)	198
	2-Chloro-17α-methylandrost-1-en-17β-ol-3-one	$t\text{-}C_4H_9OH$, AcOH/—/—	2-Chloro-17α-methylandrosta-1,4-dien-17β-ol-3-one (—)	208
	6-Dehydroestrone acetate	AcOH/10–15 min/ reflux	Equilenin acetate (95)	463, 464
	9α-Fluoro-17α-methylandrostane-11β,17β-diol-3-one	$t\text{-}C_4H_9OH$, C_5H_5N/48 hr/reflux	9α-Fluoro-17α-methyl-1-androstene-11β, 17β-diol-3-one (—)	465
	9α-Fluoro-17α-methylandrostene-17β-ol-3,11-dione	—/—/—	9α-Fluoro-17α-methyl-1-androsten-17β-ol-3,11-dione (—)	465

9α-Fluoro-20-spirox-4-en-11β-ol	—/—/—	9α-Fluoro-20-spiroxa-1,4-dien-11β-ol-3-one (—)	466
4-Methyltestosterone	t-C₄H₉OH, AcOH/24 hr/reflux	4-Methyl-1-dehydrotestosterone (29)	467
17α-Methyltestosterone	t-C₄H₉OH, AcOH/—/reflux	17α-Methyl-1-dehydrotestosterone (80)	67, 226
20-Spiroxa-4,6-diene-3,11-dione	—/—/—	20-Spiroxa-1,4,6-triene-3,11-dione (—)	466
20-Spiroxa-4,6-dien-11β-ol-3-one	—/—/—	20-Spiroxa-1,4,6-trien-11β-ol-3-one (—)	466
20-Spiroxa-4,6-dien-3-one	t-C₅H₁₁OH/20 hr/reflux, HgO	20-Spiroxa-1,4,6-trien-3-one (—)	466
20-Spirox-4-ene-3,11-dione	—/—/—	20-Spiroxa-1,4-diene-3,11-dione (—)	466
20-Spirox-4-en-11β-ol-3-one	—/—/—	20-Spiroxa-1,4-dien-11β-ol-3-one (—)	466
17β-Acetoxy-3β-fluoro-5-androstene	Dioxane, HOAc/1 hr/110	4β,17β-Diacetoxy-3β-fluoro-5-androstene (9)	468

21

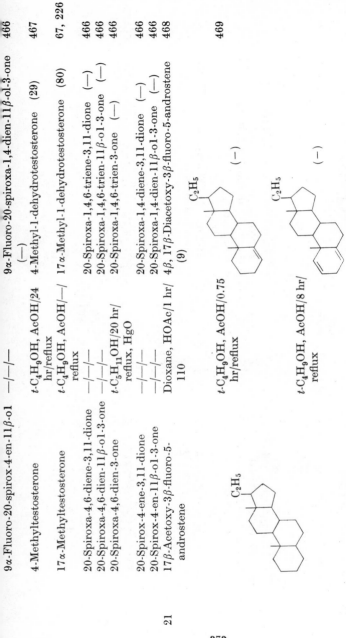

a. 17α,21-Dihydroxy-3,11,20-trioxo- 21-acetate	t-C₄H₉OH, AcOH/0.75 hr/reflux	a. 17α,21-Dihydroxy-3,11,20-trioxo-21-acetate (—)	469
b. Allo isomer of a.		b. Allo isomer of a.	
c. 11β,17α,21-Trihydroxy-3,20-dioxo-21-acetate	t-C₄H₉OH, AcOH/8 hr/reflux	c. 11β,17α,21-Trihydroxy-3,20-dioxo-21-acetate (—)	

373

Note: References 249–634 are on pp. 407–415.

TABLE XI STEROIDS—(Continued)

No. of C Atoms	Reactant	Solvent/Time/ Temperature (°C)	Product(s) and Yield(s) (%)	Refs.
21 (contd.)	a. 17α,21-Dihydroxy-3,11,20-trioxo-21-acetate b. 11β,17α-Trihydroxy-3,20-dioxo-21-acetate c. 3,17-Dioxo- d. 9α-Fluoro-11β,17α,21-trihydroxy-3,20-dioxo-21-acetate	C₆H₅CH₃, AcOH/3 hr/reflux	(—) a. 17α,21-Dihydroxy-3,11,20-trioxo-21-acetate b. 11β,17α-Trihydroxy-3,20-dioxo-21-acetate c. 3,17-Dioxo- d. 9α-Fluoro-11β,17β,21-trihydroxy-3,20-dioxo-21-acetate	470
	4-Androsten-17β-ol-3-one acetate	AcOH/1 hr/reflux	Seleno-1,4-androstadien-17β-ol-3-one acetate (40)	78
	9α-Chlorohydrocortisone	—/—/—	9α-Chloro-1-dehydrohydrocortisone (—)	231
		t-C₄H₉OH, AcOH/42 hr/reflux	(—)	471

374

X = 17β-methoxy, Y = 6-Cl
X = 17β-methoxy, Y = 6-F
X = 17β-ol acetate, Y = 6-F
X = 17β-ol, Y = 6-CH₃

X = 17β-methoxy, Y = 6-Cl
X = 17β-methoxy, Y = 6-F
X = 17β-ol acetate, Y = 6-F
X = 17β-ol, Y = 6-CH₃

471

X = 17β-ol acetate, Y = H
X = 17β-methoxy, Y = H
X = 17β-methoxy, Y = 6α-F
X = 17β-ol acetate, Y = 6α-Cl
X = 17β-ol acetate, Y = 6α-F
X = 17β-ol, Y = 6α-CH₃

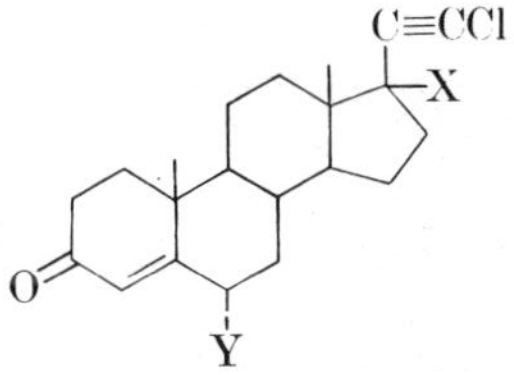

X = 17β-ol acetate, Y = H
X = 17β-methoxy, Y = H
X = 17β-methoxy, Y = 6α-F
X = 17β-ol acetate, Y = 6α-Cl
X = 17β-ol acetate, Y = 6α-F
X = 17β-ol, Y = 6α-CH₃

t-C₄H₉OH, AcOH/42 hr/reflux

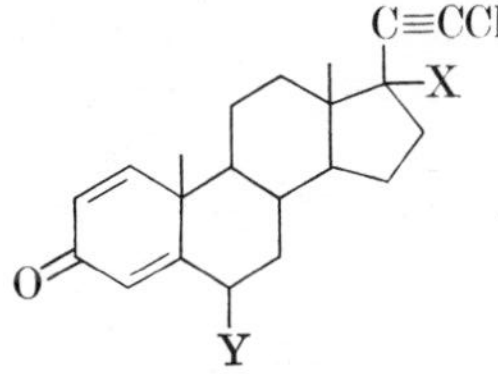

Reactant	Conditions	Product (yield)	Ref.
Cortisone	t-C₄H₉OH, AcOH/—/reflux	1-Dehydrocortisone (80)	67
17α-Ethynyl-6α-fluorotestosterone	t-C₄H₉OH, C₅H₅N/24 hr/reflux	17α-Ethynyl-6α-fluoro-1,4-androstadien-17β-ol-3-one (19)	472
17α-Ethynyltestosterone	t-C₄H₉OH, AcOH/—/reflux	17α-Ethynyl-1-dehydrotestosterone (80)	67
6α-Fluoro-11-ketoprogesterone	t-C₄H₉OH, AcOH/7 hr/reflux	6α-Fluoro-1-dehydro-11-ketoprogesterone (—)	473
9α-Fluoro-2α-methyl-20-spirox-4-en-11β-ol	—/—/—	9α-Fluoro-2α-methyl-20-spiroxa-1,4-dien-11β-ol-3-one (—)	466
1-Methyl-6-dehydrosterone acetate	AcOH/10 min/reflux	1-Methylequilenin acetate (60)	474

Note: References 249–634 are on pp. 407–415.

375

STEROIDS—*(Continued)*

No. of C Atoms	Reactant	Solvent/Time/ Temperature (°C)	Product(s) and Yield(s) (%)	Refs.
21 *(contd.)*	2α-Methyl-20-spirox-4-en-3-one	—/—/—	2α-Methyl-20-spiroxa-1,4-dien-3-one (—)	466
	Progesterone	t-C₄H₉OH, AcOH/—/ reflux	1-Dehydroprogesterone (80)	67
	Testosterone acetate	t-C₄H₉OH, AcOH/48 hr/reflux	1-Dehydrotestosterone (—)	226
		C₄H₉OCOCH₃, AcOH, Fe, HOSeO₂CH₃/ 8 hr/reflux	1-Dehydrotestosterone acetate (59)	192
		Dioxane, AcOH/1 hr/ reflux	Seleno-1-dehydrotestosterone acetate (—)	78
22	3-Acetoxy-12-ketoetiocholanic acid	AcOH/10 hr/reflux	3-Acetoxy-12-keto-9,11-etiocholenic acid (—)	66
	17β-Acetoxymethyl-17α-hydroxy-4-androsten-3-one	t-C₅H₁₁OH/—/—	17β-Acetoxymethyl-17α-hydroxy-1,4-androstadien-3-one (—)	475
	6-Dehydroestradiol 3,17-diacetate	AcOH/8 min/reflux	17-Dihydroequilenin-17β 3,17-diacetate (80)	476
	3α,17β-Diacetoxy-5(10)-estrene	AcOH/3 hr/60	3α,17β-Diacetoxy-5(10),9(11)-estradiene (40)	198
	17α,21-Dihydroxy-3,11,20-trioxo-C-norallopregnane	t-C₄H₉OH, C₅H₅N/ 24 hr/reflux	17α,21-Dihydroxy-3,11,20-trioxo-C-nor-1,4-pregnadiene 21-acetate (—)	477
	6α-Fluoro-16-methyl-4,9(11), 16-pregnatriene-3,20-dione	t-C₄H₉OH, AcOH/ 48 hr/reflux	6α-Fluoro-16-methyl-1,4,9(11),16-pregnatetraene-3,20-dione (56)	478
	3-(17β-Hydroxy-3-oxo-4,6-androstadien-17α-yl)propionic acid lactone	t-C₄H₉OH, AcOH/ —/—	3-(17β-Hydroxy-3-oxo-1,4,6-androstatrien-17α-yl)propionic acid lactone (15)	209
	3-(17β-Hydroxy-3-oxo-4-androsten-17α-yl)propionic acid lactone	t-C₄H₉OH, AcOH/21 hr/reflux	3-(17β-Hydroxy-3-oxo-1,4-androstadien-17α-yl)propionic acid lactone (33)	209

376

Substrate	Conditions	Product	Ref.
Methyl 3,11-diketo-(Z)-4,17(20)-pregnadien-21-oate	t-C$_4$H$_9$OH, AcOH/24 hr/reflux	Methyl 3,11-diketo-16α-hydroxy-(Z)-1,4,17-(20)-pregnatrien-21-oate (16), methyl 3,11-diketo-16β-hydroxy-(E)-1,4,17-(20)-pregnatrien-21-oate (22)	204
	THF/5 hr/reflux	Methyl 3,11-diketo-16α-hydroxy-(Z)-4,17(20)-pregnadien-21-oate (18), methyl 3,11-diketo-16β-hydroxy-(E)-4,17(20)-pregnadien-21-oate (30)	204
Methyl 3,11-diketo-(E)-4,17(20)-pregnadien-21-oate	t-C$_4$H$_9$OH, AcOH/18 hr/reflux	Methyl 3,11-diketo-16α-hydroxy-(E)-1,4,17-(20)-pregnatrien-21-oate (29)	204
2α-Methylpregn-4-ene-11α,20β-diol-3-one	t-C$_4$H$_9$OH, AcOH/24 hr/reflux/Hg	2-Methylpregna-1,4-diene-11α,20β-diol-3-one (—)	210
4-Methyltestosterone acetate	t-C$_4$H$_9$OH, AcOH/24 hr/reflux	4-Methyl-1-dehydrotestosterone acetate (21)	467
3-Oxo-10β,17β-diacetoxyestr-4-ene	t-C$_4$H$_9$OH, AcOH/6 hr/reflux	3-Oxo-10β,17β-diacetoxyestra-1,4-diene (—)	479
17α-Acetoxy-6-chloro-6-dehydroprogesterone	—/—/—	17α-Acetoxy-1,6-bisdehydro-6-chloroprogesterone (—)	480
17α-Acetoxy-6α-chloroprogesterone	—/—/—	17α-Acetoxy-6α-chloro-1-dehydroprogesterone (—)	480
17α-Acetoxy-6α-fluoroprogesterone	t-C$_4$H$_9$OH, C$_5$H$_5$N/18 hr/reflux	6α-Fluoro-1,4-pregnadien-17α-o1-3,20-dione 17-acetate (90)	481
21-Acetoxy-17α-hydroxy-4-pregnene-3,20-dione	C$_6$H$_5$OC$_2$H$_5$, C$_5$H$_5$N/48 hr/reflux	1,4-Pregnadiene-3,20-dione-17α,21-diol 21-acetate (60)	231
Allodihydrocortisone 21-acetate	t-C$_4$H$_9$OH, resin/—/—	Prednisone 21-acetate (68)	212
17,20;20,21-Bismethylenedioxy-6α-fluoromethyl-11β-hydroxy-5α-pregnan-3-one	t-C$_4$H$_9$OH, AcOH/48 hr/reflux	17,20;20,21-Bismethylenedioxy-6α-fluoromethylprednisolone (34)	482
17,20;20,21-Bismethylenedioxy-7β-methylpregnane-3,11-dione	t-C$_4$H$_9$OH, AcOH/16 hr/reflux	17,20;20,21-Bismethylenedioxy-7β-methylprednisone (38)	207

Note: References 249–634 are on pp. 407–415.

STEROIDS—*(Continued)*

No. of C Atoms	Reactant	Solvent/Time/Temperature (°C)	Product(s) and Yield(s) (%)	Refs.
23 (contd.)	Corticosterone 21-acetate	t-C_4H_9OH, AcOH/—/ reflux	1-Dehydrocorticosterone (80)	67
	Cortisone 21-acetate	t-C_4H_9OH, AcOH/4 hr/reflux	1-Dehydrocortisone 21-acetate (I) (90)	67, 68
		t-C_4H_9OH, resin/—/—	I (85)	212
		t-$C_5H_{11}OH$, AcOH, catalysts/—/reflux	I (—)	30
		$C_4H_9OCOCH_3$, AcOH, Fe, $HOSeO_2CH_3$/ 8.5 hr/reflux	I (69)	192
		$(C_4H_9OCH_2)_2$/1.5 hr/ 175	I (—)	231
	11-Dehydrocorticosterone 21-acetate	t-C_4H_9OH, AcOH/—/ reflux	1,11-Bisdehydrocorticosterone 21-acetate (80)	67
	3-Dehydrodigoxigenin	t-C_4H_9OH/2 hr/ reflux	12-β-Hydroxy-4,5-dehydrodigitoxigenone (17)	483
	1-Dehydro-4,5-dihydro-5α-cortisone 21-acetate	t-C_4H_9OH, AcOH/—/ reflux	1,4-Bisdehydro-5α-cortisone 21-acetate (80)	67
	6-Dehydro-6,16α-dimethyl-progesterone	t-C_4H_9OH, AcOH/16 hr/reflux	6,16α-Dimethyl-1,4,6-pregnatriene-3,20-dione (28)	484
	6-Dehydro-1-methylestradiol 3,17-diacetate	AcOH/10 min/reflux	1-Methyl-17-dihydroequilenin-17β 3,17-diacetate (—)	474
	Dehydronorcholene	Ac_2O, C_6H_6/2 hr/ reflux	Dehydronorcholadiene (36)	485
	11-Deoxycorticosterone 21-acetate	t-C_4H_9OH, AcOH/72 hr/70	21-Acetoxy-1,4-pregnadiene-3,20-dione (—)	67,68
	11-Deoxy-17α-hydroxycorticosterone 21-acetate	t-C_4H_9OH, AcOH/—/ reflux	1-Dehydro-11-deoxy-17α-hydroxy-corticosterone 21-acetate (80)	67

Starting material	Conditions	Product (% yield)	Ref.
9α,11β-Dichloro-4-pregnene-17α,21-diol-3,20-dione 21-acetate	t-C$_4$H$_9$OH, C$_5$H$_5$N/48 hr/reflux	9α,11β-Dichloro-1,4-pregnadiene-17α,21-diol-3,20-dione 21-acetate (86)	486
6α,16α-Difluorodihydrocortisone acetate	t-C$_4$H$_9$OH, AcOH/23 hr/reflux	6α,16α-Difluoroprednisolone acetate (47)	487
6α,9α-Difluorohydrocortisone acetate	t-C$_4$H$_9$OH, C$_5$H$_5$N/40 hr/reflux	6α,9α-Difluoroprednisolone acetate (30)	235
6α-Difluoromethyl-16β-methyl-9α,11β-dichloro-4-pregnen-17α-ol-3,20-dione	t-C$_4$H$_9$OH, C$_5$H$_5$N/48 hr/reflux	6α-Difluoromethyl-16β-methyl-9α,11β-dichloro-1,4-pregnadien-17α-ol-3,20-dione (—)	488
Dihydrocortisone 21-acetate	t-C$_4$H$_9$OH, AcOH/1 hr/reflux	Cortisone 21-acetate (—), prednisone 21-acetate (—)	67,68
17α,21-Dihydroxy-3,11,20-trioxoallopregane 21-acetate	t-C$_4$H$_9$OH, AcOH/0.75 hr/reflux	17α,21-Dihydroxy-3,11,20-trioxo-1-allopregnene 21-acetate (—), 17α,21-dihydroxy-3,11,20-trioxo-1,4-allopregnadiene 21-acetate (—)	203
17,21-Dihydroxy-3,11,20-trioxo-4,6-pregnadiene 21-acetate	t-C$_4$H$_9$OH, resin/—/—	17,21-Dihydroxy-3,11,20-trioxo-1,4,6-pregnatriene 21-acetate (80)	212
17α,21-Dihydroxy-3,11,20-trioxopregnane 21-acetate	t-C$_4$H$_9$OH, AcOH/0.75 hr/reflux	17α,21-Dihydroxy-3,11,20-trioxo-4-pregnene 21-acetate (—)	203
	t-C$_4$H$_9$OH, AcOH, add SeO$_2$/8 hr/reflux	17α,21-Dihydroxy-3,11,20-trioxo-1,4-pregnadiene 21-acetate (—)	203
6α,16α-Dimethylprogesterone	t-C$_4$H$_9$OH, AcOH/16 hr/reflux	6α,16α-Dimethyl-1,4-pregnadiene-3,20-dione (11)	484
16,17-Epoxy-4-pregnen-21-ol-3,20-dione acetate	—/—/—	16,17-Epoxy-1,4-pregnadien-21-ol-3,20-dione acetate (—)	489
6α-Fluorohydrocortisone acetate	t-C$_4$H$_9$OH, C$_5$H$_5$N/24 hr/reflux	6α-Fluoroprednisolone 21-acetate (95)	235
9α-Fluorohydrocortisone acetate	AcOH/0.5 hr/reflux	9α-Fluoroprednisolone 21-acetate (I) (9)	78

Note: References 249–634 are on pp. 407–415.

No. of C Atoms	Reactant	Solvent/Time/Temperature (°C)	Product(s) and Yield(s) (%)	Refs.
23 (contd.)	6-Fluoro-4,6-pregnadien-17α-ol-3,20-dione acetate	t-C_4H_9OH, AcOH/24 hr/70	I (—)	67
		t-C_4H_9OH, C_5H_5N/36 hr/reflux	6-Fluoro-1,4,6-pregnatrien-17α-ol-2,20-dione acetate (65)	481
	Hydrocortisone acetate	AcOH/1 hr/reflux	Prednisolone 21-acetate (I) (8), selenoprednisolone 21-acetate (—)	78
		AcOH/—/—, $^{75}SeO_2$ $C_4H_9OCOCH_3$, AcOH, Fe, $HOSeO_2CH_3$/8.5 hr/reflux	^{75}Se selenoprednisolone 21-acetate (—), I (69)	490, 192
		t-C_4H_9OH, AcOH/24 hr/reflux	I (80)	67
	3-(17β-Hydroxy-6α-methyl-3-oxo-4-androsten-17α-yl)-propionic acid lactone	t-C_4H_9OH, AcOH/13 hr/reflux	3-(17β-Hydroxy-6α-methyl-3-oxo-1,4-androstadien-17α-yl)propionic acid lactone (13)	491
	19-Hydroxytestosterone diacetate	t-C_4H_9OH, AcOH/29 hr/reflux	1-Dehydro-19-hydroxytestosterone diacetate (77)	492
	Methyl 3,11-dioxo-(Z)-4,17-(20)-pregnadien-21-oate	t-C_4H_9OH, AcOH/24 hr/reflux	Methyl 16β-hydroxy-3,11-dioxo-(E)-1,4,17(20)-pregnadien-21-oate (I) (24), 16α-hydroxy-(Z)-isomer (II) (17)	211
	(E)-isomer	THF/4 hr/reflux	I (23), II (12)	211
		t-C_4H_9OH, AcOH/18 hr/reflux	Methyl 3,11-dioxo-(E)-1,4,17-(20)-pregnatrien-21-oate (25), methyl 16α-hydroxy-3,11-dioxo-(E)-1,4,17(20)-pregnatrien-21-oate (36)	211

	Reactant	Conditions	Products	Ref.
	Methyl 5α-hydroxy-6β-fluoro-3,11-dioxo-(Z)-17(20)-pregnen-21-oate	THF/6 hr/reflux	Methyl 5α,16β-dihydroxy-6β-fluoro-3,11-dioxo-(E)-17(20)-pregnen-21-oate (42), 5α,16α-dihydroxy-6β-fluoro-3,11-dioxo-(Z)-17(20)-pregnen-21-oic acid (20)	211
	Methyl 5α-hydroxy-6β-methyl-3-oxo-5α-pregn-17(20)-(E)-en-21-oate	THF/4.5 hr/reflux	Methyl 5α,16β-dihydroxy-6β-methyl-3-oxo-5α-pregn-17(20)-(E)-en-21-oate (—)	493
	4,6-Pregnadiene-11β,17α,21-triol-3,20-dione acetate	—/—/—	1,4,6-Pregnatriene-11β,17α,21-triol-3,20-dione acetate (—)	494
	5α-Pregnane-17α,21-diol-3,11,20-trione acetate	C₄H₉OCOCH₃, AcOH, Fe, HOSeO₂CH₃/ 8.5 hr/reflux	Prednisone acetate (59)	192
	11β,17α,21-Trihydroxypregnane-3,20-dione 21-acetate	—/—/—	11β,17α,21-Trihydroxy-4-pregnene-3,20-21-acetate (—), 11β,17α,21-trihydroxy-1,4-pregnadiene-3,20-dione 21-acetate (—)	203
		—/—/—	(—)	495
24	3β-Acetoxy-17a,17a-dimethyl-D-homoandrostan-17-one	t-C₄H₉OH, AcOH/24 hr/reflux	3β,16-Dihydroxy-17a,17a-dimethyl-D-homoandrost-15-en-17-one 16-acetate (—), 17a,17a-dimethyl-3β-hydroxy-D-homoandrost-15-en-17-one (—), 16-bis(17a,17a-dimethyl-3β-hydroxy-D-homoandrost-15-en-17-one diselenide (—)	81

Note: References 249–634 are on pp. 407–415.

381

No. of C Atoms	Reactant	Solvent/Time/Temperature (°C)	Product(s) and Yield(s) (%)	Refs.
24 (contd.)	3-Acetoxy-12-ketobisnorcholanic acid	AcOH/5 hr/reflux	3-Hydroxy-12-keto-9,11-bisnorcholenic acid (after saponification) (—)	66
	17α-Acetoxy-6α-methyl-progesterone	t-C₄H₉OH, C₅H₅N/30 hr/reflux	6α-Methyl-1,4-pregnadien-17α-ol-3,20-dione acetate (60)	496
	17α-Acetoxy-16α-methyl-progesterone	t-C₄H₉OH, C₅H₅N/24 hr/reflux	16α-Methyl-1,4-pregnadien-17α-ol-3,20-dione acetate (30)	497
	17α-Acetoxy-6α-trifluoromethyl-progesterone	—/—/—	6α-Trifluoromethyl-1,4-pregnadien-17α-ol-3,20-dione acetate (—)	498
	17,20,20,21-Bismethylenedioxy-6α-difluoromethyl-5α-pregnan-11β-ol-3-one	t-C₄H₉OH, AcOH/44 hr/reflux	17,20,20,21-Bismethylenedioxy-6α-difluoromethyl-1,4-pregnadien-11β-ol-3-one (23)	499, 500
	17α,20,20,21-Bisemethylenedioxy-2α-methylpregn-4-en-11α-ol-3-one	t-C₄H₉OH, AcOH/—/reflux/Hg	17α,20,20,21-Bisemethylenedioxy-2-methylpregna-1,4-dien-11α-ol-3-one (—)	210
	17,20,20,21-Bismethylenedioxy-6α-methyl-4-pregnen-11β-ol-3-one	t-C₄H₉OH, AcOH/87 hr/reflux	17α,20,20,21-Bismethylenedioxy-1,4-pregnadien-11β-ol-3-one (27)	219
	16-Chloromethylene-6-fluoro-4,6-pregnadiene-17α,21-diol-3,11-20-trione 21-acetate	C₄H₉OH, AcOH/48 hr/reflux	16-Chloromethylene-6-fluoro-1,4,6-pregnatriene-17α,21-diol-3,11,20-trione 21-acetate (—)	501
	16α-Chloro-6α-methylhydro-cortisone acetate	t-C₄H₉OH, AcOH/21 hr/reflux	16α-Chloro-6α-methylprednisolone (—)	211
	16α-Chloromethyl-5α-pregnane-17α,21-diol-3,20-dione 21-acetate	t-C₄H₉OH, AcOH/48 hr/reflux	16α-Chloromethyl-1,4-pregnadiene-17α,21-diol-3,20-dione 21-acetate (—)	502
	6α-Difluoromethyl-4,5α-dihydro-cortisone 21-acetate	t-C₄H₉OH, AcOH/42 hr/reflux	6α-Difluoromethylprednisolone 21-acetate (8)	499
	6α,9α-Difluoro-16α-methylhy-drocortisone acetate	—/—/—	6α,9α-Difluoro-16α-methylprednisolone acetate (—)	503
	6α,9α-Difluoro-16α-methyl-4-pregnene-11β,17α,21-triol-3,20-	t-C₄H₉OH, C₅H₅N/62 hr/reflux	6α,9α-Difluoro-16α-methyl-1,4-pregnadiene-11β,17α,21-triol-3,20-dione 21-acetate	504

382

6α-Difluoromethyl-11β,17α,21-trihydroxy-5α-pregnan-3-one 21-acetate	t-C₄H₉OH, AcOH/42 hr/reflux	6α-Difluoromethyl-11β,17α,21-trihydroxy-1,4-pregnadiene-3,20-dione 21-acetate (—) 500
6α-Fluoro-16α-methylhydrocortisone 21-acetate	t-C₄H₉OH, C₅H₅N/60 hr/reflux	6α-Fluoro-16α-methylprednisolone 21-acetate (94) 504
16α-Methoxycortisone acetate	t-C₄H₉OH, C₅H₅N/28 hr/reflux	16α-Methoxyprednisone acetate (64) 505
6α-Methoxyhydrocortisone acetate	t-C₄H₉OH/24 hr/reflux	6α-Methoxyprednisolone acetate (24) 506
16α-Methoxyhydrocortisone acetate	t-C₄H₉OH, C₅H₅N/28 hr/reflux	16α-Methoxyprednisolone acetate (55) 505
Methyl 6β-acetoxy-3α,5α-cyclopregn-17(20)-(Z)-en-21-oate	THF/4.5 hr/reflux	Methyl 6β-acetoxy-16β-hydroxy-3α,5α-cyclopregn-17(20)-(E)-en-21-oate (—) 493
16-Methyl-15-allopregnene-17α,21-diol-3,11,20-trione 21-acetate	t-C₅H₁₁OH, C₅H₅N/24 hr/reflux	16-Methyl-1,15-allopregnadiene-17α,21-diol-3,11,20-trione 21-acetate (20), 16-methyl-1,4,15-pregnatriene-17α,21-diol-3,11,20-trione 21-acetate (10) 497
Methyl 3-ethylenedioxy-5α-hydroxy-6β-fluoro-11-keto-(Z)-17(20)-pregnen-21-oate	THF/6 hr/reflux	Methyl 5α,16α-dihydroxy-3-ethylenedioxy-6β-fluoro-11-keto-(Z)-17(20)-pregnen-21-oate (23), methyl 5α,16β-dihydroxy-3-ethylene-dioxy-6β-fluoro-11-keto-(E)-17(20)-pregnen-21-oate (41) 204
16β-Methylhydrocortisone 21-acetate	—/—/—	16β-Methylprednisolone acetate (—) 507
6α-Methyl-17-hydroxyprogesterone acetate	t-C₄H₉OH, C₅H₅N/72 hr/reflux	6α-Methyl-1,4-pregnadien-17α-ol-3,20-dione acetate (—) 508
6-Methyl 4,6-pregnadien-17α-ol-3,20-dione acetate	t-C₄H₉OH, C₅H₅N/30 hr/reflux	6-Methyl-1,4,6-pregnatrien-17α-ol-3,20-dione acetate (54) 496
16α-Methyl-4-pregnene-17α,21-diol-3,11,20-trione 21-acetate	—/—/—	16α-Methyl-1,4-pregnadiene-17α,21-diol-3,11,20-trione 21-acetate (—) 509

Note: References 249–634 are on pp. 407–415.

383

STEROIDS—(Continued)

No. of C Atoms	Reactant	Solvent/Time/Temperature (°C)	Product(s) and Yield(s) (%)	Refs.
25	3-Acetoxy-12-ketonorcholanic acid	AcOH/5 hr/reflux	3-Hydroxy-12-keto-9,11-norcholenic acid (after saponification) (—)	66
	17α-Acetoxyspiro[-4-pregnene-6,1'-cyclopropane]-3,20-dione	t-C_4H_9OH, AcOH/24 hr/75	17-Acetoxyspiro[-1,4-pregnadiene-6,1'-cyclopropane]-3,20-dione (—)	510
	10α-Androst-1-ene-2,5ξ,17β-triol triacetate	Dioxane/20 hr/reflux	2ξ,17β-Diacetoxy-10α-androst-4-en-3-one (14)	511
	6-Dehydroestrone 3-benzyl ether	AcOH/70 min/100–110; then Zn/1 hr/90–100	Equilenin 3-benzyl ether (50)	224
	16α,21-Diacetoxy-11β,17α-dihydroxy-9α-fluoro-4,6-pregnadiene-3,20-dione	t-C_4H_9OH, AcOH/23 hr/reflux	16α,21-Diacetoxy-11β,17α-dihydroxy-9α-fluoro-1,4,6-pregnatriene-3,20-dione (12), 9α-fluoro-17α-hydroxy-11β,16α,21-triacetoxy-1,4,6-pregnatriene-3,20-dione (14)	512
	16α,21-Diacetoxy-11β,17α-dihydroxy-9α-fluoro-4-pregnene-3,20-dione	t-C_4H_9OH, AcOH/48 hr/70	16α,21-Diacetoxy-11β,17α-dihydroxy-9α-fluoro-1,4-pregnadiene-3,20-dione (15)	513
	16β,21-Diacetoxy-11β,17α-dihydroxy-4-pregnene-3,20-dione	t-C_4H_9OH, AcOH/23 hr/reflux	16β,21-Diacetoxy-11β,17α-dihydroxy-9α-fluoro-1,4-pregnadiene-3,20-dione (14)	222
	16α,21-Diacetoxy-11β,17α-dihydroxy-4-pregnene-3,20-dione	C_6H_6/65 hr/reflux	16α,21-Diacetoxy-11β,17α-dihydroxy-1,4-pregnadiene-3,20-dione (3)	513
	3β,20α-Diacetoxy-5α-pregnan-12-one	AcOH, HCl/20 hr/reflux	3β,20α-Diacetoxy-5α-pregn-9(11)-en-12-one (68)	514
	Digitoxigenin 3-acetate	Dioxane/16 hr/reflux	17α-Hydroxydigitoxigenin 3-acetate (60)	130, 193
	6α-Fluorocortisone diacetate	t-C_4H_9OH, C_5H_5N/24 hr/reflux	6α-Fluoroprednisone diacetate (45)	235
	6α-Hydroxyhydrocortisone 6,21-diacetate	t-C_4H_9OH, AcOH/18 hr/reflux	6α-Hydroxyprednisolone 6,21-diacetate (47)	515, 516

Methyl 3-cholenate	AcOH/35 hr/25	Methyl 3α-hydroxy-4-cholenate (—), methyl 3β-hydroxy-4-cholenate (—) methyl 3,5-choladienoate (—)	218
Methyl 3,11-dioxo-(Z)-5,17(20)-pregnadien-21-oate 3-ethylene ketal	THF/3.5 hr/reflux	Methyl 3,11-dioxo-16β-hydroxy-(E)-5,17(20)-pregnadien-21-oate 3-ethylene ketal (—), 16α-hydroxy-(Z)-isomer (—), methyl 3,11-dioxo-7,16-dihydroxy-5,17(20)-pregnadien-21-oate 3-ethylene ketal (—)	211
Methyl 3,11-dioxo-5α-(Z)-17(20)-pregnen-21-oate ketal	t-C$_4$H$_9$OH, AcONa/—/—	Methyl 3,11-dioxo-16β-hydroxy-5α-(E)-17(20)-pregnen-21-oate 3-ethylene ketal (28), 16α-hydroxy-(Z)-isomer (24)	211
Methyl 3-ethylenedioxy-6β-fluoro-5α-hydroxy-11-keto-(Z)-17(20)-pregnen-21-oate	THF/6 hr/reflux	Methyl 3-ethylenedioxy-5α,16α-dihydroxy-6β-fluoro-11-keto-(Z)-17(20)-pregnen-21-oate (23), methyl 3-ethylenedioxy-5α,16β-dihydroxy-6β-fluoro-11-keto-(E)-17(20)-pregnen-21-oate (41)	204
Methyl 3β-hydroxy-11-oxo-5α-(Z)-17(20)-pregnen-21-oate acetate	t-C$_4$H$_9$OH, AcOH/18 hr/reflux	Methyl 3β,16β-dihydroxy-11-oxo-(E)-17(20)-pregnen-21-oate 3-acetate (38), 16α-hydroxy-(Z)-isomer (13)	211
Strophanthidin acetate	Dioxane/25 hr/reflux	17α-Hydroxystrophanthidin acetate (48)	517
3-Acetoxy-12-oxocholanic acid	AcOH/8 hr/reflux	3-Hydroxy-12-oxochol-9(11)enic acid (—) (after saponification)	66
Allopregnane-11β,12β,17α,21-tetrol-3,20-dione 11β,12β-acetonide 21-acetate	t-C$_4$H$_9$OH, C$_5$H$_5$N/48 hr/reflux	12β-Hydroxyprednisolone 11β,12β-acetonide 21-acetate (22)	518
Cortisone 21-trimethylacetate	t-C$_4$H$_9$OH, AcOH/—/reflux	1-Dehydrocortisone trimethylacetate (80)	67
11-Deoxycorticosterone 21-trimethylacetate	t-C$_4$H$_9$OH, AcOH/—/reflux	21-Trimethylacetoxy-1,4-pregnadiene-3,20-dione (80)	67

26

Note: References 249–634 are on pp. 407–415.

No. of C Atoms	Reactant	Solvent/Time/ Temperature (°C)	Product(s) and Yield(s) (%)	Refs.
26 (contd.)	16α,21-Diacetoxy-11β,17α-dihydroxy-2α-methyl-4-pregnene-3,20-dione	t-C₄H₉OH, AcOH/48 hr/reflux	16α,21-Diacetoxy-11β,17α-dihydroxy-2-methyl-1,4-pregnadiene-3,20-dione (23)	512
	16α,21-Diacetoxy-11β,17α-dihydroxy-6α-methyl-4-pregnene-3,20-dione	t-C₄H₉OH, AcOH/20 hr/reflux	16α,21-Diacetoxy-11β,17α-dihydroxy-6α-methyl-1,4-pregnadiene-3,20-dione (38)	519
	16α,21-Diacetoxy-11β,17α-dihydroxy-9α-fluoro-2α-methyl-4-pregnene-3,20-dione	t-C₄H₉OH, AcOH/96 hr/reflux	9α-Fluoro-2-methyl-11β,16α,17α,21-tetrahydroxy-1,4-pregnadiene-3,20-dione 16,21-diacetate (35)	512, 520
	6α-Fluoro-16α-hydroxyhydrocortisone 16,17-acetonide 21-acetate	—/—/—	6α-Fluoro-16α-hydroxyprednisolone 16,17-acetonide 21-acetate (—)	521
	17α-Hydroxycorticosterone 21-trimethylacetate	t-C₄H₉OH, AcOH/—/reflux	1-Dehydro-17α-hydroxycorticosterone 21-trimethylacetate (80)	67
	Methyl 3-acetoxynorallochol-20(22)-enate	AcOH, Ac₂O/—/reflux	3-Acetoxy-21-hydroxynorallochol-20(22)-enic acid lactone (—)	522, 523
	Methyl 3-acetoxynorchol-20(22)-enate	Ac₂O/—/—	3-Acetoxy-21-hydroxynorchol-20(22)-enic acid lactone (—)	230
	Methyl 3α-acetoxy-11-oxo-16α-methyl-(Z)-17(20)-pregnen-21-oate	t-C₄H₉OH, AcOH/72 hr/reflux	Methyl 3α-acetoxy-11-oxo-16α-hydroxy-16β-methyl-(E)-17(20)-pregnen-21-oate (—), methyl 3α-acetoxy-11-oxo-16-hydroxy-16-methyl-(Z)-17(20)-pregnen-21-oate (—)	211
	Methyl 5α-hydroxy-6β-methyl-3,11-dioxo-(Z)-17(20)-pregnen-21-oate 3-ethylene ketal	THF/6 hr/reflux	Methyl 5α,16β-dihydroxy-6β-methyl-3,11-dioxo-(E)-17(20)-pregnen-21-oate 3-ethylene ketal (55), 16α-hydroxy-(Z)-isomer (25)	211

27	20-(Carbethoxymethylene)-3β,14β-dihydroxy-5β-pregnane 3-acetate	C₆H₆/10 hr/reflux	Digitoxigenin acetate (30)	131, 132
	Cholestane-3,6-dione	AcOH/—/30	(Rate of oxidation)	524
	Cholest-7-en-3β-ol	AcOH/—/25	7α-Acetoxycholest-8(14)-en-3β-ol (—)	525
	5α-Cholest-8-en-3β-ol	C₂H₅OH/1 hr/reflux	8,14-Cholestadien-3β-ol (60)	248
	4-Cholesten-3-one	t-C₄H₉OH, AcOH/48 hr/reflux	1,4-Cholestadien-3-one (I) (75–95)	226
		C₆H₅OC₂H₅, AcOH/22 hr/reflux	I (—)	231
	Cholesterol	AcOH/1 hr/reflux	Seleno-1,4-cholestadien-3-one (17)	77, 78
		C₆H₆, AcOH/1 hr/reflux	5-Cholestene-3β,4β-diol (55)	468
	2β,3β-Dihydroxy-5α-cholest-7-en-6-one	Dioxane/30 min/80	2β,3β,14-Trihydroxy-5α,14α-cholest-7-en-6-one (38)	526
	2β,3β-Dihydroxy-5β-cholest-7-en-6-one	Dioxane/1 hr/80	2β,3β,14-Trihydroxy-5β,14α-cholest-7-en-6-one (25)	526
	Diosgenone	t-C₅H₁₁OH, AcOH/4 hr/reflux	20α,22β,25D-Spirosta-1,4-dien-3-one (4)	527
	3β-Fluoro-5-cholestene	C₆H₆, AcOH/1 hr/75	4β-Acetoxy-3β-fluoro-5-cholestene (83)	468
	Methyl 3-acetoxyallochol-20(22)-enate	Ac₂O, AcOH/—/reflux	3-Acetoxy-21-hydroxyallochol-20(22)-enic acid lactone (—)	523
	Methyl 3α-acetoxychol-7-enate	(C₂H₅)₂O, AcOH/12–14 hr/25	Methyl 3α-acetoxy-7ξ,15ξ-dihydroxychol-8(14) enate (15)	156
	Methyl (20S)-2β,3β-diacetoxy-6-oxo-5α-pregn-7-ene-20-carboxylate	Dioxane/30 min/90	Methyl (20S)-2β,3β-diacetoxy-14-hydroxy-6-oxo-5α,14α-pregn-7-ene-20-carboxylate (80)	526 (95)
	11-Oxo-3-tigogenone	t-C₄H₉OH, AcOH/—/reflux	1-Dehydro-11-oxo-3-tigogenone (80)	67

Note: References 249–634 are on pp. 407–415.

387

STEROIDS—*(Continued)*

No. of C Atoms	Reactant	Solvent/Time/ Temperature (°C)	Product(s) and Yield(s) (%)	Refs.
27 *(contd.)*	25D,5β-Spirost-9(11)-en-3-one	t-C_4H_9OH, AcOH/50 hr/reflux	25D,5β-Spirosta-1,4,9(11)-trien-3-one (—)	528
28	4-Dehydrotigogenone	—/—/	1,4-Bisdehydrotigogenone (—)	489
	3-(β-Carboxypropionyloxy)-12-ketocholanic acid	AcOH/5 hr/reflux	3-(β-Carboxypropionyloxy)-12-ketochol-9(11)-enic acid (—)	66
	5,6-Dihydroergosterol	C_6H_6, C_2H_5OH/20 hr/35	7α-Ethoxy-3β-hydroxyergosterol (as 3,5-dinitrobenzoate) (77)	21
	4,4-Dimethyl-3-oxo-A(1)-norcholest-5-ene	AcOH/4 hr/reflux	4,4-Dimethyl-2,3-dioxo-A(1)-norchlest-5-ene (92)	529
	4,7,22-Ergostatrien-3-one	$C_6H_5OC_2H_5$, AcOH/22 hr/reflux	1,4,7,22-Ergostatetraen-3-one (—)	231
	Methyl 3-acetoxyhomochol-20(22)-enate	Ac_2O, AcOH/—/reflux	3-Acetoxy-21-hydroxyhomochol-20(22)-enic acid lactone (—)	523
	2-Methyl-5α-cholest-2-ene	C_6H_6, C_2H_5OH/30 hr/reflux	5α-Cholest-2-ene-2-carboxaldehyde (15), 2-methyl-5α-cholest-2-en-1α-ol (20)	19
	3-Methyl-5α-cholest-2-ene	C_6H_6, C_2H_5OH/30 hr/reflux	5α-Cholest-2-ene-3-carboxaldehyde (9), 3-methyl-5α-cholest-2-en-4α-ol (30), 3-hydroxymethyl-5α-cholest-2-ene (9)	19
29	3β-Acetoxycholestan-6-one	Ac_2O, AcOH/2 hr/reflux	3β-Acetoxy-5α-hydroxycholestan-6-one (13)	206
	3-Acetoxy-1-methyl-19-nor-1,3,5,6-cholestatetraene	AcOH/0.5 hr/reflux	3-Acetoxy-1-methyl-19-nor-1,3,5,8-cholestapentaene (83)	530
	Botogenin acetate	t-C_4H_9OH, C_5H_5N/96 hr/reflux	9(11)-Dehydrobotogenin acetate (70)	531
	7-Cholestenol acetate	C_6H_6, AcOH/16 hr/0-5	Cholest-8(14)ene-3β,7α-diol diacetate (I) (55)	21
		$(C_2H_5)_2O$, AcOH/8 hr/25	I (45)	21

388

C	Substrate	Conditions	Product (yield %)	Refs
25	Correllogenin acetate	$t\text{-}C_4H_9OH$, C_5H_5N/96 hr/reflux	cholesta-7,14-dienol(B_3) acetate (—)	531
	γ-Diosgenin acetate	C_6H_6, AcOH/15 hr/0–5	9(11)-Dehydrocorrellogenin acetate (64)	21
	Hecogenin acetate	$t\text{-}C_4H_9OH$, C_5H_5N/96 hr/reflux	Mixture of 7,9(11)- and 7,14-dienes (80)	229, 532
	22-Isospirosta-7,9(11)-dien-3β-ol acetate	C_6H_6, AcOH/17 hr/0–5	22-Isoallospirost-9(11)-en-3β-ol-12-one 3-acetate (—)	21
		AcOH/18 hr/reflux	22-Isoprosta-7,9(11)-diene-3,14-diol 3-acetate (26)	533
	Methyl 3α,7α-diacetoxy-12-oxocholanate	$t\text{-}C_4H_9OH$, AcOH/20 hr/reflux	Methyl 3α,7α-diacetoxy-12-oxochol-9(11)-enate (76)	528
	Nogiragenone acetate	Dioxane/—/—	11α-Hydroxy-25D-spirosta-1,4-dien-3-one acetate (—)	534
	6-Oxo-2β,3β,5α-triacetoxy-23,24-bisnorchol-7-en-22-oic acid methyl ester	$t\text{-}C_4H_9OH$, C_5H_5N/66 hr/reflux	14α-Hydroxy-6-oxo-2β,3β,5α-triacetoxy-23,24-bisnorchol-7-en-22-oic acid methyl ester (—)	229
	25D-Spirost-5-en-3β-ol-12-one acetate (drawn structure; C_8H_{17}, CH_3O_2C, H)	Dioxane/—/reflux	22,25α-Spirosta-5,9(11)-dien-3β-ol-12-one acetate (50)	535
30	(drawn structure; CH_3O_2C)	AcOH/—/30	(drawn structure; CH_3O_2C, CHO) (70)	524
	3β-Acetoxyergost-22-ene-7,11-dione	Dioxane/3 hr/80	(Rate of oxidation)	220
	3β-Acetoxy-5α-hydroxy-6-oxo-7,22-ergostadiene	AcOH/5 hr/reflux	3β-Acetoxy-5α,14ξ-dihydroxy-6-oxo-7,22-ergostadiene (50)	536
	5,6-Dihydroergosterol acetate		Ergosterol-D acetate (—), ergosterol-B_3 acetate (—), 3β-acetoxy-9ξ-hydroxy-7,22-ergostadiene (—)	

Note: References 249–634 are on pp. 407–415.

No. of C Atoms	Reactant	Solvent/Time/Temperature (°C)	Product(s) and Yield(s) (%)	Refs.
30 (contd.)	7,11-Dioxoergostan-3β-yl acetate	AcOH/3 hr/reflux	7,11-Dioxoergost-5-en-3β-yl acetate (—)	537
	7,11-Dioxo-8α-ergost-22-en-3β-yl acetate	C₂H₅OH/30 min/reflux	7,11-Dioxoergosta-8,22-dien-3β-yl acetate (—)	537
	Ethyl 3,12-diacetoxy-7-oxo-cholanate	—/—/—	Ethyl 3,12-diacetoxy-6,7-dioxocholanate (—), ethyl 12-acetoxy-6,7-dioxo-3-cholenate (—)	538
		Dioxane/—/reflux	(—)	535
31	2β,3β-Diacetoxy-5α-cholest-7-en-6-one	Dioxane/1 hr/80	2β,3β-Diacetoxy-14-hydroxy-5α,14α-cholest-7-en-6-one (—)	526
	(25R)-3β,16β-Diacetoxy-22,26-imino-5α-cholest-22(N)-ene	Dioxane/2 hr/70	(25R)-3β,16β-Diacetoxy-22,26-imino-5α-cholest-22(N)-en-23-one (60)	539
	(25S)-3β,16β-Diacetoxy-22,26-imino-5α-cholest-22(N)-ene	Dioxane/2 hr/70	(25S)-3β,16β-Diacetoxy-22,26-imino-5α-cholest-22(N)-en-23-one (30)	539

32	Methyl 3α-benzoxy-12-oxo-cholanate	C_6H_5Cl, AcOH, HCl/72 hr/reflux	3α-Hydroxy-12-oxo-chol-9(11)-enic acid (after hydrolysis) (—)	540
	(22R)-25-(Tetrahydropyran-2-yloxy)-2β,3β,22-trihydroxy-5β-cholest-7-en-6-one	Dioxane/1.5 hr/90	Ecdyson (30)	541
34	3β-Benzoyloxycholestan-6-one	C_6H_5Cl, AcOH, HCl/70 hr/reflux	3β-Benzoyloxy-5α-hydroxycholestan-6-one (—)	206
	Cholest-7-enyl benzoate	C_6H_6, AcOH/19 hr/25	Cholestadienyl benzoates (—)	21
	2β-Tosyloxy-25D,5β-spirost-9(11)-en-3-one	t-C_4H_9OH, AcOH/48 hr/reflux	25D-Spirosta-1,4,9(11)-trien-3-one (—)	528
35	Testosterone 17-stearate	t-C_4H_9OH, AcOH/24 hr/reflux	1-Dehydrotestosterone 17-stearate (—)	80

3,9-Epoxy-22-hydroxy-11-ketobisnorchol-anyldiphenylethylene   (—)] 542

| 36 | 3,9-Epoxy-12-ketobisnorchol-anyldiphenylethylene] | —/—/— |

391

Note: References 249–634 are on pp. 407–415.

TABLE XII. SULFUR COMPOUNDS

No. of C Atoms	Reactant	Solvent/Time/Temperature (°C)	Product(s) and Yield(s) (%)	Refs.
		A. Aliphatic		
1	Thiourea	HCl/—/reflux	Urea (—), formamidine-C-sulfonic acid (—)	543–545
3	Cysteine	—/—/—	Cystine (—), selenium dicysteine (—)	104
4	*meso*-2,3-Dimercaptosuccinic acid	CH$_3$OH/—/<40	$\left[\begin{array}{c} HO_2CCH-S \\ HO_2CCH-S \end{array}Se\right]_2$ (—)	106
5	N,N-Diethyldithiocarbamic acid	Xylene/—/reflux	Tetraethylthiuram disulfide (—)	188
10	Glutathione	—/—/—	Selenium diglutathione (—), oxidized glutathione (—)	105
		B. Aromatic		
6	2-NH$_2$-5-RC$_6$H$_3$SH R = Cl, CH$_3$O	AcOH/—/—	Isolated as: R = Cl, CH$_3$O	546

No.	Substrate	Conditions	Product (yield)	Ref.
7	$p\text{-}CH_3C_6H_4SO_2NHNHR$ R = cyclohexyl, cycloheptyl, $(CH_2)_nCH_3$ n = 13, 15	THF/2 hr/25	$p\text{-}CH_3C_6H_4SO_2CH_2R$ (40–80), $RCH = CH_2$ (14–18); $p\text{-}CH_3C_6H_4SO_3H$ (—) R = cyclohexyl, cycloheptyl, $(CH_2)_nCH_3$, n = 13, 15	547
13	m-Tolyl benzenesulfonate	None/40 min/250	$m\text{-}OHCC_6H_4OSO_2C_6H_5$ (30)	374
	p-Tolyl benzenesulfonate	None/40 min/250	$p\text{-}OHCC_6H_4OSO_2C_6H_5$ (23)	374
19	3-Amino-4-benzoylaminophenyl 2-aminophenyl sulfide	—/—/—	6-(2-Aminophenylthio)-1',2,3'-benzoselenoimidazole (—)	548

C. Heterocyclic

No.	Substrate	Conditions	Product (yield)	Ref.
6	2-Acetylthiophene	C_5H_5N/2 hr/90	α-(2-Thienyl)glyoxylic acid (63)	549
7	2-Mercaptobenzimidazole	AcOH/0.5 hr/reflux	Benzimidazole (—)	165
8	5,5-Diethyl-2-mercaptobarbituric acid	AcOH/0.5 hr/reflux	5,5-Diethylbarbituric acid (—)	165
9	2-Acetyl-5-(trimethylsilyl)thiophene	Dioxane/2 hr/reflux	(5-Trimethylsilyl-2-thienyl)glyoxal (73)	447
10	2-Mercapto-4-methylquinoline	AcOH/0.5 hr/reflux	2-Hydroxy-4-methylquinoline (—)	165
		AcOH/48 hr/25	Bis-(4-methylquinol-2-yl) disulfide (—)	165
	2-Mercapto-4-phenyluracil	AcOH/0.5 hr/reflux	4-Phenyluracil (—)	165
	4-Methylquinoline-2-sulfonic acid	AcOH/0.5 hr/reflux	2-Hydroxy-4-methylquinoline (—)	165
11	4,6-Dimethyl-2-mercaptoquinoline	AcOH/0.5 hr/reflux	4,6-Dimethyl-2-hydroxyquinoline (—)	165
		AcOH/48 hr/25	Bis-(4,6-dimethylquinol-2-yl) disulfide (—)	165
12	4-Carbomethoxymethyl-2-(p-chlorophenyl)thiazole	—/—/—	Methyl 2-(p-chlorophenyl)thiazolylglyoxylate (—)	550
13	9-Mercaptoacridine	AcOH/0.5 hr/reflux	Acridone (—)	165

Note: References 249–634 are on pp. 407–415.

TABLE XIII. TERPENES

No. of C Atoms	Reactant	Solvent/Time/Temperature (°C)	Product(s) and Yield(s) (%)	Refs.
10	3-Bromocamphor	None/6 hr/150	Camphorquinone (55)	213
		Ac₂O/10 hr/reflux	No reaction	213
	10-Bromocamphor	Ac₂O/—/—	10-Bromocamphorquinone (—)	551
	Camphene	Ac₂O/—/reflux	Camphenilone (—), tricyclol (—), 2,2-dimethyl-carbocamphenilone (—), bicyclo[3.2.1]-octane-3,4-dione (—), campheneglycol carbonate (—), camphenilanaldehyde enol acetate (—)	552–555
	Camphor	C₂H₅OH/8 hr/reflux	Camphorquinone (I) (73)	213, 556
		C₆H₅CH₃/8 hr/120	I (88)	213
		Xylene/8 hr/140	I (90)	213
		Ac₂O/4 hr/reflux	I (90)	213
	Camphor-¹⁸O	Ac₂O/3 hr/145	Camphorquinone-¹⁸O (I) (30)	327
		Toluene/15.5 hr/reflux	I (3)	327
	3-Carene	C₂H₅OH/—/—	p-Isopropenyltoluene (—), p-mentha-1,5-dien-8-ol (—), p-mentha-1,5-dien-8-ol ethyl ether (—), 4,(7)-caren-5-one (—), 2,8-epoxy-p-mentha-1(7),5-diene (—), 2-p-tolyl-2-propanol (—)	557
				558, 559
	(+)-3-Carene	—/—/60	(+)-3-Caren-7-ol (—)	560
	(+)-Carvone	C₂H₅OH/1.25 hr/reflux	4-Hydroxy-p-mentha-6,8-dien-2-one (70), dehydrocarvacrol (8), 4-methyl-α-methylene-5-oxo-3-cyclohexen-1-acetaldehyde (10)	62, 561
		AcOH/—/95	3,6-Dimethyl-1-selenanaphthene-4,7-quinone (—)	562

Substrate	Conditions	Products	Refs.
D-(+)-(trans)-Carvotanacetol	Dioxane/1 hr/reflux	D-(+)-Carvotanacetone (—)	45
α-Chlorocamphor	None/3 hr/150	Camphorquinone (32)	213
Citronellal	—/—/—	2,6-Dimethyl-2-octenedial (—)	561, 563
β-Cyclocitral	—/—/—	Safranal (2)	564
(—)-Dihydrocarveol	t-C_4H_9OH/4-20 hr/15-50	Dihydrocarvone (—), p-meth-8(9)-en-2-ol-10-al (—), p-menth-8(9)-en-2,10-diol (—)	565
1,2,3,4,5,6-Hexahydroazulene	Dioxane/24 hr/25; 2 hr/90	6-Oxo-1,2,3,6-tetrahydroazulene (19)	227
3-Hydroxycamphor	C_2H_5OH/2 hr/reflux; None/15 min/150-160	Camphorquinone (I) (40); I (85)	213; 213
trans-8-Hydroxycamphor	Ac_2O/—/—	π-Acetoxycamphorquinone (—)	551
Isonitrosocamphor	None/—/—	Camphoric anhydride (I) (27), camphoric mononitrile (II) (23)	213
	C_2H_5OH/5 hr/reflux	I (12), II (20)	213
	$C_6H_5CH_3$/5 hr/reflux	I (36), II (36)	213
Isopulegol	Ac_2O/4 hr/reflux	p-Menth-8-ene-3,4-diol diacetate (35)	119
	t-C_4H_9OH/—/50	Isopulegylselenious acid (41), cis-(—)-p-menth-8(9)-ene-3,4-diol (1.8), (—)-p-menth-8(9)-en-3-ol-10-al (1.2)	566
(+)-Limonene	Ac_2O/1 hr/80-90	Carveyl acetate (40), p-mentha-1,8-dien-10-yl acetate (30), trans-p-mentha-1(7),8-dien-2-yl acetate (20)	43
	C_2H_5OH/2 hr/95-96	(+)-p-Mentha-1,8-dien-4-ol (I) (12), (+)-p-mentha-1,8-dien-10-ol (II) (6), carveol (—)	43, 63, 64
	C_2H_5OH/—/reflux	I (11), II (—), (+)-trans-carveol (—), cis-carveol (—), carvone (—)	36, 567, 568
D-(+)-p-Menth-1-ene	C_2H_5OH/20 hr/reflux	D-(+)-Carvotanacetone (I) (75), D-(+)-phellandral (10)	47

Note: References 249-634 are on pp. 407-415.

No. of C Atoms	Reactant	Solvent/Time/ Temperature (°C)	Product(s) and Yield(s) (%)	Refs.
10 (contd.)		Dioxane/12–20 hr/ reflux	I (1.5), D-(+)-carvotanacetol (cis, 3.3, trans, 7.7), D-(+)-phellandrol (3.8)	45, 569
		Ac₂O, AcOH/10 hr/ 25	Carvotanacetol acetates (15)	45
	(+)-p-Menth-3-(and-2-)ene	Ac₂O, AcOH/40 hr/50	p-Menth-3-en-5-ol (I) (—), p-menth-3-en-5-one (II) (—), p-menth-3-en-5-yl acetate (—)	570
	p-Menth-3-ene	C₂H₅OH/4 hr/80	I (—), II (—), menthene selenide (—)	571
	Myrcene	C₂H₅OH/1 hr/reflux	(E)-2-Methyl-6-methylen-2,7-octadienal (15), (E)-2-methyl-6-methylen-2,7-octadien-1-ol (46)	58
	1,2,3,4,5,6,7,8-Octahydroazulene	Dioxane/24 hr/25; 2 hr/90	1-Oxo-1,2,3,4,5,6,7,8-octahydroazulene (23)	227
	4-Oxo-1,2,3,4,5,6,7,8-octa-hydroazulene	C₂H₅OH/2 hr/reflux	4,5-Dioxo-1,2,3,4,5,6,7,8-octahydroazulene (48)	227
	α-Pinene	Ac₂O/—/—	Myrtenol (I) (23), pinol (4.4)	572
		C₂H₅OH/24 hr/reflux	Myrtenal (II) (55)	573
		AcOH/—/—	I (—), II (—)	574, 575
	β-Pinene	C₂H₅OH/4 hr/reflux	Pinocarveol (53–62)	576, 577
	Sabinene	C₂H₅OH/5 hr/reflux	Dihydrocumaldehyde (—)	578
	(+)-Sabinol	C₂H₅OH/4 hr/reflux	Sabinone dimer (19)	579–582
	β-Terpineol	AcOH/2 hr/95–105	p-Menth-8(9)-en-1-ol-9-al (I) (—), p-mentha-3,8(9)dien-1-ol acetate (II) (—)	583
		t-C₄H₉OH, C₆H₆/—/ 25	I (—), II (—), trans-p-menth-8(9)-ene-1,4-diol (—)	583

Terpinolene	C₂H₅OH/—/60	p-Cymene (—), 1-methyl-4-isopropenyl-benzene (—), 4-isopropenylbenzyl alcohol (—)	584
α-Thujene	C₂H₅OH/—/—	p-Cymene (—), 4-isopropyl-1,3-cyclohexadiene-4-carboxaldehyde (—)	585
12			
exo-5-Acetoxycamphor	Dioxane/5 hr/155	exo-5-Acetoxycamphorquinone (69)	556
trans-8-Acetoxycamphor	Ac₂O/—/—	8-Acetoxycamphorquinone (—)	551
10-Acetoxycamphor	Ac₂O/—/—	10-Acetoxycamphorquinone (—)	551
(—)-cis-Carvyl acetate	t-C₄H₉OH/—/25	p-Mentha-6,8(9)-diene-2,4-diol cis and trans (—)	586
(—)-Dihydrocarvyl acetate	t-C₄H₉OH/4-20 hr/15-50	trans-p-Menth-8(9)-ene-2,4-diol monoacetate (—), trans-p-menth-8(9)-en-2-ol-10-al acetate (—), p-mentha-3,8(9)-dien-2-ol acetate (—)	565
1,4-Dimethyl-6-oxo-1,2,6,7,8,9-hexahydroazulene	Dioxane/1 week/25; 2 hr/90	6-Methyl-4,5-(1-methylcyclopentyl)-tropolone (7)	227
3-Ethylcamphor	None/2 hr/180-190	Camphorquinone (12), ethylidenecamphor (?) (—)	213
Ethyl α-safranate	AcOH/15 min/100-110	Ethyl 2,3,6-trimethylbenzoate (63)	147
	Dioxane/30 min/reflux	5-Carboethoxy-4,6,6-trimethyl-2,4-cyclohexadienone (40)	147
Geranyl acetate	C₂H₅OH/—/reflux	OHC⟶ ⟶OAc (54)	123
Isopulegyl acetate	t-C₄H₉OH/—/50	(—)-3-Acetoxy-p-menth-8(9)-en-4-ol (14), 3-acetoxy-p-menth-8(9)-en-10-al (24), trans-p-menth-8(9)-ene-3,4-diol (—)	566, 587

Note: References 249-634 are on pp. 407-415.

397

No. of C Atoms	Reactant	Solvent/Time/Temperature	Product(s) and Yield(s) (%)	Refs.
12 (*contd.*)	Ketoisobornyl acetate	—/—/—	5,6-Diketoisobornyl acetate (29)	588
	Linalyl acetate	Dioxane/—/—	CHO, (38)	126, 589, 590
	α-Terpinyl acetate	Ac$_2$O/3 hr/reflux	p-Menth-1-ene-6,8-diol diacetate (37)	119
	β-Terpinyl acetate	Ac$_2$O, AcOH/15 hr/25	trans-p-Menth-8(9)-ene-1,4-diol (—), p-mentha-3,8(9)-dien-1-ol (after saponification) (—)	591
13	Pseudoionone	Ac$_2$O/17 hr/reflux	11-Acetoxy-6,10-dimethyl-3,5,9-undecatrien-2-one (22)	592
14	(+,−)-Deoxy-11-norsantonic acid	AcOH/6 hr/reflux	(+,−)-11-Norsantonin (10)	593
15	4,11-Epoxy-cis-eudesmane	—/18 hr/300	Eudalene (—)	594
	Tricyclic diketone, C$_{14}$H$_{20}$O$_2$	AcOH/1 hr/reflux	Tricyclic diketone, C$_{14}$H$_{18}$O$_2$ (—)	368
	Cadinene	C$_2$H$_5$OH/9 hr/80–87	Cadinene dimer (?) (—)	595
	Cedrene	C$_4$H$_9$OH/3 hr/reflux	Cedrenal (75)	578
	β-Cedrene	—/—/—	trans-β-Cedranol (—)	596
	Costunolide	C$_6$H$_6$/—/25	(—)	597
	α-Cyclodihydrocostunolide	—/—/—	(30)	135

398

β-Cyclodihydrocostunolide	[structure]	—/—/—	(—)	136
Dihydrocostunolide	[structure]	C_6H_6/—/25	(—)	597
Dihydroselinene		C_2H_5OH/4 hr/reflux	Dihydrocostal (—), dihydrocostol (—)	563
Guaiazulene		$(CH_3)_2CO$/4 hr/25; 4 hr/reflux	Diguaiazulenyl ether (—), 7-isopropyl-4-methylazulene-1-carboxyaldehyde (—), 3,3'-diguaiazulenylacetone (—), 3-formyl-guaiazulene (—)	598, 599
1-Hydroxy-4,4,8-trimethyl-tricyclo[6.3.1.0²·⁵]-9-dodecanone		C_2H_5OH/2 hr/reflux	1,4-Dihydroxy-4,4,8-trimethyltricyclo[6.3.1.0²·⁵]-10-dodecen-9-one (—)	600
Methyl 3-oxo-11-noreusanton-4-enate		AcOH/45 min/reflux	Methyl 3-oxo-11-noreusantona-1,4-dienate (33)	601
(—)-(3-Oxo-11α(H)-eudesma-1,4-dien-13-oic acid		—/—/—	(—)-β-Santonin (—)	602
3-Oxoeusanton-4-enonitrile		AcOH/30 min/reflux	3-Oxoeusantonadienonitrile (50)	601
3-Oxoeusantonin-1,4-dieno-nitrile		AcOH/3 hr/reflux	(+—)α-Santonin (—), (+—)β-santonin (—)	601
α-Santalene		C_2H_5OH/4 hr/reflux	α-Santalol (—), α-santalal (—)	563
β-Santalene		C_2H_5OH/4 hr/reflux	β-Santalol (—), β-santalal (—)	563

Note: References 249–634 are on pp. 407–415.

TERPENES—(Continued)

No. of C Atoms	Reactant	Solvent/Time/ Temperature (°C)	Product(s) and Yield(s) (%)	Refs.
	1(10)-Tetrahydrocostunolide	C_6H_6/—/25	(—)	597
	Tetrahydro-β-elemene	C_2H_5OH/6 hr/reflux	Tetrahydroelemol (—), tetrahydroelemal (—)	563
		—/—/—	(—)	603
		C_2H_5OH/4 hr/reflux		604

400

16	11-Cyano-3-oxoeusantona-1,4-dienoic acid	AcOH/5 hr/reflux	11-Cyano-6α-hydroxy-3-oxoeusantona-1,4-dienic acid lactone (—)	605
	12-Methoxydihydrocostunolide	C_6H_6/—/25	CH_2OCH_3 (—)	597
	Methyl β-3,6-dioxoeudesmanoate	AcOH/90 min/reflux	Methyl 3,6-dioxoeudesm-4-enoate (30)	524
	Methyl 3,6-dioxoeudesmanoates (5-stereoisomers)	AcOH/—/30	Methyl 3,6-dioxoeudesm-4-enoates (—)	524
	Methyl 3-oxo-11-epieusanton-4-enic acid	AcOH/30 min/reflux	Methyl 3-oxo-11-epieusantona-1,4-dienate (42)	601
		C_6H_6/10 hr/reflux	(—),	606
	Methyl α-ionylidenacetate	C_2H_5OH/—/—	(—), CO_2CH_3 (—)	607

Note: References 249–634 are on pp. 407–415.

401

No. of C Atoms	Reactant	Solvent/Time/Temperature (°C)	Product(s) and Yield(s) (%)	Refs.
16 (contd.)			[structure] CO_2CH_3 (—), [structure with OH]	
			[structure] CO_2CH_3, OH, HO (—)	
17	3-Benzylcamphor	None/8 hr/200	3-Benzylidenecamphor (95)	213
	Camphor enol benzoate	C_6H_6/3.3 hr/150–160 SeO_2, 31.25% ^{18}O	Camphorquinone-^{18}O (60)	327
	Methyl 11-cyano-3-oxoeusanton-4-enic acid	AcOH/45 min/reflux	Methyl 11-cyano-3-oxoeusantona-1,4-dienate (36)	601
18	11-Carbethoxy-6α-hydroxy-3-oxoeusanton-4-enic acid lactone	AcOH/30 min/reflux	1-Carbethoxy-6α-hydroxy-3-oxoeusantone-1,4-dienic acid lactone (30)	601
20	Ethyl 11-carbethoxy-3-oxoeusanton-4-enate	AcOH/45 min/reflux	$C_{20}H_{26}O_5Se$ (—)	601
	Geranyl mesitoate	C_2H_5OH/—/reflux	[structure] OHC···OCO··· $C_6H_2(CH_3)_3$-2,4,6 (43)	124
29	A(1)-Norallobetul-3-one	AcOH/24 hr/reflux	A(1)-Norallobetulane-2,3-dione (87)	529
	A(1)-Norfriedelan-3-one	AcOH/30 min/reflux	A(1)-Norfriedel-4(23)-en-3-one (70)	608
	A(1)-Norfriedel-4(23)-en-3-one	Dioxane/16 hr/200 (sealed tube)	A(1)-Norfriedela-2(10),4(23)-dien-3-one (18)	608

	2-Lupene	AcOH/—/— Dioxane/6 hr/160 CH$_3$CH$_2$CO$_2$H/1 hr/ reflux	Fema-7,9(11)-diene (—) 2-Lupen-4-one (—) C$_{33}$H$_{54}$O$_2$ (—), C$_{30}$H$_{50}$O (—)	609 610 610
31		—/—/—	(—)	611
32	3-Acetoxycoriaceolide	AcOH/15 hr/reflux	Acetoxy-12,19-dioxo-Δ$^{19(11),13.18}$-coriaceolide (64), acetoxy-12,19-seleno-Δ$^{9(11),12,18}$-coriaceolide (5)	612
	3β-Acetoxyeuphane-7,11-dione	AcOH/—/30	(Rate of oxidation)	524
	3β-Acetoxylanostane-7,11-dione	AcOH/—/30	(Rate of oxidation)	524
	Acetoxylanostanone	Dioxane/4 hr/180	(60)	613
	Acetoxylanostenedione	Ac$_2$O, AcOH/3.5 hr/ reflux	Acetoxylanostadienedione (60)	614
	Dihydroeuphyl acetate	—/—/—	Not isolated	615
	Dihydrolanosteryl acetate	AcOH/4 hr/reflux	Dihydroagnosteryl acetate (85)	194
	Diketodihydrolanosteryl acetate	Ac$_2$O, AcOH/4 hr/ reflux	Triketodihydrolanosterol acetate (33)	194
	Diketolanostanyl acetate	Ac$_2$O, AcOH/3 hr/ reflux	Dehydrodiketolanostanyl acetate (36)	194

Note: References 249–634 are on pp. 407–415.

403

Terpenes—(Continued)

No. of C Atoms	Reactant	Solvent/Time/Temperature (°C)	Product(s) and Yield(s) (%)	Refs.
32 (contd.)	Diketolanostenyl acetate	Ac$_2$O, AcOH/3.5 hr/reflux	Diketolanostadienyl acetate (40)	195
	7,11-Dioxoeuphanyl acetate	—/—/—	7,11-Dioxoeuphan-8-enyl acetate (—)	616
	Glutinol acetate	—/—/—	A dienol acetate (—)	617
	7-Ketolanosta-5,8-dien-3-βyl acetate	AcOH/3 hr/reflux	7-Ketolanosta-5,8,11-trien-3β-yl acetate (—)	618
	7-Ketolanostan-3β-yl acetate	AcOH/4 hr/reflux	7-Ketolanost-5-en-3β-yl acetate (80)	618, 619
	Ketolanostenyl acetate	AcOH/4 hr/reflux	Ketolanostatrienyl acetate (70)	195
	7-Oxoeuphan-8-enyl acetate	—/—/—	7-Oxoeupha-5,8,11-trienyl acetate (—)	616
	Taraxasteryl acetate	AcOH/—/reflux	Acetoxy diene, C$_{32}$H$_{50}$O$_2$ (40)	620
		—/—/—	30-Oxotaraxast-20-ene-3β-yl acetate (—)	621
	(structure, AcO—)	—/—/—	*(structure, AcO—)* (—)	622
33	7,11-Dioxo-6a-aza-B-homo-8α,9α-lanostan-3β-yl acetate	AcOH/3 hr/reflux	7,11,12-Trioxo-6a-aza-B-homolanost-8-en-3β-yl acetate (—)	537
	7,11-Dioxo-6a-aza-B-homo-8β,9α-lanostan-3β-yl acetate	AcOH/2 hr/reflux	7,11,12-Trioxo-6a-aza-B-homo-8β,9α-lanostan-3β-yl acetate (—)	537
	Methyl 3β-acetoxy-12,19-dioxo-18a-oleanan-28-oate (cis and trans)	AcOH/—/30	(Rate of oxidation)	524

404

	Methyl 2-acetoxy-12-oxoursan-28-oate	AcOH/18 hr/reflux	Acetoxy lactone, $C_{32}H_{46}O_5$ (—)	625
	Methyl 2-acetoxy-12-oxo-$\Delta^{10.11}$-ursen-28-oate	AcOH/18 hr/reflux	Acetoxy lactone, $C_{32}H_{46}O_5$ (—)	623
34	Daturadiol diacetate	AcOH/3 hr/reflux	(—)	624
		AcOH/15 hr/reflux	(—)	624
35	2,24-Diacetoxy-$\Delta^{13.18.x,y}$ oleandiene	Dioxane/4 hr/190	2,24-Diacetoxy-$\Delta^{12,13,18.19.x.y}$-oleantriene (—)	625
	Methyl 3-benzoxy-7,11-dioxo-trisnoreuphanate	—/—/—	Methyl 3-benzoxy-7,11-dioxotris-noreuphan-8-enate (—)	616
	Methyl diacetoxymachaerinate	AcOH/22 hr/reflux	Methyl diacetoxy-$\Delta^{11,13(18)}$-machaerinate (12.5), methyl diacetoxy-12,19-dioxo-$\Delta^{9(11),12(18)}$-machaerinate (—)	612
36	2,24,x-Triacetoxy-$\Delta^{13.18}$-oleanene	Dioxane/24 hr/200	2,24,x-Triacetoxy-12,19-dioxo-$\Delta^{10.11.13,18}$-oleandiene (—)	625

Note: References 249–634 are on pp. 407–415.

TERPENES—(Continued)

No. of C Atoms	Reactant	Solvent/Time/Temperature (°C)	Product(s) and Yield(s) (%)	Refs.
	(structure) AcO····OAc, CR₂OAc; R = H, D	—/—/—	(structure) AcO····O, O, OAc, CR₂OAc; R = H, D (—)	626

TABLE XIV. MISCELLANEOUS COMPOUNDS AND MIXTURES

No. of C Atoms	Reactant	Solvent/Time/Temperature (°C)	Product(s) and Yield(s) (%)	Refs.
3	Reductone	H₂O/15 hr/25	Mesoxaldehyde (—)	250
5	5,5-Dimethyl-2-oxo-1,3,2-dioxaphosphorinane	—/—/—	Bis-(5,5-dimethyl-2-oxo-1,3,2-dioxaphosphorinan-2-yl) selenide (25)	627
6	L-Ascorbic acid	C₂H₅OH/20 hr/25	Dehydro-L-ascorbic acid (I) (—)	250
		H₂O, HCl/—/—	I (100)	628
7	2-Propioselenophene	Dioxane/4 hr/reflux	1-(2-Selenophenyl)-1,2-propanedione (22)	629
8	2-Methylbenzoselenazole	m-Xylene/1 hr/reflux	2-Benzoselenazolecarboxaldehyde (23)	630
21	Tri-m-tolylphosphine	C₆H₆/—/reflux	Tri-m-tolylphosphine oxide (—)	166
21	Tri-p-tolylphosphine	C₆H₆/—/reflux	Tri-p-tolylphosphine oxide (—)	166
22	Alkylidenetriphenylphosphoranes; (C₆H₅)₃P=CHCOR, R = OC₂H₅, aryl	Dioxane/—/—	RCOCH=CHCOR (75–87)	167
31	Fluorenylidenetriphenylphosphorane	Dioxane/—/—	9,9'-Bifluorenylidene (73)	167
	Dithiols	—/—/—	Polymers	631
	Linseed oil	C₂H₅OH/3 hr/reflux	Hydroxylated and dehydrated products	632, 633
	Selenochromenes	C₅H₅N/—/reflux	2-Formylbenzo[b]selenophenes	634

REFERENCES TO TABLES II-XIV

[249] R. B. Thompson and J. A. Chenicek, *J. Amer. Chem. Soc.*, **69**, 2563 (1947).

[250] J. R. Holker, *J. Chem. Soc.*, **1955**, 579.

[251] A. Tubul-Peretz, E. Ucciani, and M. Naudet, *Bull. Soc. Chim. Fr.*, **1966**, 2331.

[252] J. H. Fried, S. Heim, S. H. Etheredge, P. Sunder-Plassmann, T. S. Santhanakrishman, J. Himizu, and C. H. Lin, *Chem. Commun.*, **1968**, 634.

[253] Y. Ohtsuka, *Kagaku Keisatsu Kenkyusho Hokoku*, **24**, 61 (1971) [*C.A.*, **77**, 87238v (1972)].

[254] F. Bigler, P. Quitte, M. Vecchi, and W. Vetter, *Arzneim.Forsch.*, **22**, 2191 (1972).

[255] S. Raymond, *J. Amer. Chem. Soc.*, **72**, 3296 (1950).

[256] J. N. Marx, J. H. Cox, and L. R. Norman, *J. Org. Chem.*, **37**, 4489 (1972); J. N. Marx and L. R. Norman, *ibid.*, **40**, 1602 (1975).

[257] K. Sato, S. Suzuki, and Y. Kojima, *J. Org. Chem.*, **32**, 339 (1967).

[258] J. N. Marx, J. C. Argyle, and L. R. Norman, *J. Amer. Chem. Soc.*, **96**, 2121 (1974).

[259] H. Böhme and H. Schneider, *Chem. Ber.*, **91**, 988 (1958).

[260] D. Caine, P. F. Brake, J. F. DeBardelen, Jr., and J. B. Dawson, *J. Org. Chem.*, **38**, 967 (1973).

[261] H. Rodé-Gowal, H. L. Dao, and H. Dahn, *Helv. Chem. Acta*, **57**, 2209 (1974).

[262] V. D. Azatyan and R. S. Gyuli-Kevkhyan, *Dokl. Akad. Nauk Arm. SSR*, **21**, 209 (1955) [*C.A.*, **50**, 11257h (1956)].

[263] H. J. E. Loewenthal, *J. Chem. Soc.*, **1961**, 1421.

[264] F. Bohlman and E. Inhoffen, *Chem. Ber.*, **89**, 1276 (1956).

[265] C. Descoins, C. A. Henrick, and J. B. Siddall, *Tetrahedron Lett.*, **1972**, 3777.

[266] E. Trommsdorf and G. Able, Ger. Pat. 803,959 [*C.A.*, **45**, 5972b (1951)].

[267] M. Elliott, N. F. Janes, and D. A. Pulman, *J. Chem. Soc.*, *Perkin Trans*, I, **1974**, 2470.

[268] M. Matsui and Y. Yamada, *Agr. Biol. Chem.* (Tokyo) **29**, 956 (1965) [*C.A.*, **64**, 3605g (1966)].

[269] Y. Watanabe, *J. Sci. Hiroshima Univ.*, Ser. A. **21**, 151 (1957) [*C.A.*, **52**, 16191d (1958)].

[270] H. Achenbach and H. Huth, *Tetrahedron Lett.*, **1974**, 119.

[271] K. Takaoka and Y. Toyama, *Nippon Kagaku Zasshi*, **89**, 405, 618 (1968) [*C.A.*, **69**, 76559j (1968)].

[272] F. Dallacker, W. Imoehl, and M. Pauling-Walther, *Ann. Chem.*, **681**, 11 (1965).

[273] J. F. Eastham and D. J. Feeney, *J. Org. Chem.*, **23**, 1826 (1958).

[274] T. Weiss, W. Nitsche, F. Boehnke, and G. Klar, *Ann. Chem.*, **1973**, 1418.

[275] K. Kariyone and T. Yazawa, Jap. Pat. 74 11,202 [*C.A.*, **81**, 120227y (1974)].

[276] J. B. Bredenberg, G. A. Nyman, P. Mahonen, and E. Rautoma, *Kem. Teollisuus*, **27**, 903 (1970) [*C.A.*, **74**, 87141w (1971)].

[277] M. Mousseron and R. Jacquier, *Bull. Soc. Chim. Fr.*, **1952**, 467.

[278] M. Mousseron, R. Jacquier, and F. Winternitz, *C. R. Acad. Sci.*, **224**, 1230 (1947).

[279] A. F. Plate and E. M. Mil'vitskaya, *Uchenye Zapiski Moskov. Gosudarst. Univ. im. M.V. Lomonosova No. 132, Org. Khim.*, **7**, 248 (1950) [*C.A.*, **49**, 3835i (1955)].

[280] A. Byers and W. J. Hickinbottom, *J. Chem. Soc.*, **1948**, 1328.

[281] W. J. Hickinbottom, *J. Chem. Soc.* **1948**, 1331.

[282] C. W. Jefford and A. F. Boschung, *Helv. Chem. Acta*, **57**, 2242 (1974).

[283] J. P. Schaefer and B. Horvath, *Tetrahedron Lett.*, **1964**, 2023.

[284] S. Tsutsumi and N. Sonoda, Jap. Pat. (71) 18,979 [*C.A.*, **75**, 76420f (1971)].

[285] I. Iwai and Y. Okajima, Jap. Pat. 11,828-9 (1960) [*C.A.*, **55**, 11367b (1961)].

[286] Z. Eckstein, A. Sacha, T. Urbański, and H. Wojnowska-Makaruk, *J. Chem. Soc.*, **1959**, 2941.

[287] K. Alder, F. H. Flock, and P. Janssen, *Chem. Ber.*, **89**, 2689 (1956).

[288] T. J. Katz, M. Rosenberger, and R. K. O'Hara, *J. Amer. Chem. Soc.*, **86**, 249 (1964).

[289] J. W. Cook, G. T. Dickson, and J. D. Loudon, *J. Chem. Soc.*, **1947**, 746.

[289a] S. I. Goldberg and R. L. Matteson, *J. Org. Chem.*, **33**, 2926 (1968).

[290] H. J. E. Loewenthal, *J. Chem. Soc.*, **1958**, 1367.

408 ORGANIC REACTIONS

[291] H. J. E. Loewenthal and P. Rona, *J. Chem. Soc.*, **1961**, 1429; *Proc. Chem. Soc.*, **1958**, 114.

[292] D. G. Lindsay and C. B. Reese, *Tetrahedron*, **21**, 1673 (1965).

[293] J. B. Lambert, A. P. Jovanovich, J. W. Hamersma, F. R. Koeng, and S. S. Oliver, *J. Amer. Chem. Soc.*, **95**, 1570 (1973).

[294] T. Kobayashi, J. Furukawa, and N. Hagihara, *Yuki Gosei Kagaku Kyokai Shi*, **20**, 551 (1962) [*C.A.*, **58**, 4436d (1963)].

[295] L. A. Paquette and J. S. Ward, *J. Org. Chem.* **37**, 3569 (1972).

[296] M. Rosenblum, *J. Amer. Chem. Soc.*, **79**, 3179 (1957); P. Wilder, Jr., A. R. Portis, Jr., G. W. Wright, and J. M. Shepherd, *J. Org. Chem.*, **39**, 1636 (1974).

[297] R. B. Woodward and T. J. Katz, *Tetrahedron*, **5**, 70 (1959).

[298] K. Alder and F. H. Flock, *Chem. Ber.*, **87**, 1916 (1954).

[299] M. N. Azidlewicz, *Rocz. Chem.*, **42**, 437 (1968) [*C.A.*, **69**, 35516z (1968)].

[300] M. Zaidlewicz, A. Uzarewicz, and W. Zacharewicz, *Rocz. Chem.* **40**, 437 (1966) [*C.A.*, **65**, 3783g (1966)].

[301] A. Uzarewicz and W. Zacharewicz, *Rocz. Chem.*, **35**, 541 (1961) [*C.A.*, **55**, 23378d (1961)].

[302] A. Uzarewicz and W. Zacharewicz, *Rocz. Chem.*, **35**, 887 (1961) [*C.A.*, **56**, 443d (1962)].

[303] W. Zacharewicz and A. Uzarewicz, *Rocz. Chem.*, **31**, 721, 729 (1957) [*C.A.*, **52**, 5312b (1958)].

[304] W. Zacharewicz and A. Uzarewicz, *Rocz. Chem.*, **34**, 413 (1960) [*C.A.*, **55**, 420f (1961)].

[305] L. J. Altman, L. Ash, and S. Marson, *Synthesis*, **1974**, 129.

[306] A. S. Sultanov, V. M. Rodionov, and M. M. Shemyakin, *J. Gen. Chem. USSR*, **16**, 2072 (1946) [*C.A.*, **42**, 880i (1948)].

[307] L. Prajer, *Rocz. Chem.*, **28**, 55 (1954).

[308] E. Gazis and P. Heim, *Tetrahedron Lett.*, **1967**, 1185.

[309] H. J. Bestmann and D. Ruppert, *Angew. Chem.*, *Int. Ed.*, **7**, 637 (1968).

[310] E. Clar, *J. Chem. Soc.*, **1949**, 2013.

[311] R. K. Eruenlue, *Chem. Ber.*, **100**, 533 (1967).

[312] N. P. Greco, U.S. Pat. 3,679,753 [*C.A.*, **77**, 100915k (1972)].

[313] J. Meinwald, C B. Jensen, A. Lewis, and C. Swithenbank, *J. Org. Chem.* **29**, 3469 (1964).

[314] R. L. Cargill and T. Y. King, *Tetrahedron Lett.*, **1970**, 409.

[315] K. B. Wiberg and R. W. Ubersax, *Tetrahedron Lett.*, **1968**, 3063.

[316] I. N. Nazarov and I. V. Torgov, *Zh. Obshch. Khim.*, **19**, 1766 (1949) [*C.A.*, **44**, 8906i (1950)].

[317] J. D. Chanley, *J. Amer. Chem. Soc.*, **70**, 244 (1948).

[318] M. Covello, F. De Simone, and A. Dini, *Rend. Accad. Sci. Fis. Mat. Naples*, **35**, 298 (1968) [*C.A.*, **74**, 141695v (1971)].

[319] N. J. Leonard and G. C. Robinson, *J. Amer. Chem. Soc.*, **75**, 2714 (1953).

[320] W. Logemann, G. Cavagna, and G. Tosolini, *Chem. Ber.*, **96**, 2248 (1963).

[321] G. Fodor and O. Kovács, Jr., Hung. Pat. 139,554 [*C.A.*, **44**, 4034d (1950)].

[322] T. Sato and M. Ohto, *Bull. Chem. Soc. Jap.* **28**, 480 (1955).

[323] C. Musante and V. Parrini, *Gazz. Chim. Ital.*, **81**, 451 (1951).

[324] G. Fodor and O. Kovács, *J. Amer. Chem. Soc.*, **71**, 1045 (1949).

[325] W. C. M. C. Kokke and F. A. Varkevisser, *J. Org. Chem.*, **39**, 1653 (1974).

[326] W. C. M. C. Kokke and L. J. Oosterhoff, *J. Amer. Chem. Soc.*, **94**, 7583 (1972).

[327] W. C. M. C. Kokke, *J. Org. Chem.*, **38**, 2989 (1973).

[328] R. D. Haworth and J. D. Hobson, *Chem. Ind.* (London), **1950**, 441.

[329] R. D. Haworth and J. D. Hobson, *J. Chem. Soc.*, **1951**, 561.

[330] R. D. Miller and D. L. Dolce, *Tetrahedron Lett.*, **1974**, 3813.

[331] P. V. Chatfield, Fr. Demande 2,160,647 [*C.A.*, **79**, 136807g (1973)].

[332] K. Schank, *Chem. Ber.*, **103**, 3087 (1970).

[333] I. Tabushi, Z. Yoshida, and Y. Aoyama, *Chem. Lett.*, **1973**, 123.

[334] F. Weygand and I. Frank, *Chem. Ber.*, **84**, 591 (1951).

[335] T. Nozoe, Y. Kitahara, and S. Ito, *Proc. Jap. Acad.*, **26**, 47 (1950) [*C.A.*, **45**, 7099 (1951)].

[336] T. Nozoe, S. Seto, K. Kikuchi, T. Mukai, S. Matsumoto, and M. Murase, *Proc. Jap. Acad.*, **26**, 43 (1950) [*C.A.*, **45**, 7099g (1951)].

337 T. Nozoe, S. Seto, K. Kikuchi, and H. Takeda, *Proc. Jap. Acad.*, **27**, 146 (1951) [*C.A.*, **46**, 4522c (1952)].

338 W. G. Dauben, C. H. Schallhorn, and D. L. Whalen, *J. Amer. Chem. Soc.*, **93**, 1446 (1961).

339 N. F. Woolsey and M. H. Khalil, *Tetrahedron Lett.*, **1974**, 4309.

340 T. Nozoe, H. Kishi, and A. Yoshikoshi, *Proc. Jap. Acad.*, **27**, 149 (1951)[*C.A.*, **46**, 4523d (1952)].

341 K. Hafner, K. P. Meinhardt, and W. Richarz, *Angew. Chem.*, **86**, 235 (1974).

342 V. V. Dhekne and B. V. Bhide, *J. Indian Chem. Soc.*, **28**, 504 (1951).

343 K. H. Schulte-Elte, M. Gadola, and G. Ohloff, *Helv. Chim. Acta*, **56**, 2028 (1973).

344 N. Rigassi and U. Schwieter, Ger Pat 2,032,919 [*C.A.*, **74**, 64323t (1971)].

345 E. L. Engelhardt and M. E. Christy, Brit. Pat. 1,265,052 [*C.A.*, **76**, 153342g (1972)]: Ger. Offen. 2,104,312 [*C.A.*, **78**, 71658s (1973)].

346 J. H. Gorvin, *Nature*, **161**, 208 (1948).

347 E. R. Bockstahler, U.S. Pat. 2,570,181 [*C.A.*, **46**, 4572g (1952)].

348 C. Musante and V. Parrini, *Gazz. Chim. Ital.*, **80**, 868 (1950).

349 D. Caine and F. N. Tuller, *J. Org. Chem.*, **38**, 3663 (1973).

350 L. Christiaens and M. Renson, *Bull. Soc. Chim. Belg.*, **79**, 133 (1970).

351 P. Yates and E. G. Lewars, *Can. J. Chem.*, **48**, 788, 796 (1970).

352 D. N. Shah, S. K. Parikh, and N. M. Shah, *J. Amer. Chem. Soc.*, **77**, 2223 (1955).

353 A. Schiavello and C. Sebastiani, *Gazz. Chim. Ital.*, **79**, 909 (1949).

354 D. R. Patel and S. R. Patel, *J. Indian Chem.*, *Soc.* **45**, 703 (1968).

355 Y. A. Rozin, V. E. Blokhin, N. M. Sokolova, Z. V. Pushkareva, and L. G. Surovtsev, *Khim. Geterotsikl. Soedin.*, **1975**, 86 [*C.A.*, **83**, 9900g (1975)].

356 T. R. Govindachari and P. C. Parthasarathy. *Tetrahedron Lett.*, **1972**, 3419.

357 F. Giarrusso and R. E. Ireland, *J. Org. Chem.*, **33**, 3560 (1968).

358 R. C. Fuson and T. Tan, *J. Amer. Chem. Soc.*, **70**, 602 (1948).

359 S. I. Burmistrov and E. I. Shilov, *J. Gen. Chem. USSR*, **17**, 1684 (1947)[*C.A.*, **42**, 2595f (1948)].

360 E. Boelema, J. Strating, and H. Wynberg, *Tetrahedron Lett.*, **1972**, 1175.

361 P. V. Radhakrishnan and A. V. R. Rao, *Indian J. Chem.*, **4**, 406 (1966).

362 S. Matsuura and T. Kunii, *J. Pharm. Soc. Jap.*, **94**, 645 (1974)[*C.A.*, **81**, 63440m (1974)].

363 F. Dayer, H. L. Dao, H. Gold, H. Rodé-Gowal and H. Dahn, *Helv. Chim. Acta*, **57**, 2201 (1974).

364 H. Musso and D. Döpp, *Chem. Ber.*, **97**, 1147 (1964).

365 S. A. Osadchii and V. A. Barkhash, *Zh. Org. Khim.*, **6**, 1815 (1970) [*C.A.*, **73**, 120381d (1970)].

366 G. Rabilloud and B. Sillion, *Bull. Soc. Chim. Fr.*, **1970**, 4052.

367 D. H. R. Barton and A. S. Lindsey, *Chem. Ind.* (London), **1951**, 313.

368 D. H. R. Barton and A. S. Lindsey, *J. Chem. Soc.*, **1951**, 2988.

369 K. Balenović, D. Cerar, and L. Filipović, *J. Org. Chem.*, **19**, 1556 (1954).

370 G. I. Eremeeva, B. K. Strelets, and L. S. Efros, *Khim. Geterotsikl. Soedin.*, **1975**, 276, *C.A.*, **82**, 156192t (1975)].

371 P. Jacquignon, G. Marechal, M. Renson, A. Ruwet, and Do Phuoc Hien, *Bull. Soc. Chim. Fr.*, **1973**, 677.

372 T. Kh. Gladysheva and M. V. Gorelik, *Khim. Geterotsikl. Soedin*, **1970**, 554[*C.A.*, **73**, 87858q (1970)].

373 M. V. Gorelik and V. I. Lomzakova, *Khim. Geterotsikl. Soedin.*, **1974**, 1275 [*C.A.*, **82**, 16755d (1975)].

374 G. Zemplén and L. Kisfaludy, *Chem. Ber.*, **93**, 1125 (1960).

375 H. Igeta, T. Tsuchiya, C. Kaneko, and S. Suzuku, *Chem. Pharm. Bull.* (Tokyo), **21**, 125 (1973).

376 V. M. Clark, B. Sklarz, and A. R. Todd, *J. Chem. Soc.*, **1959**, 2123.

377 L. Achremowicz and L. Syper, *Rocz. Chem.*, **46**, 409 (1972) [*C.A.*, 77, 101351k (1972)].

378 D. J. Cook and R. S. Yunghans, *J. Amer. Chem. Soc.*, **74**, 5515 (1952).

379 T. Slebodzinski, H. Kietczewska, and W. Biernacki, *Przem. Chem.*, **48**, 90 (1969) [*C.A.*, **71**, 38751z (1969)].

[380] S. Furukawa and Y. Kuroiwa, *Pharm. Bull.* (Japan), **3**, 232 (1955) [*C.A.*, **50**, 10092d (1956)].

[381] K. Schank, *Chem. Ber.*, **102**, 383 (1969).

[382] M. Giannella and F. Gualtieri, *Boll. Chim. Farm.*, **105**, 708 (1966) [*C.A.*, **66**, 104945r (1967)].

[383] E. S. Hand and W. W. Paudler, *J. Org. Chem.*, **40**, 2916 (1975).

[384] L. Rappen and O. Koch, *Ger. Pat.* 1,620,174 [*C.A.*, **77**, 48273h (1972)].

[385] J. Koncewicz and Z. Shrowaczewska, *Rocz. Chem.*, **42**, 1873 (1968) [*C.A.*, **70**, 114972u (1969)].

[386] L. Achremowicz, *Rocz. Chem.*, **47**, 2367 (1973). [*C.A.*, **80**, 133200p 1974)].

[387] 1. Matsumoto and J. Yoshizawa, *Jap. Pat.* 72 02,093 [*C.A.*, **76**, 126801z (1972)].

[388] R. F. C. Brown, V. M. Clark, and A. R. Todd, *J. Chem. Soc.*, **1959**, 2105.

[389] A. Matsumoto, M. Yoshida, and O. Simamura, *Bull. Chem. Soc. Jap.*, **47**, 1493 (1974) [*C.A.*, **81**, 105404k (1974)].

[390] S. Murahashi and S. Otuka, *Mem. Inst. Sci. Ind. Res. Osaka Univ.*, **7**, 127 (1950 [*C.A.*, **45**, 9054g (1951)].

[391] J. F. K. Wilshire, *Aust. J. Chem.*, **20**, 359 (1967).

[392] B. Witkop and H. Fiedker, *Ann. Chem.*, **558**, 91 (1947).

[393] M. Seyhan and S. Avan, *Rev. Fac. Sci. Univ. Istanbul*, **16A**, 30 (1951) [*C.A.*, **46**, 8090c (1952)].

[394] W. Sliwa and Z. Skrowaczewska, *Rocz. Chem.*, **44**, 1941 (1970) [*C.A.*, **75**, 20141y (1971)].

[395] E. Giovannini and P. Portmann, *Helv. Chim. Acta*, **31**, 1392 (1948).

[396] E. Tojo and K. Kurosaki, *Bull. Soc. Sci. Phot. Jap.*, **12**, 5 (1962) [*C.A.*, **59**, 8298f (1963)].

[397] H. Umezawa and T. Nagatsu, *S. African Pat.* 70 06,634 [*C.A.*, **76**, 3702k (1972)].

[398] A. M. Simonov and L. M. Sitkina, *Khim. Geterotsikl. Soedin.*, *Sb.* 1.: *Azotsoderzhashchie Geterosikly*, **1967**, 116 [*C.A.*, **70**, 8767q (1969)].

[399] M. P. Lamontagne, *J. Med. Chem.*, **16**, 68 (1973).

[400] K. C. Agrawal, B. A. Both, E. C. Moore, and A. C. Sartorelli, *J. Med. Chem.*, **15**, 1154 (1972).

[401] S. Sakai, A. Kubo, K. Katsuura, K. Mochinaga, and M. Ezaki, *Chem. Pharm. Bull.* (Tokyo), **1972**, 76.

[402] M. Seyhan, *Chem. Ber.*, **84**, 477 (1951).

[403] J. D. Johnston, *U.S. Pat.* 3,296,257 [*C.A.*, **67**, 3100b (1967)].

[404] H. E. Baumgarten and J. E. Dirks, *J. Org. Chem.*, **23**, 900 (1958).

[405] C. E. Teague, Jr. and A. Roe, *J. Amer. Chem. Soc.*, **73**, 688 (1951).

[406] V. G. Ramsey, *J. Amer. Pharm. Assoc.*, **40**, 564 (1951).

[407] E. V. Brown and N. G. Frazer, *J. Heterocycl. Chem.*, **6**, 567 (1969).

[408] C. A. Buehler, L. A. Walker, and P. Garcia, *J. Org. Chem.*, **26**, 1410 (1961).

[409] G. Heinisch, A. Jentzsch, and M. Pailer, *Monatsh. Chem.*, **105**, 648 (1974).

[410] D. J. Cook, R. W. Sears, and D. Dock. *Proc. Indiana Acad. Sci.*, **58**, 145 (1949) [*C.A.*, **44**, 4473f (1950)].

[411] M. Seyhan, *Chem. Ber.* **92**, 1480 (1959).

[412] M. Seyhan, *Chem. Ber.*, **85**, 425 (1952).

[413] M. Seyhan, *Rev. Fac. Sci. Univ. Istanbul*, **16A**, 252 (1951) [*C.A.*, **47**, 3312b (1953)].

[414] M. Seyhan, *Chem. Ber.*, **90**, 1386 (1957).

[415] C. A. Buehler, *J. Amer. Chem. Soc.*, **74**, 977 (1952).

[416] M. Seyhan and W. C. Fernelius, *Chem. Ber.*, **89**, 2212 (1956).

[417] P. Duballet, A. Godard, G. Quequiner, and P. Pastour, *J. Heterocycl. Chem.*, **10**, 1079 (1973).

[418] I. Matsumoto and K. Tomimoto, *Jap. Kokai* 74 51,282 [*C.A.*, **81**, 120482c (1974)].

[419] R. E. Lyle, S. A. Leone, H. J. Troscianiec, and G. H. Warner, *J. Org. Chem.*, **24**, 330 (1959).

[420] M. Seyhan and W. C. Fernelius, *Chem. Ber.*, **91**, 469 (1958).

[421] R. I. Fryer, G. A. Archer, B. Brust, W. Zally, and L. H. Sternbach, *J. Org. Chem.*, **30**, 1308 (1965).

[422] E. Hayashi and C. Iijima, *Yakugaku Zasshi*, **82**, 1093 (1962) [*C.A.*, **58**, 4551f (1963)].

[423] E. Hayashi and C. Iijima, *Yakugaku Zasshi*, **84**, 156 (1964) [*C.A.*, **61**, 3108c (1964)].

[424] R. S. Klein and J. J. Fox, *J. Org. Chem.*, **37**, 4381 (1972).

[425] W. Reid and W. Kunstmann, *Chem. Ber.*, **102**, 1418 (1969).

[426] R. J. Sundberg, F. X. Smith, and L.-Su Lin, *J. Org. Chem.*, **40**, 1433 (1975).

[427] W. A. Ayer, W. R. Bowman, T. C. Joseph, and P. Smith, *J. Amer. Chem. Soc.*, **90**, 1648 (1968).

[428] P. L. Julian, W. J. Karpel, A. Magnani, and E. W. Meyer, *J. Amer. Chem. Soc.*, **70**, 180 (1948).

[429] R. B. Woodward, M. P. Cava, W. D. Ollis, A. Hunger, H. U. Daeniker, and K. Schenker, *J. Amer. Chem. Soc.*, **76**, 4749 (1954); *Tetrahedron*, **19**, 247 (1963).

[430] E. P. Taylor, *J. Pharm. Pharmacol.*, **2**, 324 (1950) [*C.A.*, **44**, 6582c (1950)].

[431] A. Bertho, *Chem. Ber.*, **80**, 316 (1947).

[432] G. Tsatsas, *C.R. Acad. Sci.*, **229**, 218 (1949).

[433] S. Fatutta, *Univ. Studi Trieste, Fac. Sci., Inst. Chim.*, No. **31**, 33 (1961) [*C.A.*, **58**, 526d (1963)].

[434] A. Bertho, W. Schönberger, and L. Kaltenborn, *Ann. Chem.*, **557**, 220 (1947).

[435] G. R. Newkome and J. M. Robinson, *Tetrahedron Lett.*, **1974**, 691.

[436] N. S. Prostakov, A. V. Varlamov, and V. P. Zvolinskii, *Khim. Geterotsikl. Soedin.*, **1972**, 957 [*C.A.*, **77**, 126741a (1972)].

[437] M. Brunold and A. E. Siegrist, *Helv. Chim. Acta*, **55**, 818 (1972).

[438] A. Caplin, *J. Chem. Soc., Perkin Trans. I*, **1974**, 30.

[439] H. Meier and I. Menzel, *Tetrahedron Lett.*, **1972**, 445.

[440] L. A. Sternson and D. A. Coviello, *J. Org. Chem.*, **37**, 139 (1972).

[441] F. Venien and C. Mandrier, *C.R. Acad. Sci., C*, **270**, 845 (1970).

[442] H. Meiner, M. Layer, and A. Zetzsche, *Chem.-Ztg.*, **98**, 460 (1974) [*C.A.*, **82**, 43086t (1975)].

[443] H. D. Vogelsang and Th. Wagner-Jauregg, *Ann. Chem.*, **568**, 116 (1950).

[444] M. W. Miller, *Tetrahedron Lett.*, **1969**, 2545.

[445] E. Suzuki, R. Hamajima, and S. Inoue, *Synthesis*, **1975**, 192.

[446] I. K. Korobitsyna, Yu.K. Yur'ev, and O. I. Nefedova, *Zh. Obshch. Khim.*, **24**, 188 (1954) [*C.A.*, **49**, 3197a (1955)].

[447] R. A. Benkeser and H. Landesman, *J. Amer. Chem. Soc.*, **71**, 2493 (1949).

[448] M. Ebel and L. Legrand, *Bull. Soc. Chim. Fr.*, **1971**, 176.

[449] J. Thibault and P. Maitte, *Bull. Soc. Chim. Fr.*, **1969**, 915.

[450] A. Shafiee, *J. Heterocycl. Chem.*, **12**, 177 (1975).

[451] I. K. Korobitsyna, Yu.K. Yur'ev, Y. A. Cheburkov, and E. M. Lukina, *Zh. Obshch. Khim.*, **25**, 734 (1955) [*C.A.*, **50**, 2536f (1956)].

[452] G. Renzi and P. Perini, *Farmaco, Ed. Sci.*, **24**, 1073 (1969) [*C.A.*, **72**, 78917k 1970)].

[453] N. R. Bannerjee and T. R. Seshadri, *Current Sci.*, **25**, 143 (1956) [*C.A.*, **51**, 395a (1957)].

[454] Y. Kishi, M. Aratani, T. Fukuyama, F. Nakatsubo, T. Goto, S. Inoue, H. Tanino, S. Sugiura, and H. Kakoi, *J. Amer. Chem. Soc.*, **94**, 9217 (1972).

[455] G. Büchi, D. M. Foulkes, M. Kurono, and G. F. Mitchell, *J. Amer. Chem. Soc.*, **88**, 4534 (1966).

[456] G. Büchi, D. M. Foulkes, M. Kurono, G. F. Mitchell, and R. S. Schneider, *J. Amer. Chem. Soc.*, **89**, 6745 (1967).

[457] A. Stener, *Farmaco, Ed. Sci.*, **15**, 642 (1960) [*C.A.*, **58**, 497e (1963)].

[458] S. Inayama, A. Sawa, and E. Hosoya, *Chem. Pharm. Bull.* (Tokyo), **22**, 1519 (1974) [*C.A.*, **81**, 135887n (1974)].

[459] I. Inoue, K. Kondo, T. Oine, and K. Okumura, *Jap. Kokai* 74 45,073 [*C.A.*, **82**, 31163c (1975)].

[460] M. Davis and V. Petrow, *J. Chem. Soc.*, **1949**, 2973.

[461] E. H. Reerink, P. Westerhof, and H. F. L. Schoeler, U.S. Pat. 3,198,702 [*C.A.*, **63**, 16429h (1965)].

[462] H. J. Ringold, G. Rosenkranz, and F. Sondheimer, *J. Org. Chem.*, **21**, 239 (1956).

[463] C. Djerassi, G. Rosenkranz, St. Kaufmann, J. Pataki, and J. Romo, U.S. Pat. 3,020,294 [*C.A.*, **57**, 915a (1962)].

[464] St. Kaufmann, J. Pataki, G. Rosenkranz, J. Romo, and C. Djerassi, *J. Amer. Chem. Soc.*, **72**, 4531, 4534 (1950).

[465] G. Rosenkranz, U.S. Pat. 3,019,246 [C.A., 57, 917i (1962)].

[466] Merck and Co., Inc., Belg. Pat. 631,469 [C.A., 61, 9566h (1964)].

[467] F. Sondheimer and Y. Mazur, J. Amer. Chem. Soc., 79, 2906 (1957).

[468] M. Mousseron-Canet, C. Chavis, and A. Guida, Bull. Soc. Chim. Fr., 1971, 627.

[469] N. V. Organon, Neth. Pat. 86,368 [C.A., 53, 6295d (1959)].

[470] N. V. Organon, Neth. Pat. 85,526 [C.A., 53, 5348b (1959)].

[471] A. E. Oberster, R. E. Beyler, and L. H. Sarett, U.S. Pat. 3,211,725 [C.A., 63, 18216h (1965)].

[472] L. H. Knox, J. A. Zderic, J. P. Ruelas, and C. Djerassi, J. Amer. Chem. Soc., 82, 1230 (1960).

[473] Upjohn Co., Brit. Pat. 882,604 [C.A., 57, 1387g (1962)].

[474] C. Djerassi, G. Rosenkranz, J. Romo, J. Pataki, and St. Kaufmann, J. Amer. Chem. Soc., 72, 4540 (1950).

[475] A. Schubert and S. Schwarz, Experientia, 21, 562 (1965).

[476] C. Djerassi, G. Rosenkranz, J. Romo, St. Kaufmann, and J. Pataki, J. Amer. Chem. Soc., 72, 4534 (1950).

[477] N. V. Organon, Belg. Pat. 612,592 [C.A., 58, 3490d (1963)].

[478] E. Merck A.-G., Neth. Pat. Appl., 6,602,266 [C.A., 66, 46528u (1967)].

[479] M. Amorosa, L. Caglioti, G. Cainelli, H. Immer, J. Keller, H. Wehrli, M.Lj. Mihailovic, K. Schaffner, D. Arigoni, and O. Jeger, Helv. Chim. Acta., 45, 2674 (1962).

[480] H. J. Ringold, E. Batres, A. Bowers, J. Edwards, and J. Zderic, J. Amer. Chem. Soc., 81, 3485 (1959).

[481] A. Bowers, L. C. Ibanez, and H. J. Ringold, J. Amer. Chem. Soc., 81, 5991 (1959).

[482] P. F. Beal, R. W. Jackson, and J. E. Pike, J. Org. Chem., 27, 1752 (1962).

[483] T. Okumura, Y. Nozaki, and D. Sato, Chem. Pharm. Bull. (Tokyo), 12, 1143 (1964) [C.A., 62, 4088e (1965)].

[484] R. P. Graber, M. B. Meyers, and V. A. Langeryou, J. Org. Chem., 27, 2534 (1962).

[485] A. Butenandt and H. Dannenberg, Ann. Chem., 568, 83 (1950).

[486] A. Bowers, L. C. Ibanez, E. Denot, and R. Becerra, J. Amer. Chem. Soc., 82, 4001 (1960).

[487] G. B. Spero, J. E. Pike, F. H. Lincoln, and J. L. Thompson, Steroids, 1968, 769.

[488] A. Bowers and J. A. Edwards, U.S. Pat. 3,036,098 [C.A., 58, 6890d (1963)].

[489] T. Miki and Y. Hara, Pharm. Bull. (Tokyo), 4, 421 (1956) [C.A., 51, 8771e (1957)].

[490] V. E. M. Chambers and A. L. M. Riley, Ger. Offen. 2,364,741 [C.A., 81, 136384h (1974)].

[491] N. W. Atwater, R. W. Bible, Jr., E. A. Brown, R. R. Burtner, J. S. Mihina, L. N. Nysted, and P. B. Sollman, J. Org. Chem., 26, 3077 (1961).

[492] M. Ehrenstein and K. Otto, J. Org. Chem., 24, 2006 (1959).

[493] Upjohn Co., Brit. Pat. 1,088,160 [C.A., 69, 10623u (1968)].

[494] E. J. Agnello and G. D. Laubach, J. Amer. Chem. Soc., 79, 1257 (1957); 82, 4293 (1960).

[495] C. Casagrande, F. Ronchetti, and G. Russo, Tetrahedron Lett., 1974, 2369.

[496] H. J. Ringold, J. P. Ruelas, E. Batres, and C. Djerassi, J. Amer. Chem. Soc., 81, 3712 (1959).

[497] E. Batres, T. Gardenas, J. A. Edwards, G. Monroy, O. Mancera, C. Djerassi, and H. J. Ringold, J. Org. Chem., 26, 871 (1961).

[498] W. O. Godtfredsen and S. Vangedal, Acta Chem. Scand., 15, 1786 (1961).

[499] D. G. Martin and J. E. Pike, J. Org. Chem., 27, 4086 (1962).

[500] Upjohn Co., Brit. Pat. 997,167 [C.A., 63, 13369g (1965)].

[501] E. Merck A.-G., Neth. Pat. Appl., 295,201 [C.A., 63, 13368c (1965)].

[502] Schering A.-G., Ger. Pat. 1,122,518 [C.A., 57, 920 (1962)].

[503] J. A. Edwards, H. J. Ringold, and C. Djerassi, J. Amer. Chem. Soc., 81, 3156 (1959).

[504] J. A. Edwards, H. J. Ringold, and C. Djerassi, J. Amer. Chem. Soc., 82, 2318 (1960).

[505] G. R. Allen, Jr., and N. A. Austin, J. Org. Chem., 26, 4574 (1961).

[506] M. Heller and S. Bernstein, J. Org. Chem., 26, 3876 (1961).

[507] D. Taub, R. D. Hoffsommer, H. L. Slater, and N. L. Wendler, J. Amer. Chem. Soc., 80, 4435 (1958).

[508] H. J. Ringold and G. Rosenkranz, U.S. Pat. 3,203,965 [C.A., 63, 14945a (1965)].

[509] G. E. Arth, D. B. R. Johnson, J. Fried, W. W. Spooncer, D. R. Hoff, and L. H. Sarett, J. Amer. Chem. Soc., 80, 3160 (1958).

510 Upjohn Co., Neth. Pat. Appl. 6,603, 864 [*C.A.*, **66**, 65746e (1967)].

511 R. Wenger, H. Dutler, H. Wehrli, K. Schaffner, and O. Jeger, *Helv. Chim. Acta*, **45**, 2420 (1962).

512 S. Bernstein, M. Heller, R. Littell, S. M. Stolar, R. H. Lenhard, W. S. Allen, and I. Ringler, *J. Amer. Chem. Soc.*, **81**, 1696 (1959).

513 S. Bernstein, R. H. Lenhard, W. S. Allen, M. Heller, R. Littell, S. M. Stolar, L. I. Feldman, and R. H. Blank, *J. Amer. Chem. Soc.*, **81**, 1689 (1959); S. Bernstein and R. H. Lenhard, *ibid.*, **82**, 3680 (1960).

514 C. R. Engle, S. Rakhit, and W. W. Huculak, *Can. J. Chem.*, **40**, 921 (1962).

515 S. Bernstein and R. Littell, *J. Org. Chem.*, **25**, 313 (1960).

516 R. Littell and S. Bernstein, *J. Org. Chem.*, **27**, 2544 (1962).

517 R. Kh. Ruzieva, M. B. Gorovits, and N. K. Abubakirov, *Khim. Prir. Soedin.*, **1968**, 57 [*C.A.*, **69**, 10616u (1968)].

518 J. A. Zderic, H. Carpio, and C. Djerassi, *J. Amer. Chem. Soc.*, **82**, 446 (1960)].

519 S. Bernstein and R. Littell, *J. Amer. Chem. Soc.*, **82**, 1235 (1960).

520 American Cyanamid Co., Brit. Pat., 880,071 [*C.A.*, **57**, 918h (1962)].

521 J. S. Mills, A. Bowers, C. C. Campillo, C. Djerassi, and H. J. Ringold, *J. Amer. Chem. Soc.*, **81**, 1264 (1959).

522 Ciba Ltd., Swiss Pat. 242,833 [*C.A.*, **43**, 7976d (1949)].

523 Soc. pour l'Ind. Chim. à Bâle, Brit. Pat. 587,030 [*C.A.*, **42**, 609a (1948)].

524 J. C. Banerji, D. H. R. Barton, and R. C. Cookson, *J. Chem. Soc.*, **1957**, 5041.

525 L. F. Fieser and K. L. Williamson, *Organic Experiments*, 3rd ed., D. C. Heath and Co., Lexington, Mass., 1975, p. 110.

526 A. Furlenmeier, A. Fürst, L. Langemann, G. Waldvogel, U. Kerb, P. Hocks, and R. Wiechert, *Helv. Chim. Acta*, **49**, 1591 (1966).

527 A. L. Nussbaum, F. E. Carlon, D. Gould, E. P. Oliveto, E. B. Hershberg, M. L. Gilmore, and W. Charney, *J. Amer. Chem. Soc.*, **81**, 5230 (1959).

528 K. Hamamoto, K. Horiki, A. Ikegami, and K. Takeda, *Yakugaku Zasshi*, **86**, 558 (1966) [*C.A.*, **65**, 15452c (1966)].

529 R. Hanna and G. Ourisson, *Bull. Soc. Chim. Fr.*, **1961**, 1945.

530 J. Romo, C. Djerassi, and G. Rosenkranz, *J. Org. Chem.*, **15**, 896 (1950).

531 A. Bowers, E. Denot, M. B. Sanchez, F. Neumann, and C. Djerassi, *J. Chem. Soc.*, **1961**, 1859.

532 C. Djerassi and A. Bowers, U.S. Pat. 3,257, 386 [*C.A.*, **65**, 15463f (1966)].

533 L. F. Fieser, S. Rajagopalan, E. Wilson, and M. Tishler, *J. Amer. Chem. Soc.*, **73**, 4133 (1951).

534 J. B. Siddall, J. P. Marshall, A. Bowers, A. D. Cross, J. A. Edwards, and J. H. Fried, *J. Amer. Chem. Soc.*, **88**, 379 (1966).

535 R. Kazlauskas, J. T. Pinhey, J. J. H. Simes, and T. G. Waston, *J. Chem. Soc.*, D, **1969**, 945.

536 G. Saucy, P. Geistlich, R. Helbling, and H. Heusser, *Helv. Chim. Acta*, **37**, 250 (1954).

537 C. S. Barnes and D. H. R. Barton, *J. Chem. Soc.*, **1953**, 1419.

538 K. Sasaki, *Hiroshima J. Med. Sci.*, **3**, 43 (1954) [*C.A.*, **49**, 10334g (1955)].

539 K. Schreiber and H. Ripperger, *Chem. Ber.*, **96**, 3094 (1963).

540 B. F. McKenzie, V. R. Mattox, L. L. Engel, and E. C. Kendall, *J. Biol. Chem.* **173**, 271 (1948).

541 U. Kerb, P. Hocks, R. Wiechert, A. Furlenmeier, A. Fürst, A. Langemann, and G. Waldvogel, *Tetrahedron Lett.*, **1966**, 1387; *Helv. Chim. Acta*, **49**, 1601 (1966).

542 E. C. Kendall, U.S. Pat. 2,541,074 [*C.A.*, **45**, 8564f (1951)].

543 M. K. Joshi, *Collect. Czech. Chem. Commun.*, **21**, 1108 (1956) [*C.A.*, **51**, 11914i (1957)]; *Chem. Listy*, **50**, 1928 (1956) [*C.A.*, **51**, 4195c (1957)].

544 E. N. Ovsepyan, G. N. Shaposhnikova, and N. G. Galfayan, *Zh. Neorg. Khim.*, **12**, 2411 (1967) [*C.A.*, **67**, 120531d (1967)].

545 E. N. Ovsepyan, V. M. Tarayan, and G. N. Shaposhnikova, *Izv. Akad. Nauk Arm. SSR, Khim. Nauki*, **18**, 225 (1965) [*C.A.*, **63**, 14357e (1965)].

414 ORGANIC REACTIONS

546 Yu. I. Akulin, B. Kh. Strelets, and L. S. Efros, *Khim. Geterotsikl. Soedin.*, **1974**, 138 [*C.A.*, **80**, 95832m (1974)].

547 O. Attanasi and L. Caglioti, *J. Chem. Soc.*, *Chem. Commun.*, **1974**, 138.

548 D. P. Sevbo and O. F. Ginzburg, *Zh. Org. Chim.*, **4**, 1064 (1968) [*C.A.*, **69**, 51777r (1968)].

549 L. B. Crast, Jr., U.S. Pat. 3,422,099 [*C.A.*, **70**, 68388h (1969)].

550 R. Howe, R. H. Moore, B. S. Rao, and A. H. Wood, *J. Med. Chem.*, **15**, 1040 (1972).

551 T. Isshiki, *J. Pharm. Soc. Jap.* **64**, No. 7A, 6 (1944) [*C.A.*, **45**, 5662g (1951)].

552 P. Hirsjärvi, *Suom. Kemistilehti*, **29B**, 145 (1956) [*C.A.*, **51**, 8042e (1957)].

553 P. Hirsjärvi and V. P. Hirsjärvi, *Suom. Kemistilethi*, **38B**, 290b (1965) [*C.A.*, **64**, 12726b (1966)].

554 P. Hirsjärvi, M. Hirsjärvi, and J. O. W. Kaila, *Suom. Kemistilethi*, **30B**, 72 (1957) [*C.A.*, **53**, 16194g (1959).

555 P. Hirsjärvi, D. Klenberg, M. Patala, and P. Eenila, *Suom. Kemistilethi*, **34B**, 152 (1961) [*C.A.*, **57**, 16662g (1962)].

556 A. Marquet, M. Dvolaitzky, and D. Arigoni, *Bull. Soc. Chim. Fr.*, **1966**, 2956.

557 B. A. Arbuzov, Z. G. Isaeva, and V. V. Ratner, *Dokl. Akad. Nauk. SSSR*, **164**, 1289 (1965) [*C.A.*, **64**, 3608e (1966)]; *Zh. Org. Khim.*, **2**, 1401 (1966) [*C.A.*, **66**, 46491b (1967)].

558 R. O. Hutchins and D. Koharski, *J. Org. Chem.*, **34**, 2771 (1969).

559 Z. G. Isaeva, B. A. Arbuzov, and V. V. Ratner, *Izv. Akad. Nauk SSSR, Ser. Khim.*, **1965**, 475 [*C.A.*, **63**, 633g (1965)].

560 W. Zacharewicz, J. Krupowicz, and L. Borowiecki, *Rocz. Chem.*, **31**, 739 (1957) [*C.A.*, **52**, 5312b (1958)]; **33**, 87 (1959) [*C.A.*, **53**, 16194h (1959)].

561 K. K. Chakravarti and S. C. Bhattacharyya, *Perfum. Essent. Oil Rec.*, **46**, 341 (1951) [*C.A.*, **50**, 4462d (1956)].

562 J. Schmitt and J. Seilert, *Ann. Chem.*, **562**, 15 (1949).

563 V. M. Sathe, K. K. Chakravarti, M. V. Kadival, and S. C. Bhattacharyya, *Indian J. Chem.*, **4**, 393 (1966).

564 W. M. B. Könst, L. M. van der Linde, and H. Boelens, *Tetrahedron Lett.*, **1974**, 3175.

565 Y. Sakuda, *Bull. Chem. Soc. Jap.*, **34**, 514 (1961) [*C.A.*, **56**, 7358i (1962)].

566 Y. Sakuda, *J. Sci. Hiroshima Univ.*, *Ser.* A-II, **25**, 207 (1961) [*C.A.*, **57**, 7313g (1962)].

567 H. Schmidt, *Chem. Ber.*, **83**, 193 (1950).

568 W. Zacharewicz, *Rocz. Chem.*, **23**, 301 (1949) [*C.A.*, **45**, 5661f (1951)].

569 S. P. Baniukiewicz, *Diss. Abstr.*, *Int.B.*, **34**, 1935 (1973).

570 T. Suga, M. Sugimoto, and T. Matsuura, *Bull. Chem. Soc. Jap.* **36**, 1363 (1963).

571 W. Zacharewicz, *Rocz. Chem.*, **22**, 68 (1948) [*C.A.*, **43**, 2976a (1949)].

572 T. Matsuura and K. Fujita, *J. Sci. Hiroshima Univ.*, *Ser. A*, **15**, 277 (1951) [*C.A.*, **48**, 3932e (1954)].

573 A. J. Baretta, C. W. Jefford, and B. Waegell, *Bull. Soc. Chim. Fr.*, **1970**, 3985.

574 A. Kergomard, *Ann. Chim.* (Paris), **8**, 153 (1953).

575 J. B. Lee and M. J. Price, *Tetrahedron Lett.*, **1962**, 1155; *Tetrahedron*, **20**, 1017 (1964).

576 J. M. Quinn, *J. Chem. Eng. Data*, **9**, 389 (1964).

577 V. Garsky, D. F. Koster, and R. T. Arnold, *J. Amer. Chem. Soc.*, **96**, 4207 (1974).

578 M. I. Goryaev and G. A. Tolstikov, *Izv. Akad. Nauk Kaz. SSR, Ser. Khim.*, **1962**, 72 [*C.A.*, **59**, 6443g (1963)].

579 R. E. Klinck, P. DeMayo, and J. B. Strothers, *Chem. Ind.* (London), **1961**, 471.

580 J. Kovář and F. Petrů, *Collect. Czech. Chem. Commun.*, **25**, 604 (1960) [*C.A.*, **54**, 12185a (1960)].

581 F. Petrů and J. Kovář, *Chem. Listy*, **45**, 458 (1951) [*C.A.*, **46**, 7545d (1952)].

582 F. Petrů and J. Kovář, *Collect. Czech. Chem. Commun.*, **24**, 2079 (1959) [*C.A.*, **53**, 20119h (1959)].

583 Y. Sakuda, *Nippon Kagaku Zasshi*, **82**, 117 (1961) [*C.A.*, **56**, 8752c (1962)].

584 L. Tomaszewska and W. Zacharewicz, *Rocz. Chem.*, **35**, 1597 (1961) [*C.A.*, **57**, 9884d (1962)].

585 F. Petrů and J. Kovář, *Collect. Czech. Chem. Commun.*, **15**, 478 (1950) [*C.A.*, **45**, 9008i (1951)].

586 Y. Sakuda, *Bull. Chem. Soc. Japan*, **42**, 475 (1969) [*C.A.*, **70**, 96967q (1969)].

587 V. R. Tadwalkar and A. S. Rao, *Indian J. Chem.*, **9**, 1416 (1971).

588 N. J. Toivonen and A. Halonen, *Suom. Kemistilehti*, **19B**, 1 (1946) [*C.A.*, **41**, 5487i (1947)].

589 A. F. Thomas and M. Ozainne, *Helv. Chim. Acta*, **57**, 2062 (1974).

590 T. Murakami, I. Ichimoto, and C. Tatsumi, *J. Agr. Chem. Soc. Jap.*, **47**, 699 (1973).

591 Y. Sakuda, *Nippon Kagaku Zasshi*, **81**, 1891 (1960) [*C.A.*, **56**, 2473h (1962)].

592 F. Bohlmann and H. J. Bax, *Chem. Ber.*, **107**, 1773 (1974).

593 T. Miki, *J. Pharm. Soc. Jap.*, **75**, 410 (1955) [*C.A.*, **50**, 2520d (1956)].

594 L. J. Wadhams, R. Baker, and P. E. Howse, *Tetrahedron Lett.*, **1974**, 1697.

595 J. Wang, *Formosan Sci.*, **2**, 62 (1948) [*C.A.*, **48**, 7853f (1954)].

596 G. Lucius and C. Schaefer, *Z. Chem.*, **2**, No. 1, 29 (1962).

597 B. V. Bapat and G. H. Kulkarni, *Indian J. Chem.*, **9**, 608 (1971) [*C.A.*, **75**, 88779t (1971)].

598 K. Kohara, *Bull. Chem. Soc. Jap.*, **42**, 3229 (1969).

599 W. Treibs, *Chem. Ber.*, **90**, 761 (1957).

600 D. H. R. Barton, T. Bruun, and A. S. Lindsey, *J. Chem. Soc.*, **1952**, 2210.

601 T. Miki, *J. Pharm. Soc. Jap.*, **75**, 403 (1955) [*C.A.*, **50**, 2519e (1956)].

602 M. Nakazaki and K. Naemura, *Chem. Ind.*, **1964**, 1708.

603 S. P. Pathak, B. V. Bapat, and G. H. Kulkarni, *Indian J. Chem.*, **8**, 1147 (1970) [*C.A.*, **74**, 142086j (1971)].

604 V. Viswanatha and G. S. Krishnarao, *Tetrahedron Lett.*, **1974**, 247.

605 T. Miki, *J. Pharm. Soc. Jap.*, **75**, 407 (1955) [*C.A.*, **50**, 2519i (1956)].

606 D. A. Evans and C. L. Sims, *Tetrahedron Lett.*, **1973**, 4691.

607 T. Oritani and K. Yamashita, *Agr. Biol. Chem.* (Tokyo), **38**, 801 (1974).

608 G. Brownlie, F. S. Spring, and R. Stevenson, *J. Chem. Soc.*, **1959**, 216.

609 H. Ageta, K. Iwata, and S. Natori, *Tetrahedron Lett.*, **1964**, 3413.

610 O. Jeger, M. Montavon, R. Nowak, and L. Ruzicka, *Helv. Chim. Acta*, **30**, 1869 (1947).

611 A. Milliet and F. Khuong-Huu, *Tetrahedron Lett.*, **1974**, 1939.

612 B. Tursch, D. Daloze, and G. Chiurdoglu, *Bull. Soc. Chim. Belg.* **75**, 734 (1966).

613 W. Voser, Hs. H. Günthard, H. Heusser, O. Jeger, and L. Ruzicka, *Helv. Chim. Acta*, **35**, 2065 (1952).

614 W. Voser, M. Montavon, H. H. Günthard, O. Jeger, and L. Ruzicka, *Helv. Chim. Acta*, **33**, 1893 (1950).

615 A. G. Gonzales, A. Calero, and A. H. Toste, *An. Real Soc. Espan. Fis. Quim.*, **47B**, 287 (1951) [*C.A.*, **46**, 2527f (1952)].

616 S. A. Knight and J. F. McGhie, *Chem. Ind.* (London) **1953**, 920; *ibid.*, **1954**, 24.

617 M. S. Chapon-Monteil, *Bull. Soc. Chim. Fr.*, **1955**, 1076.

618 D. H. R. Barton and B. R. Thomas, *J. Chem. Soc.*, **1953**, 1842.

619 D. H. R. Barton and B. R. Thomas, *Chem. Ind.* (London), **1953**, 172.

620 E. Koller, A. Hiestand, P. Dietrich, and O. Jeger, *Helv. Chim. Acta*, **33**, 1050 (1950).

621 S. K. Talapatra, M. Bhattacharya, and B. Talapatra, *Indian J. Chem.*, **11**, 977 (1973).

622 A. Meyer, O. Jeger, V. Prelog, and L. Ruzicka, *Helv. Chim. Acta*, **34**, 747 (1951).

623 J. Dreiding, O. Jeger, and L. Ruzicka, *Helv. Chim. Acta*, **33**, 1325 (1950).

624 M. Kocór, J. St. Pyrek, C. K. Atal, K. L. Bedi, and B. R. Sharma, *J. Org. Chem.*, **38**, 3685 (1973).

625 A. Meyer, O. Jeger, and L. Ruzicka, *Helv. Chim. Acta*, **33**, 672, 1835 (1950).

626 D. Daloze, B. Tursch, and G. Chiurdoglu, *Tetrahedron Lett.*, **1967**, 1247.

627 D. S. Rycroft and R. F. M. White, *J. Chem. Soc., Chem. Commun.*, **1974**, 444.

628 G. S. Deshmukh and M. G. Bapat, *Chem. Ber.*, **88**, 1121 (1955).

629 Yu. K. Yur'ev, N. N. Magdesieva, and A. T. Monakhova, *Zh. Obshch. Khim.*, **35**, 68 (1965) [*C.A.*, **62**, 13114g (1965)].

630 M. Seyhan, *Chem. Ber.*, **86**, 888 (1953).

631 J. P. Allison and C. S. Marvel, *J. Polymer Sci.*, **1965**, 137.

632 F. Armitage and J. A. Cottrell, *Paint Technol.*, **13**, 353 (1948) [*C.A.*, **43**, 4025g (1949)]

633 A. Turk, U.S. Pat. 2,469,059 [*C.A.*, **43**, 5792c (1949)].

634 A. Ruwet, J. Meessen, and M. Renson, *Bull. Soc. Chim. Belg.* **78**, 459 (1969).

AUTHOR INDEX, VOLUMES 1–24

417

CHAPTER AND TOPIC INDEX, VOLUMES 1–24

Many chapters contain brief discussions of reactions and comparisons of alternative synthetic methods which are related to the reaction that is the subject of the chapter. These related reactions and alternative methods are not usually listed in this index. In this index the volume number is in BOLDFACE, the chapter number in ordinary type.

SUBJECT INDEX, VOLUME 24

Since the table of contents provides a quite complete index, only those items not readily found from the contents page are listed here. Numbers in **BOLDFACE** refer to experimental procedures.